VOLUME FIVE HUNDRED AND FIFTY THREE

METHODS IN ENZYMOLOGY

Computational Methods for Understanding Riboswitches

METHODS IN ENZYMOLOGY

Editors-in-Chief

JOHN N. ABELSON and MELVIN I. SIMON
Division of Biology
California Institute of Technology
Pasadena, California

ANNA MARIE PYLE
Departments of Molecular, Cellular and Developmental Biology and Department of Chemistry Investigator
Howard Hughes Medical Institute
Yale University

Founding Editors

SIDNEY P. COLOWICK and NATHAN O. KAPLAN

VOLUME FIVE HUNDRED AND FIFTY THREE

METHODS IN ENZYMOLOGY

Computational Methods for Understanding Riboswitches

Edited by

SHI-JIE CHEN
Department of Physics, Department of Biochemistry, and the Informatics Institute, University of Missouri, USA

DONALD H. BURKE-AGUERO
Department of Molecular Microbiology & Immunology and Department of Biochemistry, University of Missouri, USA

AMSTERDAM • BOSTON • HEIDELBERG • LONDON
NEW YORK • OXFORD • PARIS • SAN DIEGO
SAN FRANCISCO • SINGAPORE • SYDNEY • TOKYO
Academic Press is an imprint of Elsevier

Academic Press is an imprint of Elsevier
225 Wyman Street, Waltham, MA 02451, USA
525 B Street, Suite 1800, San Diego, CA 92101-4495, USA
125 London Wall, London, EC2Y 5AS, UK
The Boulevard, Langford Lane, Kidlington, Oxford OX5 1GB, UK

First edition 2015

Notices

Knowledge and best practice in this field are constantly changing. As new research and experience broaden our understanding, changes in research methods, professional practices, or medical treatment may become necessary.

Practitioners and researchers must always rely on their own experience and knowledge in evaluating and using any information, methods, compounds, or experiments described herein. In using such information or methods they should be mindful of their own safety and the safety of others, including parties for whom they have a professional responsibility.

To the fullest extent of the law, neither the Publisher nor the authors, contributors, or editors, assume any liability for any injury and/or damage to persons or property as a matter of products liability, negligence or otherwise, or from any use or operation of any methods, products, instructions, or ideas contained in the material herein.

ISBN: 978-0-12-801429-5
ISSN: 0076-6879

For information on all Academic Press publications visit our website at store.elsevier.com

CONTENTS

CONTRIBUTORS

R.W. Adamiak
Department of Structural Chemistry and Biology of Nucleic Acids, Institute of Bioorganic Chemistry Polish Academy of Sciences, and European Center for Bioinformatics and Genomics, Institute of Computing Science, Poznan University of Technology, Poznan, Poland

Kirill A. Afonin
Basic Research Laboratory, Center for Cancer Research, National Cancer Institute, Frederick, Maryland, USA

M. Antczak
European Center for Bioinformatics and Genomics, Institute of Computing Science, Poznan University of Technology, Poznan, Poland

Stefan Badelt
Institute for Theoretical Chemistry, University of Vienna, Vienna, Austria

Eckart Bindewald
Basic Science Program, Leidos Biomedical Research Inc., National Cancer Institute, National Institutes of Health, Frederick, Maryland, USA

J. Blazewicz
Department of Structural Chemistry and Biology of Nucleic Acids, Institute of Bioorganic Chemistry Polish Academy of Sciences, and European Center for Bioinformatics and Genomics, Institute of Computing Science, Poznan University of Technology, Poznan, Poland

Janusz M. Bujnicki
International Institute of Molecular and Cell Biology, Warsaw, and Faculty of Biology, Institute of Molecular Biology and Biotechnology, Adam Mickiewicz University, Poznan, Poland

Giovanni Bussi
Scuola Internazionale Superiore di Studi Avanzati (SISSA), Trieste, Italy

Haoyuan Chen
Center for Integrative Proteomics Research, BioMaPS Institute and Department of Chemistry and Chemical Biology, Rutgers University, Piscataway, New Jersey, USA

Clarence Yu Cheng
Department of Biochemistry, Stanford University, Stanford, California, USA

Fang-Chieh Chou
Department of Biochemistry, Stanford University, Stanford, California, USA

P. Clote
Biology Department, Boston College, Boston, Massachusetts, USA

Francesco Colizzi
Scuola Internazionale Superiore di Studi Avanzati (SISSA), Trieste, Italy

Rhiju Das
Department of Biochemistry, and Department of Physics, Stanford University, Stanford, California, USA

Francesco Di Palma
Scuola Internazionale Superiore di Studi Avanzati (SISSA), Trieste, Italy

Thakshila Dissanayake
Center for Integrative Proteomics Research, BioMaPS Institute and Department of Chemistry and Chemical Biology, Rutgers University, Piscataway, New Jersey, USA

Nikolay V. Dokholyan
Department of Biochemistry and Biophysics, School of Medicine, University of North Carolina, Chapel Hill, North Carolina, USA

Christoph Flamm
Institute for Theoretical Chemistry, University of Vienna, Vienna, Austria

George M. Giambaşu
Center for Integrative Proteomics Research, BioMaPS Institute and Department of Chemistry and Chemical Biology, Rutgers University, Piscataway, New Jersey, USA

Holger Gohlke
Mathematisch-Naturwissenschaftliche Fakultät, Institut für Pharmazeutische und Medizinische Chemie, Heinrich-Heine-Universität Düsseldorf, Düsseldorf, Germany

Stefan Hammer
Institute for Theoretical Chemistry, and Research Group Bioinformatics and Computational Biology, University of Vienna, Vienna, Austria

Christian A. Hanke
Mathematisch-Naturwissenschaftliche Fakultät, Institut für Pharmazeutische und Medizinische Chemie, Heinrich-Heine-Universität Düsseldorf, Düsseldorf, Germany

Scott P. Hennelly
New Mexico Consortium, and Theoretical Biology and Biophysics, Theoretical Division, Los Alamos National Laboratory, Los Alamos, New Mexico, USA

Ivo L. Hofacker
Institute for Theoretical Chemistry, and Research Group Bioinformatics and Computational Biology, University of Vienna, Vienna, Austria

Changbong Hyeon
School of Computational Sciences, Korea Institute for Advanced Study, Seoul Republic of Korea

Namhee Kim
Department of Chemistry and Courant Institute of Mathematical Sciences, New York University, New York, USA

Maria Kireeva
Gene Regulation and Chromosome Biology Laboratory, Center for Cancer Research, NCI, National Cancer Institute, Frederick, Maryland, USA

Serdal Kirmizialtin
New Mexico Consortium, and Theoretical Biology and Biophysics, Theoretical Division, Los Alamos National Laboratory, Los Alamos, New Mexico, USA

Andrey Krokhotin
Department of Biochemistry and Biophysics, School of Medicine, University of North Carolina, Chapel Hill, North Carolina, USA

Erich R. Kuechler
Center for Integrative Proteomics Research, BioMaPS Institute and Department of Chemistry and Chemical Biology, Rutgers University, Piscataway, New Jersey, USA

Grzegorz Łach
International Institute of Molecular and Cell Biology, Warsaw, Poland

Tai-Sung Lee
Center for Integrative Proteomics Research, BioMaPS Institute and Department of Chemistry and Chemical Biology, Rutgers University, Piscataway, New Jersey, USA

Jong-Chin Lin
Institute for Physical Science and Technology, and Department of Chemistry and Biochemistry, University of Maryland, College Park, MD, USA

P. Lukasiak
Department of Structural Chemistry and Biology of Nucleic Acids, Institute of Bioorganic Chemistry Polish Academy of Sciences, and European Center for Bioinformatics and Genomics, Institute of Computing Science, Poznan University of Technology, Poznan, Poland

David H. Mathews
Department of Biochemistry & Biophysics, and Center for RNA Biology, University of Rochester Medical Center, Box 712, Rochester, New York, USA

Jose N. Onuchic
Center for Theoretical Biological Physics; Department of Physics and Astronomy; Department of Chemistry; Department of Biosciences, and Department of Biochemistry and Cell Biology, Rice University, Houston, Texas, USA

Maria T. Panteva
Center for Integrative Proteomics Research, BioMaPS Institute and Department of Chemistry and Chemical Biology, Rutgers University, Piscataway, New Jersey, USA

Anna Philips
European Center for Bioinformatics and Genomics, Institute of Bioorganic Chemistry, Polish Academy of Science, Poznan, Poland

M. Popenda
Department of Structural Chemistry and Biology of Nucleic Acids, Institute of Bioorganic Chemistry Polish Academy of Sciences, Poznan, Poland

K.J. Purzycka
Department of Structural Chemistry and Biology of Nucleic Acids, Institute of Bioorganic Chemistry Polish Academy of Sciences, Poznan, Poland

Brian K. Radak
Center for Integrative Proteomics Research, BioMaPS Institute and Department of Chemistry and Chemical Biology, Rutgers University, Piscataway, New Jersey, USA

Karissa Y. Sanbonmatsu
New Mexico Consortium, and Theoretical Biology and Biophysics, Theoretical Division, Los Alamos National Laboratory, Los Alamos, New Mexico, USA

Tamar Schlick
Department of Chemistry and Courant Institute of Mathematical Sciences, New York University, New York, USA

Alexander Schug
Steinbuch Centre for Computing, Karlsruhe Institute of Technology, Karlsruhe, Germany

Bruce A. Shapiro
Basic Research Laboratory, Center for Cancer Research, National Cancer Institute, Frederick, Maryland, USA

Michael F. Sloma
Department of Biochemistry & Biophysics, and Center for RNA Biology, University of Rochester Medical Center, Box 712, Rochester, New York, USA

M. Szachniuk
Department of Structural Chemistry and Biology of Nucleic Acids, Institute of Bioorganic Chemistry Polish Academy of Sciences, and European Center for Bioinformatics and Genomics, Institute of Computing Science, Poznan University of Technology, Poznan, Poland

D. Thirumalai
Institute for Physical Science and Technology, and Department of Chemistry and Biochemistry, University of Maryland, College Park, MD, USA

Jeseong Yoon
School of Computational Sciences, Korea Institute for Advanced Study, Seoul Republic of Korea

Darrin M. York
Center for Integrative Proteomics Research, BioMaPS Institute and Department of Chemistry and Chemical Biology, Rutgers University, Piscataway, New Jersey, USA

Mai Zahran
Department of Chemistry and Courant Institute of Mathematical Sciences, New York University, New York, USA

PREFACE

It is clear from the Encyclopedia of DNA Elements (ENCODE) project that a significant portion in the transcriptome does not translate into protein sequences. Many of these sequences function through the formation of specific RNA structures. Riboswitches represent an important class of noncoding RNAs. Riboswitches perform functions as genetic "switches" that regulate when and where genes are expressed. To understand the structure and function of riboswitches requires computational models that can predict stable and metastable structures, their folding stabilities, kinetic pathways, transition states, and rate constants for the conformational switches, in addition to the effects of metal ions and ligand binding on RNA folding, all from the nucleotide sequence.

This book represents a state-of-the-art collection of advanced computational methods for modeling RNA riboswitch structure, thermal stability, dynamics, and kinetics. Given the fact that riboswitches have been extensively studied experimentally, yet currently the accuracy of computational prediction for riboswitch remains generally inconsistent, this volume is particularly timely. We believe that this volume will excite future more radical advances in computational modeling for riboswitches and other noncoding RNAs.

One of the major scientific challenges for RNA modeling relates to how we use the knowledge derived from limited information about known RNA structures. Effective structure prediction methods often involve integration of knowledge-based algorithm and physics-based or experimental data-based models. Several chapters present methods along this line: Purzycka et al. and Cheng et al. developed motif library-based and fragment-based methods, respectively; Sloma and Mathews developed methods that incorporate high-throughput structure probing data into secondary structure prediction; Kirmizialtin et al. employed selective 2′-hydroxyl acylation by primer extension (SHAPE) data and developed a new force field for molecular dynamics simulation; and Krokhotin and Dokholyan incorporated SHAPE into discrete molecular dynamics to predict RNA structure.

Another major challenge for the computational modeling of riboswitches comes from conformational sampling due to the large conformational space of the molecule and the flexibility of RNA structure.

To enhance the quality of conformational sampling, Krokhotin and Dokholyan developed a three-bead coarse-grained structure model in molecular dynamics simulation and Kim et al. developed a graph-theoretic approach to efficient sampling of RNA motif structures. These methods offer very interesting new tools for modeling RNA folding.

Kinetics are intrinsic to RNA folding and conformational switching. Riboswitches can fold cotranscriptionally into the conformations that correspond to ON and OFF states. The kinetics of RNA folding and conformational switching are critical for understanding riboswitch structure and function. This volume contains several chapters that address the methods to model riboswitch kinetics. Badelt et al. developed a cotranscriptional kinetics model by integrating RNA secondary structure with the dynamic energy landscape. Lin et al. developed a three-dimensional coarse-grained self-organized polymer model to predict riboswitch dynamics under force. Using a steered molecular dynamics-based method, Di Palma et al. developed an all-atom model to predict ligand-induced riboswitch stability and structure changes at atomic resolution. Hanke and Gohlke presented a critical analysis for the performances of the different force fields, including the Mg^{2+} ion effects, in riboswitch simulations. In addition to the physical models, informatics-based approaches have been highly successful in predicting riboswitch structures and finding riboswitch genes. Clote presented a comprehensive introduction for the different computational methods with particular emphasis on informatics-based approaches for riboswitch structure and kinetics.

Conformational changes in riboswitches can be induced by ligands or ions. It is thus important to have a reliable tool to predict the binding sites and binding modes. This volume contains two chapters on this issue. Using knowledge-based scoring functions, Philips et al. developed novel methods ("LigandRNA" and "MetalionRNA") for the prediction of riboswitch binding. Panteva et al. presented a multiscale method for modeling conformational switching, metal ion binding, and enzymatic reactions. In these chapters, the related computational methods are discussed, as well as their limitations and pitfalls.

RNA conformational switching can also be induced by the presence of other RNA or DNA molecules. Afonin et al. presented a thermodynamic model for predicting association and dissociation of RNA/DNA hybrids containing a novel split functionality. The results demonstrate the great promise of using RNA structure and conformational switches in RNA nanotechnology.

Finally, we would like to thank the fabulous team of authors, who have put together high-quality chapters. Without the phenomenal works of the authors, it is simply impossible to publish such a quality volume.

Shi-Jie Chen
Donald H. Burke-Agüero

SECTION I

RNA Structure Prediction

CHAPTER ONE

Automated 3D RNA Structure Prediction Using the RNAComposer Method for Riboswitches[1]

K.J. Purzycka*, M. Popenda*, M. Szachniuk*,†, M. Antczak†, P. Lukasiak*,†, J. Blazewicz*,†, R.W. Adamiak*,†,2

*Department of Structural Chemistry and Biology of Nucleic Acids, Institute of Bioorganic Chemistry Polish Academy of Sciences, Poznan, Poland

†European Center for Bioinformatics and Genomics, Institute of Computing Science, Poznan University of Technology, Poznan, Poland

[2]Corresponding author: e-mail address: adamiakr@ibch.poznan.pl

Contents

Abstract

Understanding the numerous functions of RNAs depends critically on the knowledge of their three-dimensional (3D) structure. In contrast to the protein field, a much smaller number of RNA 3D structures have been assessed using X-ray crystallography, NMR spectroscopy, and cryomicroscopy. This has led to a great demand to obtain the RNA 3D

[1]This work is dedicated to Professor Colin B. Reese (FRS) on the occasion of his 85th birthday anniversary.

Methods in Enzymology, Volume 553
ISSN 0076-6879
http://dx.doi.org/10.1016/bs.mie.2014.10.050

structures using prediction methods. The 3D structure prediction, especially of large RNAs, still remains a significant challenge and there is still a great demand for high-resolution structure prediction methods. In this chapter, we describe RNAComposer, a method and server for the automated prediction of RNA 3D structures based on the knowledge of secondary structure. Its applications are supported by other automated servers: RNA FRABASE and RNApdbee, developed to search and analyze secondary and 3D structures. Another method, RNAlyzer, offers new way to analyze and visualize quality of RNA 3D models. Scope and limitations of RNAComposer in application for an automated prediction of riboswitches' 3D structure will be presented and discussed. Analysis of the cyclic di-GMP-II riboswitch from *Clostridium acetobutylicum* (PDB ID 3Q3Z) as an example allows for 3D structure prediction of related riboswitches from *Clostridium difficile 4*, *Bacillus halodurans 1*, and *Thermus aquaticus Y5.1* of yet unknown structures.

1. INTRODUCTION

RNAs play the numerous and important roles in all major life processes (Gesteland, Cech, & Atkins, 2005). Thorough understanding of RNA functions depends critically on informative three-dimensional (3D) structure. However, it is difficult to assess 3D structures of large RNAs experimentally. In contrast to the protein field, where nearly hundred thousand structures have been deposited in the Protein Data Bank (Rose et al., 2011), a much smaller number of RNA 3D structures have been assessed experimentally (ca. 2300 entries) using X-ray crystallography, NMR spectroscopy, and, more recently, cryomicroscopy. Prediction of the RNA secondary structure using *in silico* methods is very advanced (Puton, Kozlowski, Rother, & Bujnicki, 2013; Xu, Almudevar, & Mathews, 2012) and has been recently reinforced by incorporating restraints from chemical probing methods (Mathews et al., 2004), such as SHAPE (Merino, Wilkinson, Coughlan, & Weeks, 2005). A growing number of the reasonably accurate secondary structures of large RNAs (Pang, Elazar, Pham, & Glenn, 2011; Watts et al., 2009; Wilkinson et al., 2008) created great demand in the RNA community to predict their 3D structures using computational methods (Jossinet, Ludwig, & Westhof, 2010; Leontis & Westhof, 2012; Martinez, Maizel, & Shapiro, 2008). However, 3D structure prediction still remains a significant challenge, even applying experimental restraints data (Seetin & Mathews, 2011). This situation clearly makes any RNA structure–function relationship studies difficult.

An automated prediction of the RNA 3D structure appeared of considerable and fast-growing interest. In the last years, few web-accessible tools have been proposed for automated prediction of the RNA 3D structure. They operate on broad spectrum of input data, such as sequence, secondary structure, conformational constrains, or structural templates. Considerable differences are also visible in terms of prediction quality, which depends on the RNA strand length and topology, processor time needed, and degree of automation. Physics-based automated methods use the coarse-grained and atomic-level molecular dynamics (Cao & Chen, 2011; Jonikas, Radmer, & Altman, 2009; Jonikas, Radmer, Laederach, et al., 2009; Sharma, Ding, & Dokholyan, 2008; Xu, Zhao, & Chen, 2014), internal coordinate space dynamics (Flores & Altman, 2010; Flores, Sherman, Bruns, Eastman, & Altman, 2011), and fragment assembly (Das, Karanicolas, & Baker, 2010; Parisien & Major, 2008). Full-atomic structure predictions based on dynamics and fragment assembly are powerful tools for modeling relatively complex but small RNAs. Despite the computational cost, the coarse-grained molecular dynamics can access larger RNAs but requires demanding and not fully resolved addition of atomic details to coarse-grain models (Jonikas, Radmer, & Altman, 2009; Jonikas, Radmer, Laederach, et al., 2009). Knowledge-based comparative modeling (Rother, Rother, Puton, & Bujnicki, 2011) depends on the access to 3D structural templates and unequivocal sequence alignment. Prior to reporting on our RNAComposer method (Popenda et al., 2012), which is based on the machine translation principle and makes use of 3D fragments assembly, none of the reported methodologies has reached the stage of truly full automation, efficient access to large RNA structures, short computing time, and user-friendly performance.

To assess 3D structures of RNAs of a size too large to conduct NMR studies (see Pachulska-Wieczorek, Purzycka, & Adamiak, 2006; Purzycka, Pachulska-Wieczorek, & Adamiak, 2011), we have settled an interdisciplinary group to develop automated RNA 3D structure prediction method. Full automation of the RNA 3D structure computational prediction is a complex and very difficult task. If one considers industrial applications, robots can perform actions that man cannot or can do only with much lower precision and reproducibility. In contrast to the industrial automation, an expert scientist with extensive knowledge and experience in the field is able to perform RNA 3D modeling in much more reliable way than available automated servers. Our aim is to change this situation with a kind of the "black box" server allowing, even inexperienced users, to access RNA 3D structures very fast and with great reliability. At the beginning, based on the secondary structure, we were able to predict efficiently only 3D structures of medium-size RNA, at low-resolution level (Popenda, Bielecki, & Adamiak, 2006).

We have soon realized that an ultimate methodology must fulfill several difficult to reach criteria outlined below:

- fully automated performance,
- high prediction fidelity at atomic resolution level,
- very fast multimodel building,
- access to 3D models with assessed energy,
- application to large RNA structures,
- acceptance of experimental constraints,
- high-throughput potential,
- web accessibility, and
- user-friendly performance.

It took us nearly 6 years to accomplish most of the above criteria. Development of RNA FRABASE (Popenda, Blazewicz, Szachniuk, & Adamiak, 2008; Popenda et al., 2010) appeared to be a key step in our study and soon flourished in the RNAComposer method (Popenda et al., 2012) and servers (http://rnacomposer.ibch.poznan.pl and mirror http://rnacomposer.cs.put.poznan.pl). The RNAComposer method, supported by our allied web server tools for the RNA quality assessment (Antczak et al., 2014; Lukasiak et al., 2013), will be presented and its scopes and limitations will be discussed in this chapter. The cyclic di-GMP-II riboswitch has been selected as an application example and its 3D structure prediction will be discussed in multiple aspects.

Riboswitches are an interesting example of noncoding mRNA regions capable of binding cellular metabolites to modulate gene expression. In order to function, those RNAs must achieve high specificity. This unique metabolite sensing is possible with a diverse array of secondary and tertiary structures. However, riboswitches representing different classes adopt diverse structures, and their exceptional selectivity is encoded in conserved structural features.

Riboswitches can be divided into two general groups based on their overall special architecture: pseudoknotted and junctional (Serganov & Nudler, 2013). Pseudoknots are formed when a hairpin loop interacts with an outside region and in riboswitches usually involve a stack of two helices. Spatial structure of fluoride riboswitch is based on the small pseudoknot (Ren, Rajashankar, & Patel, 2012). Junctional riboswitches contain multihelical junction joining several helices. This group is represented by lysine (5-way junction) (Garst, Heroux, Rambo, & Batey, 2008) or TPP (3-way junction) riboswitches (Edwards & Ferre-D'Amare, 2006). For the functional architecture of several riboswitches, like c-di-GMP-II, both junction and extensive tertiary interactions are necessary (Smith, Shanahan, Moore, Simon, & Strobel, 2011). Those riboswitches represent most complex, mixed type group (Peselis & Serganov, 2014). Differing in complexity, pseudoknots and multihelical junctions are structural elements that constitute significant challenge for the prediction methods. Extensive studies are necessary to deconvolute not only their tertiary but also secondary structures.

2. RNA FRABASE—OPENING THE ROUTE TO RNAComposer

In 2006, we have presented (Popenda et al., 2006) a computer program, called 3D-RNApredict, which implements RNA secondary structure and converts various structural data (RNA fragments coordinates, experimental data) to create the input to the CYANA software (Guntert, Mumenthaler, & Wuthrich, 1997). CYANA's torsion angle dynamics algorithm provides a very fast engine for the 3D RNA structure calculation. The method was amenable to automatization. Soon we have realized that further advance of this approach needs dedicated 3D RNA fragments search engine and database. Along this line RNA FRABASE has been developed (Popenda et al., 2008, 2010). RNA FRABASE core consists of the interfaced database, the search engine, and the web interface. Its repository collects information about PDB-deposited RNA structures and their complexes, including all models of NMR-elucidated molecules. It stores RNA sequences; secondary structures encoded in the dot-bracket notation and in graphical form; atom coordinates of the unmodified and modified nucleotide and nucleoside residues; torsion and pseudotorsion angle values; sugar pucker parameters; and complete classification of base pair types, base–base parameters for base pairs, and inter-base pair parameters for dinucleotide steps. The third component has been designed with an emphasis on friendly and ergonomic interface. RNA FRABASE is a publicly available web-interfaced system, running at http://rnafrabase.cs.put.poznan.pl. RNA FRABASE is automatically updated on a monthly basis. As of June 2014, it contained 2259 RNA structures that represent about 90% of all RNAs deposited in the Protein Data Bank.

Relevant to this chapter's topic, it should be noted that within the RNA FRABASE database, one could find fragments derived from 149 PDB-deposited structures of riboswitches. They include 1119 duplexes (constituting 0.7% of all duplexes deposited in RNA FRABASE), 2106 loops (0.7%), and 71 single-stranded fragments (0.9%).

Having this valuable tool in hand we use it as the prerequisite to generate the dedicated dictionary—a key element in our RNA 3D structure prediction method.

3. RNAComposer—FROM THE RNA SECONDARY STRUCTURE TO RNA 3D STRUCTURE

3.1. General description of the RNAComposer method

The RNAComposer method for automated prediction of RNA 3D structures (Popenda et al., 2012) was founded on the machine translation concept parallel to that used in the computational linguistics. The developed workflow system allows us to progress rapidly from the RNA secondary structure to the RNA 3D structure (Fig. 1). Pivotal part of the translation system is the dedicated dictionary. This dictionary is tailored from the RNA FRABASE database (Popenda et al., 2008, 2010) on which the translation engine operates to predict RNA 3D structure. Dictionary relates the RNA secondary structure and tertiary structure elements. The RNAComposer dictionary differs from the RNA FRABASE database. In order to enable dictionary function in the translation, all the 3D structural elements collected there were attributed to a complete set of atoms, energy calculated by the CHARMM force field (XPLOR-NIH), and are characterized by a good stereochemistry and structural properties. The dictionary does not include elements with modified residues or missing heavy-atom coordinates. Since our report (Popenda et al., 2012), the volume of the dictionary has been substantially enlarged and includes 23,092 secondary structure elements (initially 14,464) and as many as 489,599 related 3D structure elements (initially 190,928). The 3D structure elements are continuously transferred from the RNA FRABASE to the RNAComposer system and are transformed into dictionary's elements. All steps of this transfer, starting from the import of the newly deposited PDB structures, are automated. As we expected, growing dictionary volume substantially increases the quality of the predicted 3D structures.

The algorithms governing the RNAComposer engine actions allow us to build the RNA 3D models automatically in the following steps exemplified on the cyclic di-GMP-II riboswitch (PDB ID 3Q3Z) (Fig. 1):

(i) *RNA secondary structure fragmentation.* The input RNA secondary structure is divided into fragments following its tree graph representation (Gan, Pasquali, & Schlick, 2003). The fragmentation algorithm provides the secondary structure elements: stems, loops (i.e., apical, bulge,

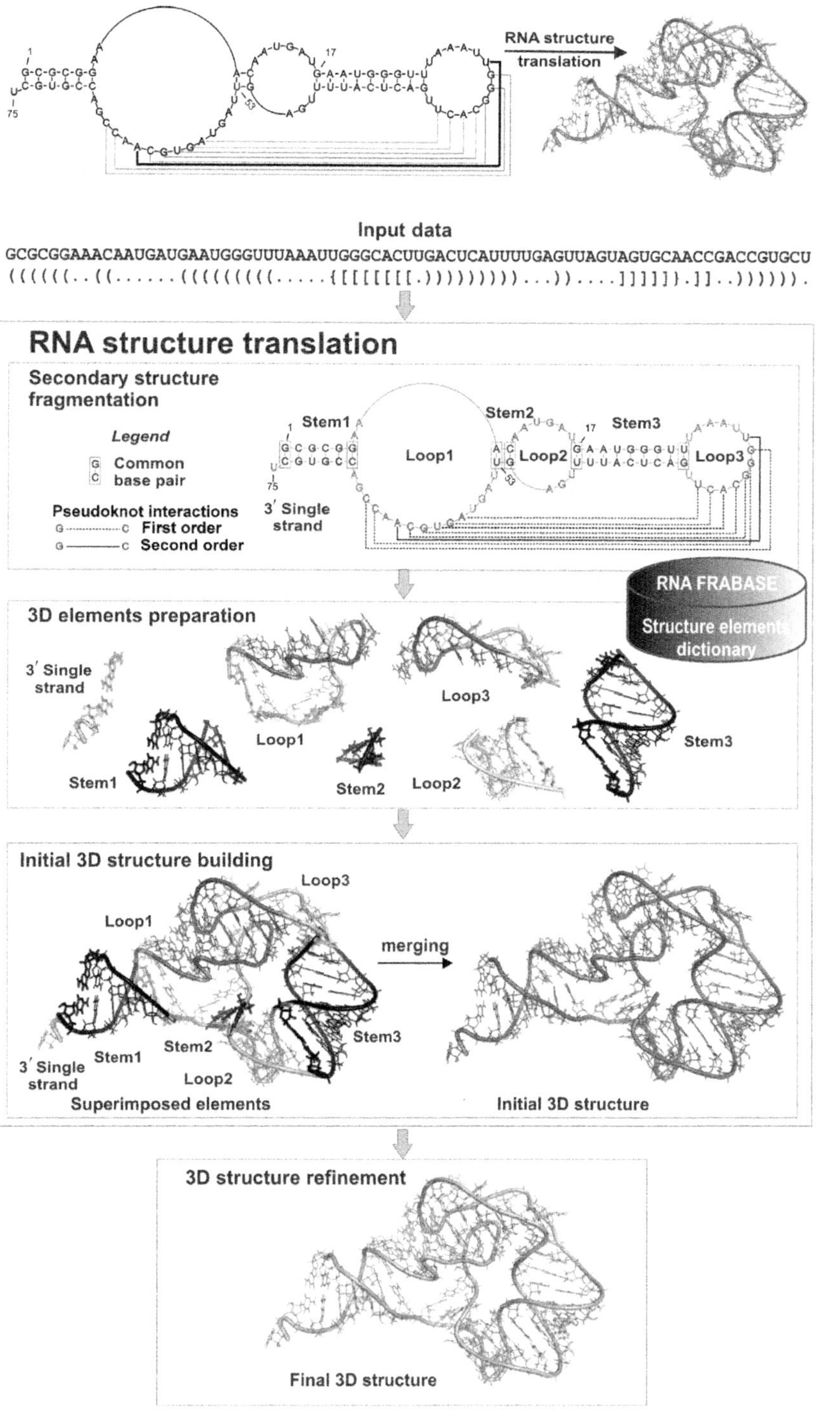

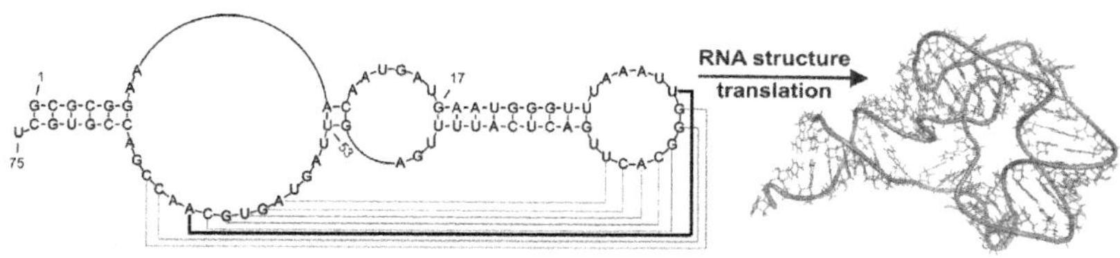

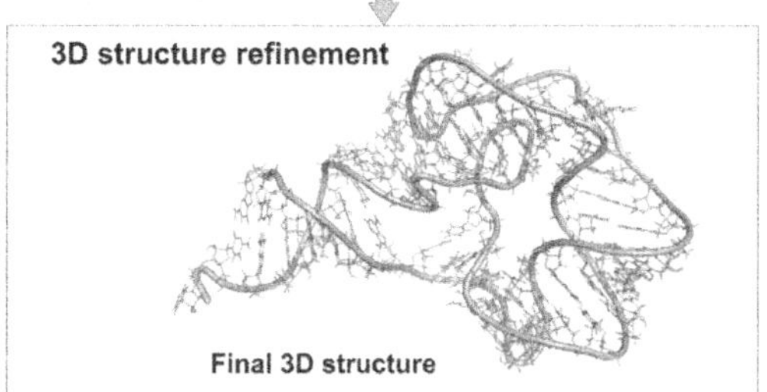

Figure 1 The RNA secondary structure to the 3D structure translation. Basic steps of RNAComposer action exemplified on the cyclic di-GMP-II riboswitch structure from *C. acetobutylicum* (PDB ID 3Q3Z). (See the color plate.)

internal, and *n*-way junctions), and single strands. Each element is closed by canonical base pair(s) according to the RNA FRABASE concept (Popenda et al., 2008). In the case of the secondary structure elements for the cyclic di-GMP-II riboswitch, three stems, three loops, and one single strand are produced by the fragmentation algorithm. At this stage pseudoknots are not considered.

(ii) *3D structure elements search.* The fragmentation constitutes input patterns (based on the RNA FRABASE notation) for an automatic search of related 3D structure elements within the dictionary. Engine makes use of only those 3D elements whose heavy-atom root mean square deviation (rmsd) is lower than 1.0 Å, when referred to the parent PDB structure.

(iii) *3D structure elements preparation.* The 3D structure elements are selected, usually from the wide spectrum of entities found in the dictionary. The selection process is crucial and is governed by an algorithm cantered on the five criteria executed in the following priority order:

- secondary structure topology,
- sequence similarity,
- pyrimidines/purines compatibility,
- source structure resolution, and
- energy.

(iv) *Initial RNA structure building.* The RNA tree graph representing secondary structure governs the building process. The 3D structure elements are superimposed with reference to the common canonical base pairs and merged to give initial, already well-shaped RNA 3D structure. Up to this step, the action of RNAComposer system is very fast. It usually takes several seconds on the single processor. For the cyclic di-GMP-II riboswitches discussed below it takes about 5 s.

(v) *3D structure refinement.* The energy minimization in the torsion angle space (Guntert et al., 1997) and subsequently in the Cartesian atom coordinate space (Schwieters, Kuszewski, Tjandra, & Clore, 2003) leads to the final, high-quality RNA 3D model. From the single secondary structure, up to 10 models can be generated in one prediction. This step takes several minutes for large RNA structures. In case of the cyclic di-GMP-II riboswitch from *Clostridium acetobutylicum*, generation of 10 models takes 2 min and 35 s.

Several important points for the users concerning RNAComposer action should be outlined. The system pipeline does not allow for changes from the secondary structure topology set as an input (in dot-bracket notation). As far as the criterion of sequence similarity (homology) is concerned, there is often departure from the full sequence homology during 3D element selection. If the RNA sequence is not found for the fragment of the required topology, the respective bases are replaced based on the pyrimidines/purines compatibility. Fragments from the highest resolution structures are preferably selected. Crystal structures are first priority, preceded by NMR and cryo-EM structures. The translation engine can generate a family of closely related 3D models in the single prediction. Up to 10 models can be generated from one secondary structure. The first model is always built using all the selection criteria described above. Other models are generated from 3D structure elements for which the criteria of structure resolution and energy are ignored. These models often include NMR-derived 3D elements. In the batch mode, RNAComposer servers (see Section 3.4) can be loaded with up to 100 secondary structures. Moreover, when launching RNAComposer, user can enter own experimental restraints, which makes the system very powerful.

During the modeling of complex RNAs, some 3D structure elements related to the certain secondary structure topology might be missing in the dictionary. For the purpose of automation, which must be considered as the process of self-reliability, the missing 3D fragments are instantly generated with dedicated algorithm. To ensure overall speed of 3D structure generation, a very fast fd_helix routine of the NAB software was chosen to generate stems and single strands. In the case of missing RNA loops being the part of hairpins, bulges, internal loops, and *n*-way junctions, molecular mechanics in the torsion angle space using CYANA (Guntert et al., 1997) is conducted.

All the details concerning major steps of the RNAComposer action are given to the user in the log.txt file accompanying respective PDB file.

In the original report (Popenda et al., 2012), RNAComposer method was carefully evaluated to estimate its scope and the quality of the predicted 3D structure models in terms of the secondary structure topology conservation, their stereochemical properties, energy, precision, and accuracy. The predicted 3D structures were validated using representative benchmark set of RNAs with the secondary structures derived from the highest resolution X-ray structures.

The RNAComposer system is based on the concept of machine translation, new in structural bioinformatics field. In the view of existing methods, its formal classification is difficult and might raise controversies.

The RNAComposer might be classified as a knowledge-based method that employs automated fragment assembly, based on the secondary structure tree graph representation and homology of structural elements.

3.2. RNAComposer input data

RNAComposer uses RNA sequence and secondary structure topology in dot-bracket notation (Vienna notation) as an input for the 3D structure prediction. In this notation, dot represents unpaired nucleotide, while bracket refers to the nucleotide involved in canonical base pairing including GU. In order to allow prediction of more complex RNA structures, square brackets are accepted to annotate first-order pseudoknots and curly brackets for higher order structures. Accomplishing the structure translation concept, RNAComposer does not change provided secondary structure topology and entire prediction depends on the input data. Correct secondary structure is therefore critical for the accurate prediction of the 3D structure. RNA secondary structure probing methods are available to support and substantially improve *in silico* RNA structure prediction, e.g., see Deigan, Li, Mathews, and Weeks (2009). Based on our own experience (Huang et al., 2013; Lusvarghi et al., 2013), we strongly encourage the user to use experimentally adjusted secondary structures to ensure more accurate 3D structure modeling with RNAComposer.

Two modes are available for the input data into RNAComposer servers (see Section 3.4). High efficiency of RNAComposer system allowed us to make an interactive mode accessible for the unregistered user. This mode allows the user to introduce only RNA sequence and choose secondary structure prediction method from three possibilities incorporated within RNAComposer system: RNAfold (Schuster, Fontana, Stadler, & Hofacker, 1994), RNAstructure (Reuter & Mathews, 2010), and Contrafold (Do, Woods, & Batzoglou, 2006). This mode is oriented toward inexperienced users and works well only for simple RNA structures, i.e., RNA for which *in silico* secondary structure prediction is reliable enough.

Batch mode, accessible upon login, requires secondary structure topology in dot-bracket notation and is oriented toward prediction of the RNA of medium size (up to 500 nts) or higher level of complexity. This allows users to introduce their own experimentally adjusted secondary structures of RNA and gives best 3D structure predictions. Several secondary structure topologies can be introduced for a single RNA sequence. Importantly, distance restraints can be entered into the RNAComposer system in the batch mode. This functionality allows users to incorporate additional data, e.g., from NMR or FRET experiments, into the 3D structure prediction pipeline. In addition, this option allows constraining noncanonical base pairs or pseudoknots.

3.3. Output data and quality control of the 3D models

The PDB and log files are provided as RNAComposer output. The PDB file contains coordinates of predicted 3D structure models. The log file describes details of the model generation process. This file includes several information allowing the user to inspect all steps of the RNAComposer action and to analyze obtained models: (i) a list of structure elements resulting from secondary structure fragmentation, (ii) a list of tertiary elements (sequence and topology) selected for the 3D structure assembly, (iii) origin (PDB ID code) of the 3D element, (iv) sequence similarity, and (v) the final energy given by the force field for the structure. In the log file, information are provided about 3D structure elements, which, due to the absence in the dictionary, were generated by the RNAComposer. Resulted RNA 3D structures containing such elements will be usually much less accurate and this should be taken into account. The energy of the final structure is a good indicator of the 3D structure models quality and should be inspected by the user (Popenda et al., 2012). Further quality control of the predicted 3D models could be accomplished using two complementary tools: RNApdbee (Antczak et al., 2014) and RNAlyzer (Lukasiak et al., 2013).

3.3.1 RNApdbee—Application in the RNA 3D structure validation

RNAComposer translation system does not allow for the departure from the input secondary structure topology. However, it is advisable to check whether the secondary structure pattern, especially in case of pseudoknots, was preserved after the refinement step. This can be done by reversing predicted 3D structure back to the secondary structure. Our recently reported method RNApdbee (Antczak et al., 2014) and publicly available server at http://rnapdbee.cs.put.poznan.pl allow the user to inspect the conservation of the input secondary structure topology in the predicted 3D structure models. RNApdbee web server has been designed to derive RNA secondary structure from the tertiary structure encoded in the PDB file or from the list of base pairs. It should be underlined that RNApdbee can process knotted and unknotted structures of large RNAs. RNApdbee supports an identification and classification of high-order pseudoknots, and their graphical visualization as well as dot-bracket encoding.

3.3.2 RNAlyzer—New tool for the 3D structure quality assessment

RNAComposer can generate a large ensemble of 3D structures within minutes. Having so many possible structures that can be generated for one sequence, one faces the problem of RNA 3D structure quality assessment. Currently, to assess 3D RNA models one can use several metrics such as rmsd, calculated for pairs of superimposed corresponding atoms, deformation index and interaction network fidelity, taking into account base–base interactions within the structures, and the deformation profile, taking into account both global and local differences (Parisien, Cruz, Westhof, & Major, 2009).

Recently, we have developed a new approach called RNAlyzer (Lukasiak et al., 2013) designed to solve the problem of comparison of RNA structure models to a reference structure. It is based on rmsd metrics and allows the user to identify the difference between them by calculating the dissimilarity between sets of atoms situated inside the series of spheres. Around selected atom of nucleotide, a sphere with defined radius is built. Selected atom plays a role of the center of the sphere that is constructed for every nucleotide of reference structure. The stage of sphere building gives sets of atoms corresponding to every sphere built on reference structure (Fig. 2).

In the next phase, for each set of the atoms identified in the previous phase, the corresponding set of atoms from the analyzed models are

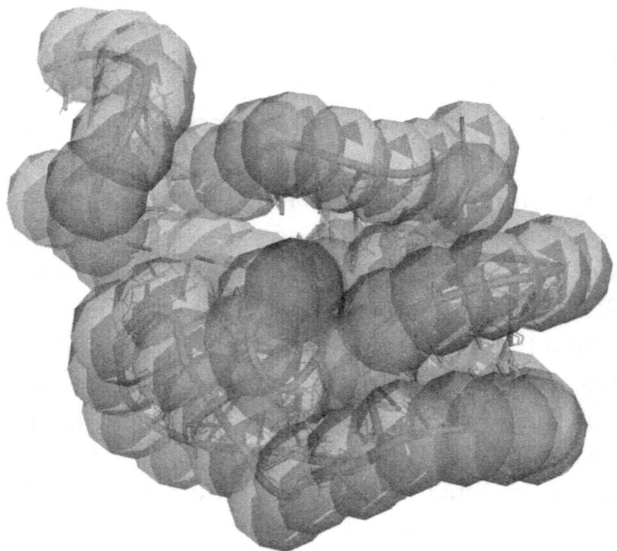

Figure 2 RNAlyzer spheres defining range of the atoms in superimposition of 3D model and reference structure, as probing structural space of every nucleotide residue. (See the color plate.)

localized. As a final stage, corresponding sets of atoms from the reference structure and the model are superimposed, and rmsd between them is calculated. The approach gives a possibility to select different radii and compare models from different levels of structural precision. Small sphere radius means the ability to analyze model from local structural point of view. With an increasing value of radius, the analysis becomes more global (lower accuracy). With sphere radius higher than radius of the analyzed molecule, the rmsd is calculated for the whole structure. The proposed quality assessment result can be analyzed and visualized on several types of 2D and 3D plots to show the model quality simultaneously for different levels of accuracy and to discriminate between correct and incorrect predictions in the entire spectrum of modeling precision. Application of RNAlyzer (http://rnalyzer.cs.put.poznan.pl) will be demonstrated on the example of cyclic di-GMP-II riboswitch from *C. acetobutylicum*.

3.4. RNAComposer web server

RNAComposer is offered to the users as a publicly available system with a web interface, running at two independent mirror sites: http://rnacomposer.ibch.poznan.pl and http://rnacomposer.cs.put.poznan.pl. Currently available version of the web server allows the user to build 3D structures of single-stranded RNAs up to 500 residues long. As a basic input, it requires a sequence and a secondary structure topology which can either be typed in directly or predicted from the sequence by one of the integrated tools (RNAstructure (Reuter & Mathews, 2010), RNAfold (Schuster et al., 1994), CONTRAfold (Do et al., 2006)) upon user choice. As already mentioned, RNAComposer works in two execution modes. An interactive mode is provided to newcomers and users processing single structure per query. Large-scale modeling is possible in a batch mode available for registered users. In this mode, one can upload a task containing even 10 RNA sequences, each associated with 1–10 secondary structures. For every secondary structure input in a batch, RNAComposer predicts up to 10 3D models. They can be generated with respect to optionally provided input atomic distance restraints. The basic output data are released in PDB-formatted files and can be immediately visualized (interactive mode), stored, and analyzed in the user workspace (batch mode) or emailed (both modes).

RNAComposer has been realized in the two-layer architecture, composed of a computational engine and a web application. The first, back-end layer hosts the translation machine (encoded in Java) and the dictionary (stored in PostgreSQL DBMS). The computational engine supports concurrent processing and adapts dynamically to available resources such as the number of processing units and RAM capacity. The front-end layer of RNAComposer provides a user-friendly web interface implemented in SpringMVC. The system is closed in the Virtual Private Network that supports effective and safe communication between its components through the integrated message brokers (Apache Active MQ).

4. PREDICTING THE TERTIARY STRUCTURE OF RIBOSWITCHES WITH RNAComposer

4.1. RNAComposer accurately predicts 3D structure of several complex riboswitches

Taking into account the predictive power of RNAComposer, we have chosen a representative benchmark set of riboswitches of the known X-ray crystal structures. While these RNAs differ in strand length and structural complexity, all contain multihelical junctions and/or pseudoknots. Their sequence and secondary structure topologies (Table 1), derived from the highest resolution X-ray structures, were inputted into RNAComposer.

It is important to note that before prediction, all the 3D structure elements comprised by the respective crystal PDB structure were excluded from the dictionary. The results are presented in Table 2 and Fig. 3. Ten models were predicted for each 2D structure and the accuracy was analyzed by calculating network fidelity parameters and the rmsd with the crystal structure. All examples, with the exception of the c-di-GMP-II riboswitch from *C. acetobutylicum* (PDB ID 3Q3Z), were characterized by high accuracy, and canonical and noncanonical base pairing and stacking were recovered, as indicated by network fidelity parameters (Table 2). Superposition of all but one predicted and crystal structures showed very good overall agreement consistent with their rmsd values (Fig. 3).

Multihelical junction conformation and orientation of the helices were correct for those examples, including 3OWW with an rmsd of 5.7 Å. However, the c-di-GMP-II riboswitch *C. acetobutylicum* (3Q3Z) was predicted with an average rmsd of 12.3 Å, and the best model showed a value of 10.1 Å. The c-di-GMP-II from *C. acetobutylicum* represents especially difficult example due to very unique structural features (see below). Since this riboswitch is represented only by one PDB entry (3Q3Z), it was chosen as a challenging application example to discuss the scope and limitations of the RNAComposer in its current version (see below).

Table 1 Sequence and secondary structure data inputted into RNAComposer for the 3D structure prediction of selected riboswitches

RNA PDB code and chain	Sequence and secondary structure topology
2YIE X+Z	GGAUCUUCGGGGCAGGGUGAAAUUCCCGACCGGUGGUAUAGUCCACGAAUCCAUCCGGAUUGAUUUGGUGAAAUUCCAAAACCGACAGUAGAGUCUGGAUGAGAGAAGAUUCG (((((((......(((.......))).[[((((((....[[))))...(((....)))....(((((.......)))))]])).(((....]])))).......))))))).
3D2V A	GGGACCAGGGGUGCUUGUUCACAGGCUGAGAAAGUCCCUUUGAACCUGAACAGGGUAAUGCCUGCGCAGGGAGUGUC ..((((((((..(((((....))))).........)))).....(((...((((......))))...)))..).)))
3GX5 A	GGCUUAUCAAGAGAGGUGGAGGGACUGGCCCGACGAAACCCGGCAACCAGAAAUGGUGCCAAUUCCUGCAGCGGAAACGUUGAAAGAUGAGCCG (((((((....(.(((...(((...[[)))......))))(((..(((....))).)))...(]]..(((((....)))))..)))))))).
4FE5 B	GGACAUAUAAUCGCGUGGAUAUGGCACGCAAGUUUCUACCGGGCACCGUAAAUGUCCGACUAUGUCC .(((((((..(.(((((.....[[)))))).)[.....)](((((((]].....))))))..))))))).
3DIL A	GGCCGACGGAGGCGCGCCCGAGAUGAGUAGGCUGUCCCAUCAGGGGAGGAAUCGGGGACGGCUGAAAGGCGAGGGCGCCGAAGGGUGCAGA GUUCCUCCCGCUCUGCAUGCCUGGGGGUAUGGGGAAUACCCAUACCACUGUCACGGAGGUCUCUCCGUGGAGAGCCGUCGGUC ((((((((..(.((((((..(.......(((((((((..(..[[[[[[....))))))))))......)..)))))))..(((((((((((.]]]]]]))))))))))).)))((.((((((((.....)))))))).)).((((((((....)))))))).....)))))))))
3PDR X	GGGCUUCGUUAGGUGAGGCUCCUGUAUGGAGAUACGCUGCUGCCCAAAAAUGUCCAAAGACGCCAAUGGGUCAACAGAAAUCAUCGACAUAAGGUGAUUUUUAAUG CAGCUGGAUGCUUGUCCUAUGCCAUACAGUGCUAAAGCUCUACGAUUGAAGCCCA (((((((((([.(((..(((..((((((((......(((((((((((....((((....))))....)))))).....(((((((((......]))))))))))..).))))).((.........))....))))))))).)))...)))..))....))))))).
3OWW A	GGCUCUGGAGAGAACCGUUUAAUCGGUCGCCGAAGGAGCAAGCUCUGCGGAAACGCAGAGUGAAACUCUCAGGCAAAAGGACAGAGUC ((((((....(.((((......)))))(((...((((...((((((((....))))))))....))).).)))......)))))))
3MXH R	GGUCACGCACAGGGCAAACCAUUCGAAAGAGUGGGACGCAAAGCCUCCGGCCUAAACCAUUGCACUCCGGUAGGUAGCGGGGUUACCGAUGG ...(.((......((...((((((....))))))..[))...(((.((((((((..((..........))))))).]))))))...))...)
3Q3Z V	**GCGCGGAAACAAUGAUGAAUGGGUUUAAAUUGGGCACUUGACUCAUUUUGAGUUAGUAGUGCAACCGACCGUGCU** **((((((..((......(((((((((.....{[[[[[[[.)))))))))...))....]]]]]}.]]..)))))).**
3SD3 A	GGAGAGUAGAUGAUUCGCGUUAAGCGUGUGUGAAUGGGAUGUCGUCACACAACGAAGCGAGAGCGCGGUGAAUCAUUGCAUCCGCUCCA ((((....(((((((((((((......(((((((...[[[[....))))))).....((....))))).))))))))))..]]]].)))).

Table 2 Characteristics and accuracy of the 3D models predicted for selected riboswitches

RFAM				PDB				Accuracy[a]		
RFAM ID	Family description	Number of sequences: Full	Number of sequences: Seed	Number of PDB entries	RNA PDB code and chain	Resolution (Å)	Strand length (nt)	rmsd (Å)	INF^{cbp}	INF^{all}
RF00050	FMN (RFN element)	4516	144	16	2YIE X, Z	2.94	113	3.2	0.98 (0.01)	0.79 (0.02)
RF00059	TPP (THI element)	11197	115	12	3D2V A	2.00	77	3.3	1.00 (0.00)	0.82 (0.02)
RF00162	SAM (S box leader)	4757	433	18	3GX5 A	2.40	94	2.7	0.99 (0.01)	0.81 (0.02)
RF00167	Purine riboswitch	2427	133	24	4FE5 B	1.32	67	2.0	1.00 (0.01)	0.85 (0.04)
RF00168	Lysine riboswitch	2422	47	14	3DIL A	1.90	174	3.0	1.00 (0.00)	0.89 (0.01)
RF00380	ykoK leader	1493	157	2	3PDR X	1.85	161	3.5	0.97 (0.01)	0.78 (0.01)
RF00504	Glycine riboswitch	6875	44	12	3OWW A	2.80	88	5.7	0.98 (0.00)	0.81 (0.03)
RF01051	Cyclic di-GMP-I	1990	155	17	3MXH R	2.30	92	3.4	0.98 (0.01)	0.79 (0.03)
RF01786	**Cyclic di-GMP-II**	**237**	**54**	**1**	**3Q3Z V**	**2.51**	**75**	**12.3**	**0.85 (0.05)**	**0.48 (0.03)**
RF01831	THF	598	98	5	3SD3 A	1.95	83	2.0	1.00 (0.01)	0.85 (0.02)

[a]Described as the average heavy-atom rmsd (in Å) between 10 individual 3D models and the crystal structure, and the average interaction network fidelity (INF) measures. INF scores range from 0.00 (worst) to 1.00 (best).

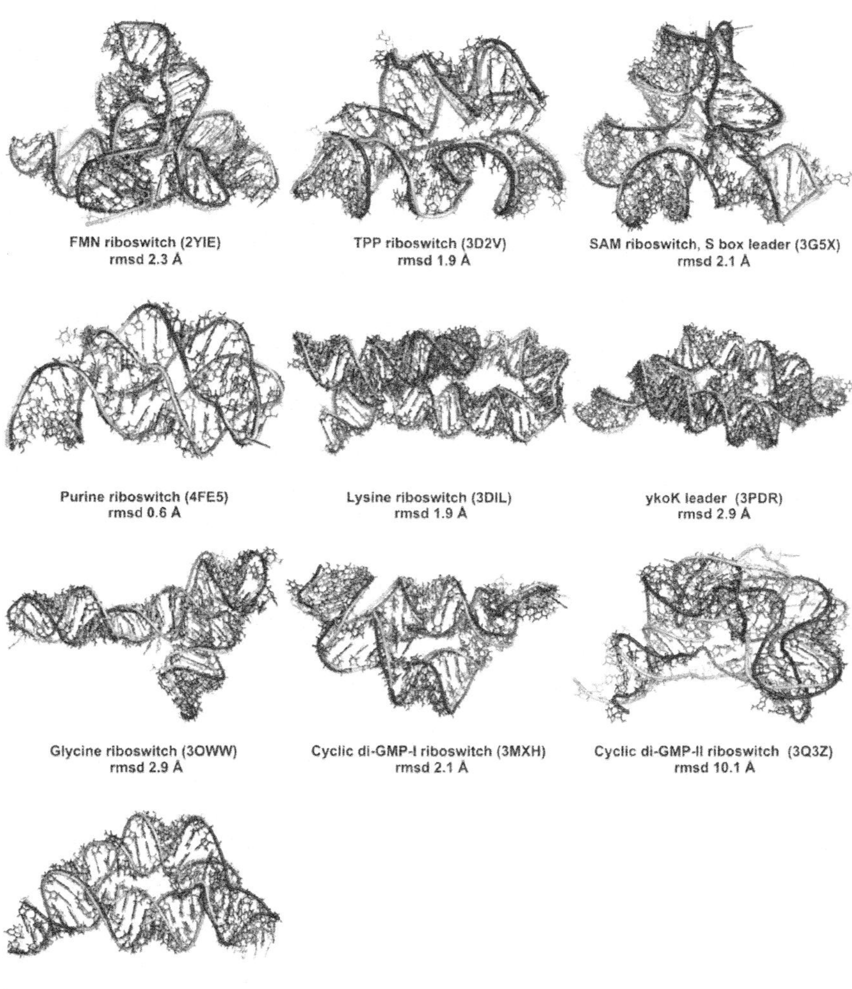

Figure 3 Superimposition of the predicted 3D models (dark blue; black in the print version) and their respective X-ray structures (cyan; light gray in the print version) for 10 selected riboswitches.

The quality of the predicted 3D models could be anticipated based on energy values calculated by RNAComposer. The energy values change linearly with the RNA strand length (Popenda et al., 2012). The models for which the calculated energy value (Table 3) was lower than the expected (Popenda et al., 2012) were in good agreement with the crystal structures.

Table 3 Energy and quality of the 3D models predicted for selected riboswitches

RNA PDB code and chain	Strand length (nt)	Energy (XPLOR) (kcal/mol)		Clash-score, all atoms		Nucleic acid geometry: Potentially incorrect: Sugar puckers		Backbone		Outlier bonds (%)		Outlier angles (%)	
		Min.	Max.	RNAComposer	X-ray	RNAComposer	X-ray	RNAComposer	X-ray	RNAComposer	X-ray	RNAComposer	X-ray
2YIE X+Z	113	−2608	−2472	7.91 (1.59)	6.26	2.70 (0.95)	1	22.00 (2.79)	12	0.00 (0.00)	0.00	0.00 (0.00)	0.00
3D2V A	77	−1941	−1852	11.87 (2.10)	5.30	1.11 (1.05)	4	16.78 (2.28)	14	0.00 (0.00)	0.00	0.00 (0.00)	0.00
3GX5 A	94	−2174	−1986	9.78 (1.43)	15.58	4.30 (1.64)	7	21.80 (2.35)	17	0.00 (0.00)	0.00	0.00 (0.01)	0.48
4FE5 B	67	−1557	−1473	12.17 (2.91)	4.66	1.00 (0.82)	1	11.70 (1.95)	3	0.00 (0.00)	0.00	0.00 (0.00)	0.00
3DIL A	174	−4628	−4462	8.87 (1.88)	0.36	1.80 (0.63)	4	26.00 (2.63)	17	0.00 (0.00)	0.00	0.00 (0.00)	0.12
3PDR X	161	−3637	−3580	10.65 (1.42)	5.07	1.50 (0.71)	2	33.30 (3.40)	18	0.00 (0.00)	0.00	0.00 (0.00)	0.32
3OWW A	88	−1985	−1859	7.28 (1.26)	1.08	1.30 (1.25)	3	15.00 (2.31)	18	0.00 (0.00)	0.00	0.00 (0.00)	0.06
3MXH R	92	−2062	−1874	11.09 (2.09)	2.72	4.50 (2.27)	1	23.50 (3.41)	10	0.00 (0.00)	0.00	0.01 (0.01)	0.09
3Q3Z V	**75**	**−1257**	**−922**	**29.26 (4.94)**	**0.43**	**3.60 (0.97)**	**5**	**26.30 (4.37)**	**10**	**0.00 (0.00)**	**0.00**	**0.08 (0.02)**	**0.29**
3SD3 A	89	−2164	−2083	8.56 (1.62)	1.76	0.20 (0.63)	0	10.80 (2.70)	1	0.00 (0.00)	0.00	0.00 (0.00)	0.00

3Q3Z riboswitch was the only example that showed higher than expected energy value, although the stereochemical quality of all models, including 3Q3Z, was high (Table 3).

4.2. Application example 1: The c-di-GMP-II riboswitch

The c-di-GMP-II riboswitch from *C. acetobutylicum* controls a gene involved in carbohydrate processing and its aptamer domain structure was solved at 2.5 Å resolution (Smith et al., 2011). This RNA assumes a compact structure comprising several structural elements that are particularly challenging for tertiary structure prediction. The structure contains not only the second-order pseudoknot, triple helix within pseudoknot major grove but also an unusual U-turn/S-turn architecture. This U-turn/S-turn motif is a unique feature not present in any other PDB-deposited structure (Smith et al., 2011). Within the predictions presented below, all structural elements derived from 3Q3Z were removed from the RNAComposer dictionary to simulate a situation in which the 3D structure of a novel RNA is predicted.

The quality of the RNAComposer-predicted 3D structure depends strongly on the input secondary structure. As the c-di-GMP-II riboswitch from *C. acetobutylicum* is a relatively small RNA (75-mer), *in silico* methods should allow prediction of the correct secondary structure. However, this is not the case. We used several web-accessible tools (Table 4) to predict secondary structure models of this riboswitch. The secondary structures obtained departed significantly from that found in the crystal structure (Table 4) as indicated by Matthews correlation coefficient (MCC) (Matthews, 1975).

The best fit was observed for the structure generated using CyloFold program (Bindewald, Kluth, & Shapiro, 2010) with MCC of 0.89 (Table 4). Secondary structures thus obtained were inputted into RNAComposer. As expected, predicted 3D structures showed little agreement with the crystal structure 3Q3Z (average rmsd of 21.17 Å). CyloFold-derived structure has rmsd of 15.85 Å despite the fact that a loop, missing in the dictionary, was generated by RNAComposer. The energy value for an RNA of this size should be lower than −1498 kcal/mol (Popenda et al., 2012) and this criterion was not fulfilled in case of the CyloFold-derived secondary structure used for 3D structure prediction (Table 4). Other structures (Table 4), with low MCC, and consequently high rmsd, showed good energy values, demonstrating that the energy criterion might not be informative if the input secondary structure topology is extensively inaccurate.

Table 4 Dependence of the 3D structure prediction of the cyclic di-GMP-II riboswitch (*C. acetobutylicum*) on the accuracy of the *in silico* predicted secondary structure

Sequence GCGCGGAAACAAUGAUGAAUGGGUUUAAAUUGGGCACUUGACUCAUUUUGAGUUAGUAGUGCAACCGACCGUGCU

Method	Secondary structure topology	MCC	Energy (XPLOR) (kcal/mol)	rmsd (Å)	Type of 3D element with lowest sequence homology
X-ray[a]	((((((..((......(((((((((..... {[[[[[[[.)))))))))...))....]]]]]}.]]..)))))).	1.00	−1257	10.1	Loop (5.26%)
CyloFold	.(((((..........(((((((((.... [[[[{{{{{.)))))))))........ .}}}}}..]]]])))))..	0.89	−1165	15.9	Loop (missing in the dictionary)
mfold, SFold	.(((((...((((..((((....)))).)))). (((((((((((.....))))))..)))))......)))))..	0.41	−1398	20.9	Junction loop (62.5%)
DotKnot	.(((((...((((........[[[[...)))). ((((((((((((...)))))))..)))))]]]]..)))))..	0.40	−1273	21.3	Loop (41.2%)
RNAfold	.(((((((((.((.....))..))))... ((((((((((((((((...)))))))..)))))..)))))))))..	0.46	−1590	21.6	Stem (64.3%)
Kinefold	.(((((((((............))))... (((((((((((((((.....))))))..)))))..)))))))))..	0.49	−1506	22.6	Loop (64.3%)
CONTRAfold	..((((....................... ((((((((((((((((...)))))))..)))))..))))))))...	0.49	−1430	28.4	Loop (missing in the dictionary)
RNAstructure, CentroidFold	.(((((....................... ((((((((((((((((...)))))))..)))))..)))))))))..	0.53	−1451	28.7	Loop (missing in the dictionary)

[a]Secondary structure topology derived from the X-ray structure (3Q3Z) using RNApdbee.

To analyze in detail how RNAComposer predicts the 3D structure of the c-di-GMP-II riboswitch from *C. acetobutylicum*, the correct secondary structure was used as an input. This was derived from the crystal structure using RNApdbee (Antczak et al., 2014) and an equivalent structure, with the exception of the U30–A62 interaction, was proposed based on the secondary structure analysis (Lee, Baker, Weinberg, Sudarsan, & Breaker, 2010). The best 3D model obtained displayed energy of −1257 kcal/mol (Table 3) and rmsd of 10.1 Å (Fig. 3). Interestingly, although one of the structural elements (loop G6-A9/U53-C69) selected by RNAComposer during assembly showed sequence homology as low as 4.8% (Table 5, first entry, exclusion of 3Q3Z from dictionary) our prediction generated a 3D model of 10.1 Å rmsd value.

This illustrates the advantage of RNAComposer over homology modeling and the importance of experimental data during input secondary structure generation. It is also important to note that the entire prediction process took about 3 min. Lack of the loop G6-A9/U53-C69 element of higher sequence homology in the RNAComposer dictionary culminated in local departure from the correct structure as visualized with RNAlyzer (Fig. 4).

Prediction of triplexes and pseudoknots as tertiary motifs annotated with the square brackets is the main limitation of RNAComposer and is currently under development. A less complex kink turn motif is well predicted; therefore, the orientation of all helices is recovered and the global structure of the c-di-GMP-II riboswitch is relatively similar to the reference structure 3Q3Z (Table 5).

As stated earlier, the quality of the predicted 3D models depends strongly on the RNAComposer dictionary content (Popenda et al., 2012). When 3D structural elements derived from 3Q3Z were returned to the dictionary (Table 5, second entry), four tertiary structure elements from the 3Q3Z were brought into play for this model building, including the G6-A9/U53-C69 loop. The 3D structure was predicted with global rmsd of 1.1 Å, indicating not only the predictive fidelity and power of the method but also the applicability of RNAComposer to model 3D structures of other members of this particular riboswitch family. Such an application case is presented below.

4.3. Application example 2: The c-di-GMP-II riboswitch relatives

As the next example we attempted to predict 3D structures of the c-di-GMP-II riboswitches, members of the family RF01786 depicted in

Table 5 3D structure prediction details for the cyclic di-GMP-II riboswitch from *C. acetobutylicum* (**3Q3Z**)

Input data

GCGCGGAAACAAUGAUGAAUGGGUUUAAAUUGGGCACUUGACUCAUUUUGAGUUAGUAGUGCAACCGACCGUGCU
((((((..((......(((((((((.....{[[[[[[[.)))))))))...))....]]]]]}.]]..)))))).

Structural element	Target secondary structure element				Source tertiary structure element					Selection criteria			
	Localization		Secondary structure		Localization			Secondary structure		Secondary structure topology	Sequence similarity (%)[a]	Pyrimidines/ purines compatibility (%)[b]	Source structure resolution (Å)
	Residues				PDB ID	Residues							
	First	Last	Sequence	Topology		First	Last	Sequence	Topology				
First entry													
Loop 1	6	9	GAAA	(..(	1VOR	959	962	UCCC	(..(	52.4	4.8	28.6	11.5
	53	69	UUAGUAGU GCAACCGAC	)....]]]]]} .]]..)		1090	1106	GGGGCUCAAGU GAUCUA	)...]....]]]].]])				
Loop 2	10	17	CAAUGAUG	(......(	3F1H	1232	1239	CGAUGAAG	(......(	Identical	76.9	84.6	3.0
	48	52	UUGAG	)...)		1259	1263	UGGAG	)...)				
Loop 3	25	40	UUAAAUUGGG CACUUG	(..... {[[[[[[[.)	4JF2	28	43	UACUUAUUUC CUUUGA	(....... [[[[[..)	81.3	25.0	37.5	2.3
Stem 1	1	6	GCGCGG	((((((	1Q2S	22	27	GCACGG	((((((	Identical	91.7	100.0	3.2
	69	74	CCGUGC	))))))		34	39	CCGUGC	))))))				
Stem 2	9	10	AC	((	2VHN	139	140	AC	((	Identical	100.0	100.0	3.7
	52	53	GU	))		640	641	GU	))				
Stem 3	17	25	GAAUGGGUU	(((((((((	3U5H	3397	3405	GGUUGCGGC	(((((((((	Identical	44.4	66.7	3.0
	40	48	GACUCAUUU	)))))))))		3508	3516	GCUGCAAUC	)))))))))				
Single strand	74	75	CU	).	3CJZ	25	26	CU	).	Identical	100.0	100.0	1.8

Second entry													
Loop 1	6	9	GAAA	(..(	3Q3Z	81	84	GAAA	(..(	Identical	100.0	100.0	2.5
	53	69	UUAGUAGUG CAACCGAC	)....]]]]]}.]]..)		128	144	UUAGUAGUG CAACCGAC	)....]]]]]}.]]..)				
Loop 2	10	17	CAAUGAUG	(......(	3Q3Z	10	17	CAAUGAUG	(......(	Identical	100.0	100.0	2.5
	48	52	UUGAG	)...)		48	52	UUGAG	)...)				
Loop 3	25	40	UUAAAUUGG GCACUUG	(..... {[[[[[[[.)	3Q3Z	25	40	UUAAAUUG GGCACUUG	(.....{[[[[[[[.)	Identical	100.0	100.0	2.5
Stem 1	1	6	GCGCGG	((((((	1Q2S	22	27	GCACGG	((((((	Identical	91.7	100.0	3.2
	69	74	CCGUGC	))))))		34	39	CCGUGC	))))))				
Stem 2	9	10	AC	((	1JZX	328	329	AC	((	Identical	100.0	100.0	3.1
	52	53	GU	))		344	345	GU	))				
Stem 3	17	25	GAAUGGGUU	(((((((((	3Q3Z	92	100	GAAUGGGUU	(((((((((	Identical	100.0	100.0	2.5
	40	48	GACUCAUUU	)))))))))		115	123	GACUCAUUU	)))))))))				
Single strand	74	75	CU	).	3CJZ	25	26	CU	).	Identical	100.0	100.0	1.8

[a]Between the target and source RNA sequence.
[b]Matching of the purine/pyrimidine residues for the target and source RNA of given sequence.

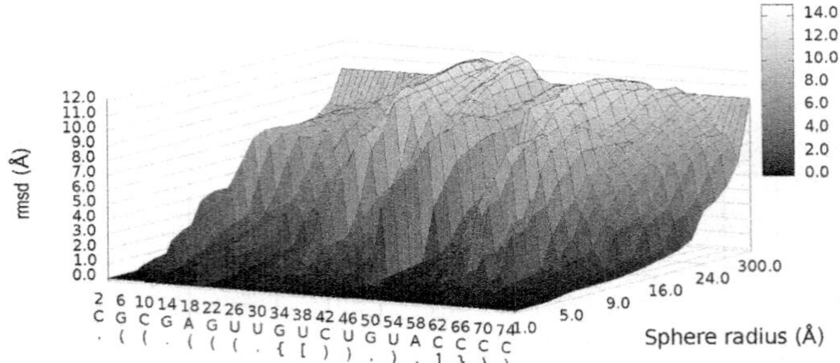

Figure 4 RNAlyzer visualization of the predicted 3D model quality for the cyclic di-GMP-II riboswitch from *C. acetobutylicum*.

the RFAM database (Burge et al., 2013). Although related to the c-di-GMP-II from *C. acetobutylicum*, these riboswitches differ in the sequence length and identity. The RF01786 family comprises 237 members, with a "seed" subgroup containing 54 entities (Table 2). Their RNA sequences are 8–15 nucleotide residues longer than c-di-GMP-II from *C. acetobutylicum*. Their consensus secondary structure, presented in the RFAM database (Stockholm notation), closely resembles that seen in the crystal structure 3Q3Z, but does not contain information about the pseudoknot. In addition, due to the length of those RNAs their graph representations differ, mostly within loops forming the pseudoknot and at the 5′- and 3′-termini. We have generated secondary structures for all RNAs present in the "seed" subgroup of the RF01786 family based on the alignment of the consensus secondary structures and respective 3Q3Z structure topology. Those structures were used for automated structure prediction using RNAComposer. One 3D model was generated for each of 54 RNA secondary structures and this ensemble was fitted into the crystal structure of the 3Q3Z using alignment of tertiary structures applying program ARTS (Dror, Nussinov, & Wolfson, 2006). Based on this, we selected three subgroup members, i.e., the riboswitches from *Clostridium difficile 4*, *Bacillus halodurans 1*, and *Thermus aquaticus Y5 1*, for which the predicted 3D structures fitted the best to the c-di-GMP-II riboswitch 3Q3Z. For each case, 10 models were generated and the best 3D structures are presented in Fig. 5.

Their alignment to 3Q3Z structure using ARTS was based on the large, consensus 3D fragment with the core size of 50–68 residues (Table 6). It appeared that within the core fragment comprising potential ligand-binding site, all three riboswitch structures very closely resemble that from *C. acetobutylicum* (rmsd less than 2 Å). Because of space limitation, 3D

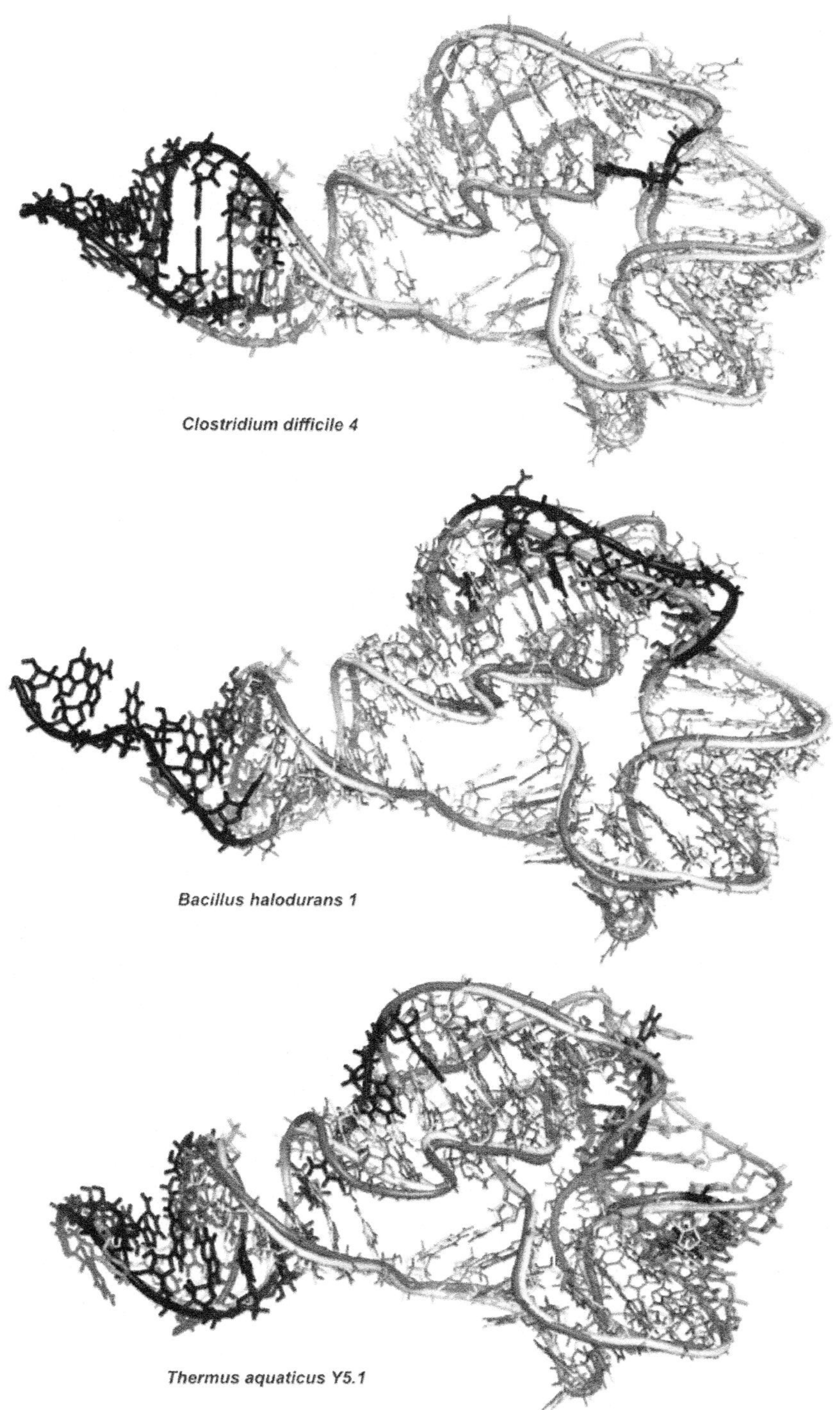

Figure 5 Pairwise superimposition of the best 3D models of the cyclic di-GMP-II riboswitch from *C. difficile 4*, *B. halodurans 1*, and *T. aquaticus Y5.1* (red and blue-core), and the X-ray structure from *C. acetobutylicum* (PDB ID 3Q3Z) (green and cyan-core). (See the color plate.)

Table 6 ARTS pairwise alignment parameters for the best 3D models of the cyclic di-GMP-II riboswitches *C. difficile 4, B. halodurans 1*, and *T. aquaticus Y5.1*, and the respective X-ray structure from *C. acetobutylicum* (3Q3Z)

	Cyclic di-GMP-II riboswitch		
ARTS parameters	***C. difficile 4***	***B. halodurans 1***	***T. aquaticus Y5.1***
Score	114	96	84
Core size (nt)	68	62	50
Number of matched base pairs	23	17	17
rmsd (Å)	1.4	1.8	1.3
Number of identical core residues	44	39	31
Structural identity	0.92	0.84	0.68
Core sequence identity	0.65	0.63	0.62
Sequential match score	66	59	46

structure prediction details are shown only for the c-di-GMP-II riboswitch from *C. difficile 4* (Table 7).

5. CONCLUSIONS AND PERSPECTIVES

We have demonstrated that our approach is not only fully automated but also characterized by very short computation time, efficiency, and user-friendly access to high-resolution 3D models. RNAComposer appeared to outperform existing automated methods (Popenda et al., 2012) and is especially suited for 3D RNA structure prediction of large RNAs, where experimentally adjusted secondary structure can be used as an input (see Huang et al., 2013; Lusvarghi et al., 2013). Since our report (Popenda et al., 2012), the accuracy of predicted 3D structures has increased considerably due to the 2.6-fold enlargement of the dictionary volume. Access to nearly 490,000 3D structure elements makes structure assembly more efficient. Both RNAComposer servers are very often visited (more than 75,000 entries worldwide for both sites). Apart from the 3D RNA predictions solely, RNAComposer method found new interesting application for deriving models coherent with experimental data from NMR (Krahenbuhl, Lukavsky, & Wider, 2014). We foresee the RNAComposer applications in fitting the 3D RNA patterns into the Cryo-EM data. Work

Table 7 3D structure prediction details for the cyclic di-GMP-II riboswitch from *C. difficile 4*

Input data

AAUAUUUUAGAAACUGAGAAGUAUAUCUUAUUAUUGGGCAUCUGGAGAUAUAUGGAGUUAGUGGUGCAACCGGCUAUGAAUAUA
.((((((((..((......(((((((((.....{[[[[[[[.).)))))))))...))....]]]]]}.]]..))).))))))).

		Target secondary structure element				Source tertiary structure element					Selection criteria			
		Localization				Localization								
	Size	Residues		Secondary structure			Residues		Secondary structure		Secondary structure	Sequence similarity	Pyrimidines/ purines compatibility	Source structure resolution
No.	(nt)	First	Last	Sequence	Topology	PDB ID	First	Last	Sequence	Topology	topology	(%)[a]	(%)[b]	(Å)
Single strands														
1	2	1	2	AA	.(	1YTU	1	2	AG	.(	Identical	50.0	100.0	2.5
Loops														
1	5	7	8	UU	((	4A1D	1351	1352	UU	((	Identical	100.0	100.0	3.5
		76	78	AUG	).)		1207	1209	AUG	).)				
2	21	10	13	GAAA	(..(	3Q3Z	6	9	GAAA	(..(	Identical	100.0	90.5	2.5
		58	74	UUAGUGGUGC AACCGGC	)....]]]]]}.]]..)		53	69	UUAGUAGUGC AACCGAC	)....]]]]]}.]]..)				
3	13	14	21	CUGAGAAG	(......(	2ZJR	1221	1228	CGGAGAAG	(......(	Identical	92.3	92.3	2.9
		53	57	UGGAG	)...)		1247	1251	UGGAG	)...)				
4	5	28	29	UU	((	2ZJR	725	716	UU	((	Identical	100.0	100.0	2.9
		44	46	GGA	).)		741	743	GGA	).)				
5	16	29	44	UAUUAUUGG GCAUCUG	(..... {[[[[[[[.)	3Q3Z	25	40	UUAAAUUGGG CACUUG	(..... {[[[[[[[.)	Identical	100.0	68.8	2.5

Continued

Table 7 3D structure prediction details for the cyclic di-GMP-II riboswitch from *C. difficile 4*—cont'd

Input data

AAUAUUUUAGAAACUGAGAAGUAUAUCUUAUUAUUGGGCAUCUGGAGAUAUAUGGAGUUAGUGGUGCAACCGGCUAUGAAUAUA
.(((((((((..((......((((((((.....{[[[[[[[.).)))))))))...))....]]]]]}.]]..))).)))))).

No.	Size (nt)	Target secondary structure element: Localization: Residues: First	Last	Secondary structure: Sequence	Topology	Source tertiary structure element: Localization: PDB ID	Residues: First	Last	Secondary structure: Sequence	Topology	Selection criteria: Secondary structure topology	Sequence similarity (%)	Pyrimidines/purines compatibility (%)	Source structure resolution (Å)
Stems														
1	12	2	7	AUAUUU	((((((	3U5B	26	31	AUAUGC	((((((	Identical	66.7	88.3	3.0
		78	83	GAAUAU	))))))		595	600	GCGUAU	))))))				
2	6	8	10	UAG	(((	1VQO	2852	2854	UAG	(((	Identical	100.0	100.0	2.2
		74	76	CUA	)))		2902	2904	CUA	)))				
3	4	13	14	AC	((	1VQO	2059	2069	AC	((	Identical	100.0	100.0	2.2
		57	58	GU	))		2074	2075	GU	))				
4	16	21	28	GUAUAUCU	((((((((	3U5B	1658	1665	GAAUGGCU	((((((((	Identical	56.3	75.0	3.0
		46	53	AGAUAUAU	))))))))		1722	1729	GGUCAUUU	))))))))				
Single strands														
1	2	83	84	UA	).	3MOJ	73	74	UA	).	Identical	100.0	100.0	2.9

[a]Between the target and source RNA sequence.
[b]Matching of the purine/pyrimidine residues for the target and source RNA of given sequence.

applying RNAComposer to predict artificial RNA 3D structures is on the way (Chworos, personal communication).

Our current perspective is to enforce the RNAComposer method with new algorithms generating missing 3D structural elements of loops and single strands more effectively. This is especially important for large *n*-way junction loops and regions involved in pseudoknots formation. More efficient energy minimization protocols will be introduced in order to enlarge applicability of the method to RNAs larger than 500 nt residues. RNAComposer servers will be equipped with new functionalities allowing user to add not only the distance restraints but also the torsion angle restraints. The user, within workspace given, will be allowed to add own 3D structural elements (e.g., obtained via molecular dynamics) to the dictionary and to force their preferred usage upon fragment assembly.

ACKNOWLEDGMENTS

This work was supported by the National Science Center, Poland [MAESTRO 2012/06/A/ST6/00384 (to R. W. A.)] and the Foundation for Polish Science [HOMING PLUS/2012-6/12 (to K. J. P.)]. Poznań Supercomputing and Networking Centre is acknowledged for hosting the RNAComposer server at http://rnacomposer.ibch.poznan.pl. R.W.A. would like to acknowledge the contribution of the COST Action CM1105.

REFERENCES

Antczak, M., Zok, T., Popenda, M., Lukasiak, P., Adamiak, R. W., Blazewicz, J., et al. (2014). RNApdbee—A webserver to derive secondary structures from pdb files of knotted and unknotted RNAs. *Nucleic Acids Research, 42*, W368–W372.

Bindewald, E., Kluth, T., & Shapiro, B. A. (2010). CyloFold: Secondary structure prediction including pseudoknots. *Nucleic Acids Research, 38*, W368–W372.

Burge, S. W., Daub, J., Eberhardt, R., Tate, J., Barquist, L., Nawrocki, E. P., et al. (2013). Rfam 11.0: 10 years of RNA families. *Nucleic Acids Research, 41*, D226–D232.

Cao, S., & Chen, S. J. (2011). Physics-based de novo prediction of RNA 3D structures. *The Journal of Physical Chemistry. B, 115*, 4216–4226.

Das, R., Karanicolas, J., & Baker, D. (2010). Atomic accuracy in predicting and designing noncanonical RNA structure. *Nature Methods, 7*, 291–294.

Deigan, K. E., Li, T. W., Mathews, D. H., & Weeks, K. M. (2009). Accurate SHAPE-directed RNA structure determination. *Proceedings of the National Academy of Sciences of the United States of America, 106*, 97–102.

Do, C. B., Woods, D. A., & Batzoglou, S. (2006). CONTRAfold: RNA secondary structure prediction without physics-based models. *Bioinformatics, 22*, e90–e98.

Dror, O., Nussinov, R., & Wolfson, H. J. (2006). The ARTS web server for aligning RNA tertiary structures. *Nucleic Acids Research, 34*, W412–W415.

Edwards, T. E., & Ferre-D'Amare, A. R. (2006). Crystal structures of the thi-box riboswitch bound to thiamine pyrophosphate analogs reveal adaptive RNA-small molecule recognition. *Structure, 14*, 1459–1468.

Flores, S. C., & Altman, R. B. (2010). Turning limited experimental information into 3D models of RNA. *RNA, 16*, 1769–1778.

Flores, S. C., Sherman, M. A., Bruns, C. M., Eastman, P., & Altman, R. B. (2011). Fast flexible modeling of RNA structure using internal coordinates. *IEEE/ACM Transactions on Computational Biology and Bioinformatics*, *8*, 1247–1257.

Gan, H. H., Pasquali, S., & Schlick, T. (2003). Exploring the repertoire of RNA secondary motifs using graph theory; implications for RNA design. *Nucleic Acids Research*, *31*, 2926–2943.

Garst, A. D., Heroux, A., Rambo, R. P., & Batey, R. T. (2008). Crystal structure of the lysine riboswitch regulatory mRNA element. *The Journal of Biological Chemistry*, *283*, 22347–22351.

Gesteland, R. F., Cech, T. R., & Atkins, J. F. (2005). *The RNA world* (3rd ed.). New York: Cold Spring Harbor Press.

Guntert, P., Mumenthaler, C., & Wuthrich, K. (1997). Torsion angle dynamics for NMR structure calculation with the new program DYANA. *Journal of Molecular Biology*, *273*, 283–298.

Huang, Q., Purzycka, K. J., Lusvarghi, S., Li, D., Legrice, S. F., & Boeke, J. D. (2013). Retrotransposon Ty1 RNA contains a 5′-terminal long-range pseudoknot required for efficient reverse transcription. *RNA*, *19*, 320–332.

Jonikas, M. A., Radmer, R. J., & Altman, R. B. (2009). Knowledge-based instantiation of full atomic detail into coarse-grain RNA 3D structural models. *Bioinformatics*, *25*, 3259–3266.

Jonikas, M. A., Radmer, R. J., Laederach, A., Das, R., Pearlman, S., Herschlag, D., et al. (2009). Coarse-grained modeling of large RNA molecules with knowledge-based potentials and structural filters. *RNA*, *15*, 189–199.

Jossinet, F., Ludwig, T. E., & Westhof, E. (2010). Assemble: An interactive graphical tool to analyze and build RNA architectures at the 2D and 3D levels. *Bioinformatics*, *26*, 2057–2059.

Krahenbuhl, B., Lukavsky, P., & Wider, G. (2014). Strategy for automated NMR resonance assignment of RNA: Application to 48-nucleotide K10. *Journal of Biomolecular NMR*, *59*, 231–240.

Lee, E. R., Baker, J. L., Weinberg, Z., Sudarsan, N., & Breaker, R. R. (2010). An allosteric self-splicing ribozyme triggered by a bacterial second messenger. *Science*, *329*, 845–848.

Leontis, N., & Westhof, E. (2012). RNA 3D structure analysis and prediction. In N. Leontis, & E. Westhof (Eds.), *Nucleic acids and molecular biology series*. *Vol. 27*. Berlin and Heidelberg: Springer-Verlag.

Lukasiak, P., Antczak, M., Ratajczak, T., Bujnicki, J. M., Szachniuk, M., Adamiak, R. W., et al. (2013). RNAlyzer—Novel approach for quality analysis of RNA structural models. *Nucleic Acids Research*, *41*, 5978–5990.

Lusvarghi, S., Sztuba-Solinska, J., Purzycka, K. J., Pauly, G. T., Rausch, J. W., & Grice, S. F. (2013). The HIV-2 Rev-response element: Determining secondary structure and defining folding intermediates. *Nucleic Acids Research*, *41*, 6637–6649.

Martinez, H. M., Maizel, J. V., Jr., & Shapiro, B. A. (2008). RNA2D3D: A program for generating, viewing, and comparing 3-dimensional models of RNA. *Journal of Biomolecular Structure & Dynamics*, *25*, 669–683.

Mathews, D. H., Disney, M. D., Childs, J. L., Schroeder, S. J., Zuker, M., & Turner, D. H. (2004). Incorporating chemical modification constraints into a dynamic programming algorithm for prediction of RNA secondary structure. *Proceedings of the National Academy of Sciences of the United States of America*, *101*, 7287–7292.

Matthews, B. W. (1975). Comparison of the predicted and observed secondary structure of T4 phage lysozyme. *Biochimica et Biophysica Acta*, *405*, 442–451.

Merino, E. J., Wilkinson, K. A., Coughlan, J. L., & Weeks, K. M. (2005). RNA structure analysis at single nucleotide resolution by selective 2′-hydroxyl acylation and primer extension (SHAPE). *Journal of the American Chemical Society*, *127*, 4223–4231.

Pachulska-Wieczorek, K., Purzycka, K. J., & Adamiak, R. W. (2006). New, extended hairpin form of the TAR-2 RNA domain points to the structural polymorphism at the 5′ end of the HIV-2 leader RNA. *Nucleic Acids Research, 34*, 2984–2997.

Pang, P. S., Elazar, M., Pham, E. A., & Glenn, J. S. (2011). Simplified RNA secondary structure mapping by automation of SHAPE data analysis. *Nucleic Acids Research, 39*, e151.

Parisien, M., Cruz, J. A., Westhof, E., & Major, F. (2009). New metrics for comparing and assessing discrepancies between RNA 3D structures and models. *RNA, 15*, 1875–1885.

Parisien, M., & Major, F. (2008). The MC-Fold and MC-Sym pipeline infers RNA structure from sequence data. *Nature, 452*, 51–55.

Peselis, A., & Serganov, A. (2014). Themes and variations in riboswitch structure and function. *Biochimica et Biophysica Acta, 1839*, 908–918.

Popenda, M., Bielecki, L., & Adamiak, R. W. (2006). High-throughput method for the prediction of low-resolution, three-dimensional RNA structures. *Nucleic Acids Symposium Series (Oxf), 50*, 67–68.

Popenda, M., Blazewicz, M., Szachniuk, M., & Adamiak, R. W. (2008). RNA FRABASE version 1.0: An engine with a database to search for the three-dimensional fragments within RNA structures. *Nucleic Acids Research, 36*, D386–D391.

Popenda, M., Szachniuk, M., Antczak, M., Purzycka, K. J., Lukasiak, P., Bartol, N., et al. (2012). Automated 3D structure composition for large RNAs. *Nucleic Acids Research, 40*, e112.

Popenda, M., Szachniuk, M., Blazewicz, M., Wasik, S., Burke, E. K., Blazewicz, J., et al. (2010). RNA FRABASE 2.0: An advanced web-accessible database with the capacity to search the three-dimensional fragments within RNA structures. *BMC Bioinformatics, 11*, 231.

Purzycka, K. J., Pachulska-Wieczorek, K., & Adamiak, R. W. (2011). The in vitro loose dimer structure and rearrangements of the HIV-2 leader RNA. *Nucleic Acids Research, 39*, 7234–7248.

Puton, T., Kozlowski, L. P., Rother, K. M., & Bujnicki, J. M. (2013). CompaRNA: A server for continuous benchmarking of automated methods for RNA secondary structure prediction. *Nucleic Acids Research, 41*, 4307–4323.

Ren, A., Rajashankar, K. R., & Patel, D. J. (2012). Fluoride ion encapsulation by Mg2+ ions and phosphates in a fluoride riboswitch. *Nature, 486*, 85–89.

Reuter, J. S., & Mathews, D. H. (2010). RNAstructure: Software for RNA secondary structure prediction and analysis. *BMC Bioinformatics, 11*, 129.

Rose, P. W., Beran, B., Bi, C., Bluhm, W. F., Dimitropoulos, D., Goodsell, D. S., et al. (2011). The RCSB Protein Data Bank: Redesigned web site and web services. *Nucleic Acids Research, 39*, D392–D401.

Rother, M., Rother, K., Puton, T., & Bujnicki, J. M. (2011). ModeRNA: A tool for comparative modeling of RNA 3D structure. *Nucleic Acids Research, 39*, 4007–4022.

Schuster, P., Fontana, W., Stadler, P. F., & Hofacker, I. L. (1994). From sequences to shapes and back: A case study in RNA secondary structures. *Proceedings of the Biological Sciences, 255*, 279–284.

Schwieters, C. D., Kuszewski, J. J., Tjandra, N., & Clore, G. M. (2003). The Xplor-NIH NMR molecular structure determination package. *Journal of Magnetic Resonance, 160*, 65–73.

Seetin, M. G., & Mathews, D. H. (2011). Automated RNA tertiary structure prediction from secondary structure and low-resolution restraints. *Journal of Computational Chemistry, 32*, 2232–2244.

Serganov, A., & Nudler, E. (2013). A decade of riboswitches. *Cell, 152*, 17–24.

Sharma, S., Ding, F., & Dokholyan, N. V. (2008). iFoldRNA: Three-dimensional RNA structure prediction and folding. *Bioinformatics, 24*, 1951–1952.

Smith, K. D., Shanahan, C. A., Moore, E. L., Simon, A. C., & Strobel, S. A. (2011). Structural basis of differential ligand recognition by two classes of bis-(3′-5′)-cyclic dimeric guanosine monophosphate-binding riboswitches. *Proceedings of the National Academy of Sciences of the United States of America, 108,* 7757–7762.

Watts, J. M., Dang, K. K., Gorelick, R. J., Leonard, C. W., Bess, J. W., Jr., Swanstrom, R., et al. (2009). Architecture and secondary structure of an entire HIV-1 RNA genome. *Nature, 460,* 711–716.

Wilkinson, K. A., Gorelick, R. J., Vasa, S. M., Guex, N., Rein, A., Mathews, D. H., et al. (2008). High-throughput SHAPE analysis reveals structures in HIV-1 genomic RNA strongly conserved across distinct biological states. *PLoS Biology, 6,* e96.

Xu, Z., Almudevar, A., & Mathews, D. H. (2012). Statistical evaluation of improvement in RNA secondary structure prediction. *Nucleic Acids Research, 40,* e26.

Xu, X. J., Zhao, P. N., & Chen, S. J. (2014). Vfold: A web server for RNA structure and folding thermodynamics prediction. *PLoS One, 9,* e107504. http://dx.doi.org/10.1371/journal.pone.0107504. eCollection 2014.

CHAPTER TWO

Modeling Complex RNA Tertiary Folds with Rosetta

Clarence Yu Cheng*, Fang-Chieh Chou*, Rhiju Das*,†,1
*Department of Biochemistry, Stanford University, Stanford, California, USA
†Department of Physics, Stanford University, Stanford, California, USA
[1]Corresponding author: e-mail address: rhiju@stanford.edu

Contents

Abstract

Reliable modeling of RNA tertiary structures is key to both understanding these structures' roles in complex biological machines and to eventually facilitating their design for molecular computing and robotics. In recent years, a concerted effort to improve computational prediction of RNA structure through the RNA-Puzzles blind prediction trials has accelerated advances in the field. Among other approaches, the versatile and expanding Rosetta molecular modeling software now permits modeling of RNAs in the 100–300 nucleotide size range at consistent subhelical (~1 nm) resolution. Our laboratory's current state-of-the-art methods for RNAs in this size range involve Fragment Assembly of RNA with Full-Atom Refinement (FARFAR), which optimizes RNA conformations in the context of a physically realistic energy function, as well as hybrid techniques that leverage experimental data to inform computational modeling. In this chapter, we give a practical guide to our current workflow for modeling RNA three-dimensional structures using FARFAR, including strategies for using data from multidimensional chemical mapping experiments to focus sampling and select accurate conformations.

Methods in Enzymology, Volume 553
ISSN 0076-6879
http://dx.doi.org/10.1016/bs.mie.2014.10.051

1. INTRODUCTION

Computational modeling of RNA structures is advancing rapidly, with recent developments improving prediction and design of both secondary and tertiary structures of RNA. Continuing improvements to secondary structure prediction algorithms (Tinoco et al., 1973), classification of RNA structural motifs (Petrov, Zirbel, & Leontis, 2013), molecular dynamics and quantum mechanical techniques (Ditzler, Otyepka, Sponer, & Walter, 2010), conformational sampling with energy scoring (Das, Karanicolas, & Baker, 2010), atomic-scale loop and motif modeling (Sripakdeevong, Kladwang, & Das, 2011), integration with conventional crystallographic (Chou, Sripakdeevong, Dibrov, Hermann, & Das, 2013) and NMR approaches (Sripakdeevong et al., 2014), and connections with recent single-molecule (Chou, Lipfert, & Das, 2014) and internet-scale videogame (Lee et al., 2014) technologies hold promise for eventually attaining confident 3D modeling and design of RNAs with high spatial resolution. An important driver of recent innovation has been the establishment of blind prediction trials, proposed during a community-wide collation of 3D RNA modeling methods in 2010 (Sripakdeevong, Beauchamp, & Das, 2012) and begun soon thereafter. The RNA-Puzzles trials (Cruz et al., 2012), modeled after the 20-year-old CASP trials in protein structure prediction, challenge participating groups to create accurate 3D models of RNAs from sequence alone; the submitted models are compared to unreleased crystallographic structures of the targets to assess the methods' predictive power. These trials provide a rigorous testing ground for current computational as well as hybrid experimental/computational structure prediction methods on RNA domains that are of strong biological interest.

This chapter describes methods from our laboratory of medium computational and experimental expense that achieve subhelix-resolution accuracy for 3D models of 100- to 300-nucleotide RNAs, a typical size range for many riboswitch and ribozyme domains and representative of RNA-Puzzles target sizes. Subhelical resolution, while not the ultimate achievable, has still been useful in guiding mutational experiments *in vitro* and *in vivo*, detecting partial structure in riboswitches without their ligands, and in revealing or illustrating evolutionary connections that are not obvious from sequence comparisons alone. The primary tools for this approach are constraints from chemical mapping experiments, which we discuss briefly here and will be described in more detail elsewhere, and computational modeling to integrate chemical mapping data into 3D portraits.

Our laboratory is developing several tools that seek to advance 3D macromolecule modeling at multiple length scales. For small RNA motifs, we leverage algorithms based on a "stepwise ansatz," which enable modeling of RNA loops and motifs with near-atomic accuracy (better than 2 Å RMSD), particularly if limited NMR or crystallographic data are available (Chou et al., 2013; Sripakdeevong et al., 2014, 2011). Unfortunately, the computational expense of those high-resolution tools is currently prohibitive for *de novo* modeling of large RNAs. Instead, our practical tools for large RNAs have largely been built on Fragment Assembly of RNA with Full-Atom Refinement (FARFAR) in the Rosetta framework, which was first introduced to model small motifs of RNAs in 2007 (Das & Baker, 2007) and was initially based on Rosetta protein structure prediction methods that we had helped in advance. Since that time, FARFAR has been progressively developed to allow for nucleotide-resolution building of not just individual RNA motifs but also more complex RNA folds involving dozens of helices. This chapter is intended to offer a practical guide to getting started with Rosetta using an up-to-date workflow from our laboratory, laid out in Fig. 1. We will illustrate this workflow below using the ligand-binding region of a tandem glycine-binding riboswitch from *F. nucleatum*, which forms a complex pseudosymmetric fold stabilized by A-minor interactions between two glycine-binding subdomains. A homolog of this domain was posed as an RNA-Puzzles challenge (Cruz et al., 2012), and crystallographic and biochemical work on this system by several RNA laboratories (Butler, Xiong, Wang, & Strobel, 2011; Cordero, Kladwang, VanLang, & Das, 2012; Erion & Strobel, 2011; Kladwang, VanLang, Cordero, & Das, 2011) have made this RNA a useful model system for calibrating and illustrating experimental and computational methodologies.

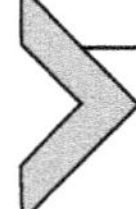

2. SETTING THE STAGE FOR 3D MODELING USING EXPERIMENTAL DATA

Several pieces of information can provide powerful constraints to help construct accurate 3D models of RNA. The most fundamental of these is the RNA's secondary structure. If phylogenetic inference of secondary structure is precluded by the lack of sequence homologs, difficulties in sequence alignment, or targeting of "alternative" states of the RNA (e.g., without ligands or in misfolded conformations), chemical mapping techniques provide useful guides to computational secondary structure prediction (Cordero, Kladwang, VanLang, & Das, 2014; Hajdin et al., 2013; Kladwang et al., 2011). In traditional "one-dimensional" (1D) chemical mapping experiments, solution-state

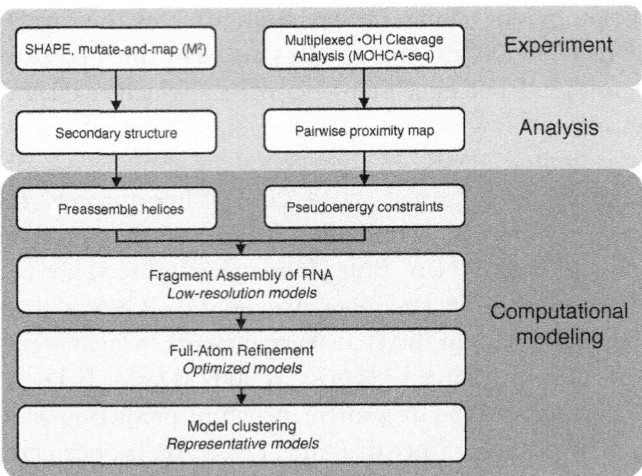

Figure 1 Workflow for modeling RNA structures in the Rosetta framework guided by experimental data. One-dimensional chemical mapping and mutate-and-map methods guide confident secondary structure prediction. To save computational expense during global modeling, secondary structure elements are separately preassembled. These ensembles of preassembled helices, along with experimental proximity mapping data from MOHCA-seq, are the inputs to global modeling by Fragment Assembly of RNA (FARNA), which generates low-resolution models. A fraction of the low-resolution models with the lowest Rosetta energy scores are then minimized using the Rosetta all-atom energy function (FARNA with Full-Atom Refinement, FARFAR) to resolve chainbreaks and unreasonable local geometries that can arise from fragment insertion. Finally, the minimized models are clustered using an RMSD threshold to collect 0.5% of the total low-resolution models in the largest cluster; this step identifies representative conformations sampled by the algorithm.

RNAs are exposed to chemical modifiers which form adducts to the backbone or nucleobases depending on backbone flexibility or base-pairing status (Fig. 2A). These modifications are traditionally detected by reverse transcription, which stops at the modified location, followed by gel or capillary electrophoresis or, more recently, deep sequencing to identify the sequence position of each modification. The reactivity of each nucleotide position to the chemical modifier can be quantified using several publically available software suites, with HiTRACE (Kim, Cordero, Das, & Yoon, 2013; Yoon et al., 2011) (https://github.com/hitrace/hitrace) and MAPseeker (Seetin et al., 2014) (https://github.com/DasLab/map_seeker) particularly optimized for high-throughput analysis of capillary electrophoresis and deep-sequencing data, respectively. Secondary structure prediction servers such as RNAstructure (Reuter & Mathews, 2010) (http://rna.urmc.rochester.edu/

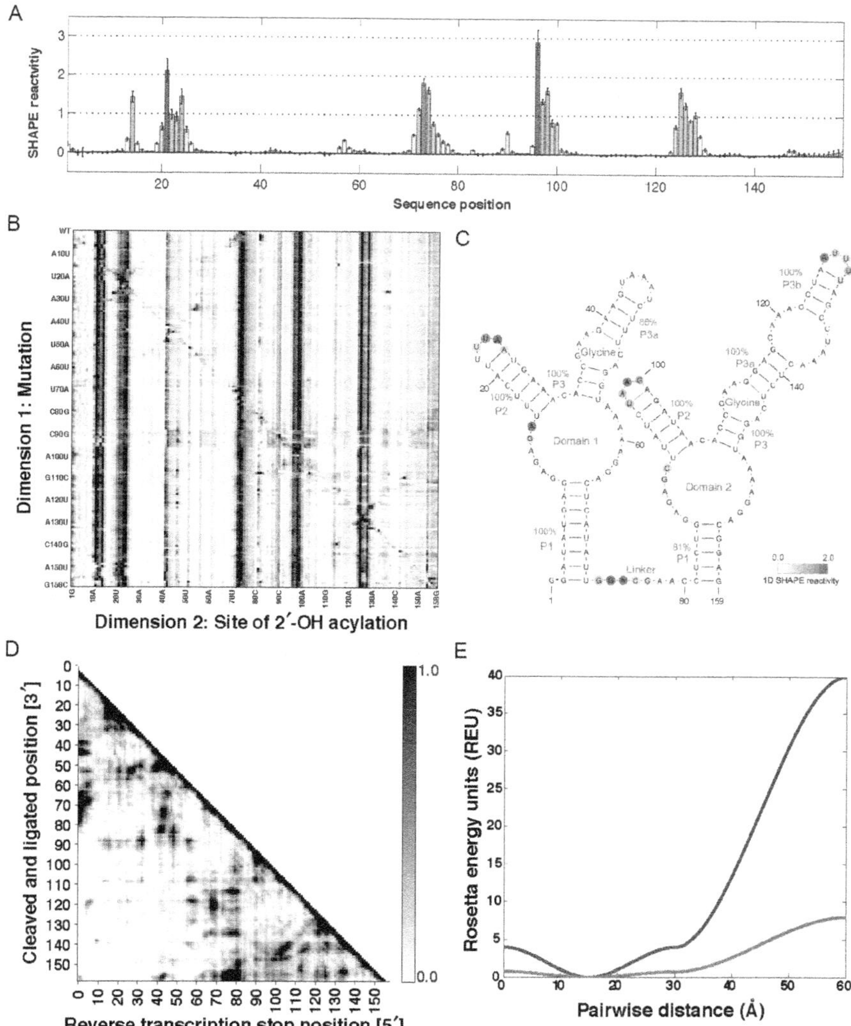

Figure 2 Rapidly acquired chemical mapping data for modeling a complex RNA fold. (A) One-dimensional SHAPE chemical mapping data for the *F. nucleatum* glycine riboswitch double ligand-binding domain in the presence of 10 m*M* glycine. Reactivities are normalized to reference hairpins (not shown) (Kladwang et al., 2014). Data are available at the RNA Mapping Database (RMDB, http://rmdb.stanford.edu) under accession code GLYCFN_1M7_0005. (B) Mutate-and-map (M^2) chemical mapping data for the glycine riboswitch in the presence of 10 m*M* glycine. Data are available at the RMDB under accession code GLYCFN_SHP_0002. (C) M^2-derived secondary structure model of the glycine riboswitch in the presence of 10 m*M* glycine, from Kladwang et al. (2011). Blue lines indicate Watson–Crick base pairs predicted in the model but not present in the crystallographic secondary structure. Red percentage values for each helix indicate confidence estimates from bootstrapping two-dimensional SHAPE chemical *(Continued)*

RNAstructureWeb) and the RNA mapping database structure server (Cordero, Lucks, & Das, 2012) (http://rmdb.stanford.edu/structureserver) can accept reactivities from chemical mapping experiments, providing additional scoring terms to guide the predictions. Nonparametric bootstrapping (Kladwang et al., 2011) can provide confidence estimates for these models.

While 1D chemical mapping experiments can provide reactivity values for every nucleotide in an RNA, the data do not directly reveal which nucleotides are base paired with which other nucleotides in the sequence, which generally limits the accuracy of the resulting models. Higher confidence secondary structures can be derived from multidimensional expansions of conventional chemical mapping. For example, the "mutate-and-map" (M^2) approach (Cordero et al., 2014) (Fig. 2B) involves systematic mutagenesis of every residue in the RNA; the suite of mutated RNAs are chemically mapped in parallel. The mutations disrupt individual Watson–Crick and noncanonical base pairs, causing the base-pairing partners of the mutated residues to increase in reactivity to the chemical modifier. Thus, M^2 can identify the base-pairing interactions throughout RNAs, which provide powerful restraints for secondary structure prediction and, in some cases, can reveal base interaction-mediated tertiary contacts (Kladwang, Chou, & Das, 2012). For the glycine riboswitch domain, M^2 was able to automatically and blindly predict the secondary structure of the domain, recovering all helices correctly and with confidence, as assessed by bootstrapping. In all cases tested to date, including blind RNA-Puzzles test cases, M^2 models achieve such accuracy; all residual errors involve helix edge base pairs (Fig. 2C). High-throughput mutation-rescue experiments read out by chemical mapping now offer the prospect of testing secondary structures at base pair resolution, and we

Figure 2—Cont'd mapping data. Nucleotides are colored according to SHAPE reactivity. (D) MOHCA-seq proximity map of the glycine riboswitch in the presence of 10 m*M* glycine, from Cheng et al. (2014). The *y*-axis represents positions that were cleaved by hydroxyl radicals, while the *x*-axis represents the locations of the radical sources from which the radicals originated. Pairwise positions are colored according to two-point correlation calculated by MAPseeker analysis (Seetin, Kladwang, Bida, & Das, 2014). Data are available at the RMDB under accession code GLYCFN_MCA_0000. (E) Pseudoenergy potential applied during modeling in Rosetta to constrain pairs of residues indicated to be in proximity by MOHCA-seq experimental data. Residue pairs showing strong MOHCA-seq signal are constrained with the blue potential and those with weaker signal are constrained with the red potential (1/5 of the blue potential). (See the color plate.)

recommend compensatory rescue tests for problems that require particularly high confidence (Tian, Cordero, Kladwang, & Das, 2014).

Another form of information that can be critical for selecting an RNA's correct 3D fold involves pairwise proximities, which reflect the topology of the tertiary structure. An experimental pipeline, Multiplexed hydroxyl radical (· OH) Cleavage Analysis by paired-end sequencing (MOHCA-seq), has been developed that can collect such pairwise proximity information, independent of traditional 3D structure determination techniques such as X-ray crystallography, cryo-EM, and NMR. In MOHCA-seq, sources of hydroxyl radicals are randomly incorporated into the RNA backbone during transcription (Cheng et al., 2014; Das et al., 2008). Activation of the sources produces localized hydroxyl radicals that diffuse outward, causing strand breaks at positions that are far away in sequence from the radical source but are brought into proximity by the 3D fold. In order to identify the locations of cleavage events and the radical sources that caused them, a DNA tail is ligated to the 3′-end of the fragmented RNAs, and reverse transcription primed on this tail stops at the radical source location. Sequencing of these complementary DNA fragments and analysis using the MAPseeker software (Seetin et al., 2014) produces pairwise proximity maps of the RNA's tertiary structure (Fig. 2D). MOHCA-seq data can be incorporated into 3D modeling via pseudoenergy terms (Cheng et al., 2014; Das et al., 2008) (Fig. 2E), as is described in further detail below.

3. MAKING MODELS OF RNA TERTIARY FOLDS

Our overall modeling pipeline still requires some manual setup of steps and has not been fully automated, mainly because it is under rapid development but also because particular steps depend on the computer cluster on which the code is tested or executed (see later). Nevertheless, it is currently fully functional without expert inspection. The following is a procedure optimized to make use of constraints from chemical mapping experiments.

3.1. Installing software and accessing computation resources

The principal framework for RNA computational modeling using our workflow is Rosetta, a collaboratively developed software suite for structure prediction and engineering of a wide range of macromolecules (https://www.rosettacommons.org/) (Leaver-Fay et al., 2011). Documentation for Rosetta can be found online (https://www.rosettacommons.org/docs/

latest/) and the modular design of the software has been described in detail (Leaver-Fay et al., 2011). Noncommercial users can install Rosetta by requesting a free license from RosettaCommons Web site, and then downloading and installing the software from the same site. Users can select which build of Rosetta to compile; we recommend that Mac users compile the `build_mac_graphics` version, which provides real-time visualization of conformational sampling and Linux users to compile the `build_release` version. General installation instructions are provided in `Rosetta/main/source/cmake/README` (see also: https://www.rosettacommons.org/docs/latest/Build-Documentation.html). Rosetta is consistently updated with weekly build releases, and the command lines referenced later in the text and given in the Appendix have been tested using a recent weekly build (weekly_releases/2014_35_57232). Beyond the core Rosetta installation, we are also developing an additional set of tools for RNA modeling, which are required for the workflow described in this chapter. The RNA tools collection is located in `Rosetta/tools/rna_tools/bin`, and documentation for setting up RNA tools is available on RosettaCommons (https://www.rosettacommons.org/docs/latest/RNA-tools.html).

The PyMOL open-source molecular visualization tool is helpful for inspecting and evaluating structural models (http://www.PyMOL.org/) (Schrodinger, 2010). Free educational subscriptions to PyMOL are available at the Web site; there is a fee for other users. Our laboratory's tools for easy visualization of RNA models in PyMOL are freely available on GitHub (https://github.com/DasLab/PyMOL_daslab). These scripts include commands to render RNAs with various levels of molecular detail, as well as to superimpose models and to color models by chemical mapping reactivities.

Most of the modeling protocols in Rosetta cannot be completed on single laptops but can be easily run on UNIX computer clusters. Sufficient computing power can be obtained from some freely available resources. For example, the Extreme Science and Engineering Discovery Environment (XSEDE, https://www.xsede.org/home) provides free startup allocations for high-performance computation. At the time of writing, 20,000 CPU hours can be acquired by research laboratories within a short time of submitting an allocation request, and this amount is more than enough to carry out several calculations. We typically carry out trial runs on local Macintosh machines and then transfer files to XSEDE or other resources for parts of the calculation that require large-scale runs.

We note that modeling of submotifs (up to 30 nucleotides) of a large RNA can also be carried out freely through the Rosetta Online Server that Includes Everyone (ROSIE, http://rosie.rosettacommons.org) (Lyskov et al., 2013), and, if desired, these submodels can be integrated into larger models (see Section 3.6). Runs on ROSIE may be useful to groups who wish to explore these tools before compiling and executing Rosetta RNA modeling on their own resources or on XSEDE.

3.2. Preassembling helices

An important principle in efficient macromolecular modeling is to not expend computation on regions of already known structure. For RNA, most helices form canonical A-form conformations. Therefore, to reduce computational expense, we preassemble the helices from high-confidence secondary structures that were predicted using chemical mapping (e.g., M^2) data.

First, we make a directory in which modeling of the target RNA will be performed. In this directory, we create a FASTA-formatted file with the name and sequence of the target RNA and a file with the secondary structure of the RNA in dot–parenthesis notation. Pseudoknots may be expressed in square brackets instead of parentheses. For example, FASTA files, secondary structure files, and UNIX command lines can be found in the Appendix and will be referenced in the text. Examples of initial FASTA and secondary structure files are given as files [F1] and [F2] in the Appendix, respectively.

To generate files containing the command lines for *de novo* RNA helix modeling in Rosetta, we run the `helix_preassemble_setup.py` script with the secondary structure and FASTA files as inputs (Appendix, command line [1]). The `helix_preassemble_setup.py` script will generate parameter and FASTA files for each helix detected in the input secondary structure, as well as a .RUN file that contains the command line for `rna_denovo`, the program that performs *de novo* RNA modeling in Rosetta. The files will be named according to order of helices in the secondary structure (e.g., `helix0.params`, `helix0.fasta`, `helix0.RUN`, `helix1.params`). The content of a `helix0.RUN` file should resemble command line [2] in the Appendix. This .RUN file can be run on a local machine in 10–20 min using `source helix0.RUN` (Appendix, command line [3]) and generates 100 FARFAR models for each helical region. The resulting models are output in compressed format (called "silent files" in Rosetta, for historical reasons) with names like `helix0.out`, etc. These files will be used as inputs for global modeling of the entire RNA. The helix models can be visualized, if desired, using the `extract_lowscore_decoys.py` script (see also below). The preassembled helices are generally nearly identical except for small variations near the ends (Fig. 3). Sampling the helices in the target RNA from these models instead of from the database of RNA fragments used for global sampling allows a greater portion of the computational effort to be spent on non-helical regions.

3.3. Defining the global fold using fragment assembly of RNA

With experimental constraints and preassembled helices in hand, the global fold of the target RNA can be tackled. At this stage, we create a set of low-resolution models using Fragment Assembly of RNA (FARNA) (Das & Baker, 2007). In FARNA, models are assembled using small RNA fragments sampled from a crystallographic database using a Monte Carlo algorithm. This heuristic allows the models to take on RNA-like conformations

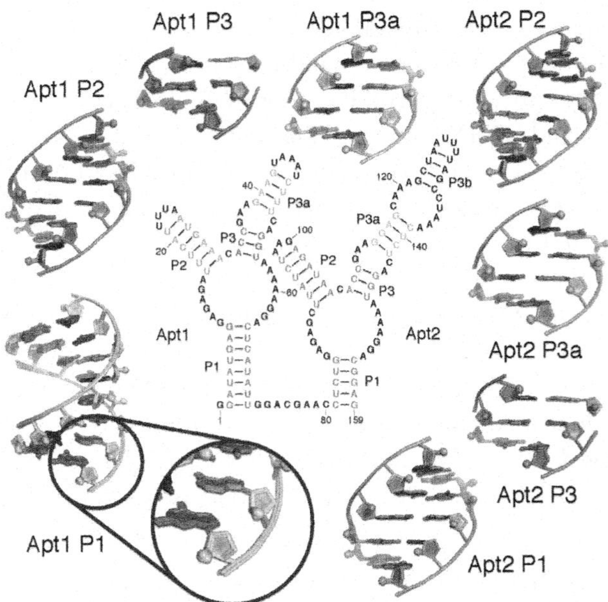

Figure 3 Preassembled helices for *F. nucleatum* double glycine riboswitch ligand-binding domain. The secondary structure is shown at center with the residues used for helix preassembly highlighted in color. Ensembles of 10 models of each helix generated by the helix preassembly protocol in Rosetta are shown at the periphery, labeled with the aptamer and helix number (e.g., Apt1 P1 for the P1 helix of aptamer 1). The magnified view of the Apt1 P1 helix highlights the slight differences in conformation between the preassembled helix models. (See the color plate.)

because the fragments are drawn from RNAs of known structure. This low-resolution modeling step does not include any refinement at the atomic level, because the all-atom energy landscape is too "rugged"; that is, it contains many energy minima that can trap the nascent model from exploring alternative conformations, and strategies for searching this landscape (Sripakdeevong et al., 2011) are currently too computationally expensive for RNA domains above 10–20 nucleotides.

For the following steps, if a comparison to a crystallographic or other reference model is desired, inputting the reference during the modeling runs will allow root mean square deviation (RMSD) values to be reported in the output silent files. To properly calculate RMSDs, reference models must have the same sequence as the construct being modeled. The `make_rna_rosetta_ready.py` command reformats PDB files with the correct sequence to be used as reference models (Appendix, command line [4]). For the glycine riboswitch example described in this chapter, the crystallographic structure includes a protein-binding loop that is not present in the construct used for experiments and modeling. To prepare the crystallographic structure for use as a reference model, we replace the protein-binding loop with a UUUA tetraloop to match the target sequence (Appendix, command lines [5] through [14]). These commands can also be used for more extensive remodeling of models and are described in detail in Section 3.6. We note that including a reference model is not required for the modeling workflow but can allow for easy visualization of modeling results through energy versus RMSD plots, such as those shown in Fig. 4.

As with the helix assembly runs above, a series of text files will record the command lines used for setup and modeling. To set up a FARNA run, we create a file called `README_SETUP`, which calls a script called `rna_denovo_setup.py` to generate the command line for low-resolution modeling. Command line [15] in the Appendix shows an example `README_SETUP` file. Special tags can be used to specify advanced options for the modeling run, including specific noncanonical base pairs (Appendix), segments of the RNA that are thought to form a tertiary contact, or soft constraints from MOHCA-seq experiments. For example, to incorporate the MOHCA-seq data into computational modeling in Rosetta, a smooth pseudoenergy potential is applied between pairs of nucleotides showing strong MOHCA-seq signal, which indicates that they are proximal in the 3D fold. Two separate pseudoenergy potential functions are used, one for strong and one for weak MOHCA-seq hits (Fig. 2E); these potentials differ

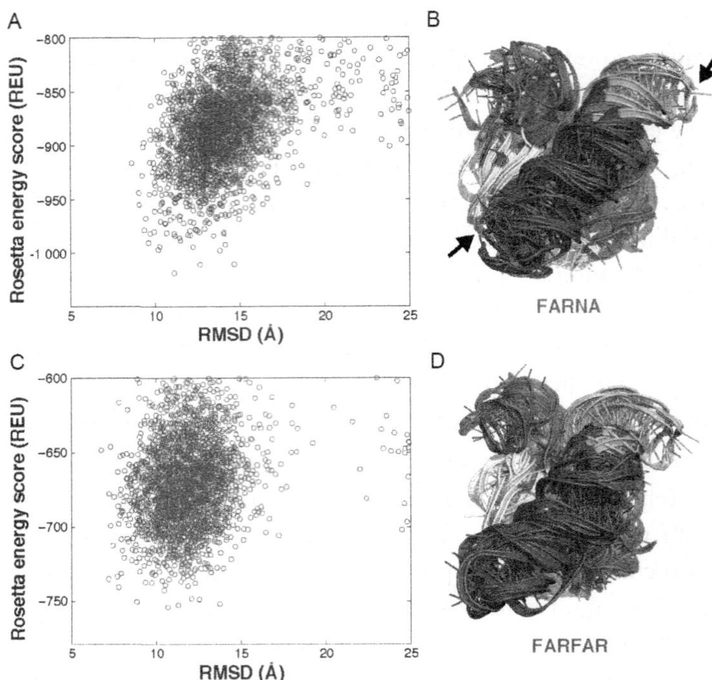

Figure 4 Low-resolution modeling and full-atom refinement using FARNA and FARFAR. (A) Rosetta energy score versus RMSD plot after low-resolution modeling using FARNA. (B) Overlaid 10 lowest-energy models after low-resolution modeling using FARNA. Chain breaks are visible in many models (arrows), and residues commonly adopt unrealistic geometries. (C) Rosetta energy score versus RMSD plot after minimization using the FARFAR algorithm. (D) Overlaid 10 lowest-energy models after minimization using the FARFAR algorithm. The models do not show any chain breaks, and poor residue geometries are greatly reduced. (See the color plate.)

only in the amplitude of the energy penalty applied for residues that are too close or too far apart. These potentials are specified in text-formatted files in Rosetta's "constraint file" format (example in Appendix, file [F3]) and can be input to `rna_denovo_setup.py`. The command `source README_SETUP` (Appendix, command line [16]) generates a file containing a command line for `rna_denovo` with the tags given in `README_SETUP`, called `README_FARFAR` (Appendix, command line [17]), as well as parameter and FASTA files.

It is a good idea to test the run locally before submitting it as a job to a cluster, in case the run is stopped by an error. To test the run, we use `source README_FARFAR` (Appendix, command line [18]) to begin a single job on a local computer and wait until sampling begins successfully (command line output similar to "`Picked Fragment Library for sequence u and sec. struct`

H... found 2308 potential fragments") before canceling the run. Then perform modeling on a computer cluster by first using the rosetta_submit.py script to generate submission files (Appendix, command line [19]) and then using source on the submission file appropriate for the cluster's queuing system (e.g., Condor, LSF, PBS, etc.). For FARNA runs, it is best to generate around 10,000–15,000 low-resolution models, from which a subset will later be minimized. The models generated by rna_denovo are by default placed in a folder named out, which is created in the modeling folder. The out folder contains individual folders for each run with a silent .out file in each that describes all of the models from that run. To collect all of the models into a single silent file, we use the easy_cat.py script (Appendix, command line [20]). This creates a single concatenated .out file with the name tag initially provided in README_SETUP.

If a reference (native) model was input during FARNA modeling, the RMSDs of the FARNA models to the reference can be compared to their Rosetta energy scores, which are all recorded in the concatenated silent file, to assess the quality of the low-resolution models. An example energy versus RMSD plot is shown in Fig. 4A. Additionally, it may be helpful to visualize the low-resolution models with the lowest—that is, most favorable—Rosetta energy scores. To do this, we extract the lowest-scoring models from the concatenated .out file using extract_lowscore_models.py (Appendix, command line [21]). These PDB-formatted models can then be loaded in PyMOL for comparison (Fig. 4B). Note that the FARNA models may contain discontinuities in the RNA backbone, which are visible in PyMOL. These chainbreaks occur because crystallographic fragments that are sampled and built into the model first may prevent a continuous backbone from being built in other regions of the RNA. Chainbreaks are not a cause for concern, however, because the following all-atom minimization step typically resolves them.

3.4. Producing and selecting models with reasonable stereochemistry using refinement

As mentioned earlier, the low-resolution models generated by FARNA may contain chainbreaks and unrealistic atomic-level geometries due to the method of sampling rigid fragments of crystallographic RNA structures. To achieve more realistic models of the RNAs, we use the rna_minimize program in Rosetta to refine the lowest-energy 1/6 of the low-resolution models (e.g., if 12,000 FARNA models were generated, minimize 2000 of them). This FARNA with Full-Atom Refinement (FARFAR) strategy optimizes the low-resolution models based on the Rosetta full-atom energy function, which accounts for physical and chemical features such as van der Waals forces, hydrogen bonding, desolvation penalties for polar groups, and RNA backbone torsion angles (Das et al., 2010; Sripakdeevong et al., 2011).

To set up refinement of the FARNA models, we create a `MINIMIZE` file similar to command line [22] in the Appendix. Running `source MINIMIZE` (Appendix, command line [23]) calls the `parallel_min_setup.py` script to generate the command lines for refinement in an output script specified in MINIMIZE (by default, `min_cmdline`). Each line in `min_cmdline` is one minimization command, and the number of lines in `min_cmdline` is the number of processors specified in MINIMIZE. As for FARNA runs, it is best to test the minimization before submitting the jobs to the cluster; here, we copy the first line from the `min_cmdline` file starting with `rna_minimize` and run it locally (Appendix, command line [24]), waiting for the output "`protocols.rna.RNA_Minimizer: Minimizing...round= 1`" before canceling the run. After confirming that the run proceeds without errors, we create submission files by running `rosetta_submit.py` on `min_cmdline` (Appendix, command line [25]), then using `source` to submit the jobs. For refinement runs, the jobs will automatically terminate after all of the specified models are minimized, which usually takes a few hours with 100 processors on a cluster. The silent files for minimized models outputted by `rna_minimize` are collected in individual folders in a folder called `min_out`, similar to the output of `rna_denovo`. Again, we use `easy_cat.py` to collect all of the minimized models into a single silent file with the tag given in `MINIMIZE` (Appendix, command line [26]).

Refinement using FARFAR improves low-resolution models by relaxing them into more realistic conformations. This generally results in better RMSDs to input reference models, as seen by energy versus RMSD plots (Fig. 4C), and more realistic models, which can be visualized using PyMOL in the same way as earlier (Fig. 4D). More base pairs are correctly formed, chainbreaks that were present in FARNA models are typically fixed, and constraints from chemical mapping and MOHCA-seq tend to be better satisfied in minimized models.

3.5. Clustering to generate final set of models

The set of refined FARFAR models often contains subsets of models that adopt similar folds to within helical resolution, especially if modeling was performed in the context of chemical mapping and MOHCA-seq data.

To select a representative set of 3D models that is likely to reflect the native fold of the RNA, we collect the largest and lowest-energy subsets of models that fall within a certain RMSD threshold of each other as described later. Such clustering suggests that the fold adopted by those models is both energetically favorable and comparatively likely to be sampled (Shortle, Simons, & Baker, 1998), and the RMSD threshold value (see later) provides an estimate of modeling precision.

First, we use the script `silent_file_sort_and_select.py` to sort the models in the silent file output by FARFAR and select the desired number of lowest-energy models, normally equal to 0.5% of the total unrefined (FARNA) models (Appendix, command line [27]). This script generates a new silent file containing only the selected lowest-energy models, usually 50–75 if 10,000–15,000 models were built by FARNA. Then, we perform clustering locally using the `cluster` application in Rosetta, which uses an RMSD threshold input by the user to sort the models in the silent file into groups that fall within the threshold (Appendix, command line [28]). Each clustering run normally takes less than a minute. The output of running `cluster` is a silent file containing the clustered models, as well as a screen output that reports how many clusters were generated and how many models were sorted into each cluster. Our standard practice is to choose an RMSD threshold that results in 1/6 of the clustered models being sorted into the largest cluster, by adjusting the input RMSD threshold over multiple clustering runs. Finally, we isolate the models in the top cluster, which is referred to as cluster0 in the output of the `cluster` application (Appendix, command line [29]). This can be done using a text editor by copying the clustered silent file, selecting the lines of the silent file comprising the cluster0 models (labeled in the silent file with `c.0.*`, where `*` is the number of the model in the cluster), and deleting the remainder. Then, we use `extract_lowscore_decoys.py` to collect these final models as PDB-formatted files (Appendix, command line [30]).

The RMSD threshold used in clustering represents an estimate of the "precision" of the final subset of FARFAR models. Because the precision captures the variation between the models, it also sets a lower bound on the accuracy of the modeling, although individual models within the cluster may have RMSDs to crystallographic models that are lower. When both chemical mapping and MOHCA-seq data are included in our pipeline, we find that the top cluster typically reflects the native fold of the target RNA, as compared to a previously or subsequently released crystal structure, to 7–15 Å RMSD (Cheng et al., 2014) (Fig. 5).

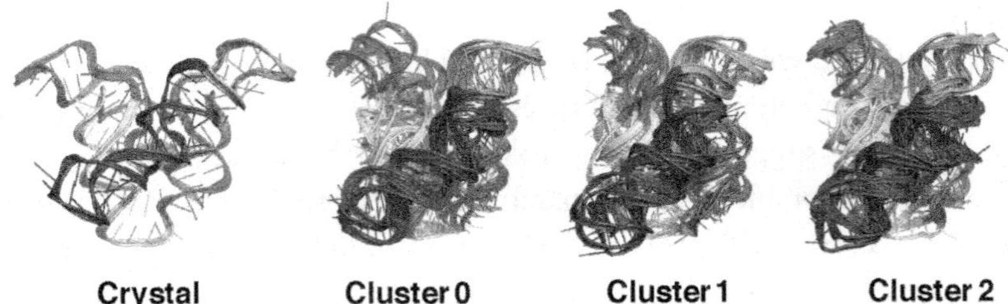

Figure 5 Clustering of minimized models to select representative models. Comparison of models generated by the experimental/computational pipeline. The crystal structure (PDB ID 3P49) is shown at left. At right, four representative models are overlaid for each of the top three model clusters. The cluster center model of cluster0 has a 7.9 Å RMSD to the crystal structure. (See the color plate.)

3.6. Advanced strategies: Building subpieces into existing models

In some cases, it may be beneficial to improve predictions of RNA structures by remodeling sections of the structure or adding additional regions to the structure. As an example, the tandem glycine-binding riboswitch, which binds two molecules of glycine using two sequentially arranged glycine aptamers, is thought to act as a cooperative sensor of glycine. However, recent studies showed that inclusion of a leader sequence abolishes cooperativity of the riboswitch, at least for the isolated ligand-binding domain. Sequence–structure alignment indicated that it likely forms a kink-turn motif (Kladwang et al., 2012; Rahrig, Petrov, Leontis, & Zirbel, 2013; Sarver, Zirbel, Stombaugh, Mokdad, & Leontis, 2008). The leader sequence was not included in prior models or crystallographic structures of the RNA (Butler et al., 2011), but modeling in Rosetta was able to automatically model the structure formed by the leader sequence when incorporated into the crystal structure (Kladwang et al., 2012) and gave support for a kink-turn conformation. Here, we will discuss how to perform this type of addition and remodeling in Rosetta.

In order to remodel a region of an RNA for which a piece is already available, e.g., in a crystallographic template, it may first be necessary to excise the desired piece from the template. This excision can be accomplished using the `pdbslice.py` command, which creates a new PDB file that contains a user-specified subset of the residues in the input PDB file (Appendix, command line [31]). In the example of the glycine riboswitch, the first nucleotide must be excised, as well as the residues comprising the linker between the two aptamers of the ligand-binding domain, which base pair with the leader sequence. The sliced model will be used as an input to FARFAR modeling so that only the nucleotides that are not present in the model will be sampled. Here, because a 5′-leader sequence must also be added to the RNA, we must also revise the FASTA and secondary structure files and renumber the input PDB, so that the sequence numbers and identities are fully consistent. The revised FASTA and secondary structure files are given as files [F4] and [F5] in the Appendix. To renumber the input PDB, we use the `renumber_pdb_in_place.py` script, providing it with the PDB to be renumbered and the desired final sequence position ranges (Appendix, command line [32]). Then, we create a new `README_SETUP` file that reads the revised FASTA and secondary structure files and includes a flag to input the sliced and renumbered input model (Appendix, command line [33]). Finally, we run the modeling as before. If only a small region of the RNA is being remodeled, fewer processors or less computational time may be necessary to reach convergence, so adjust these parameters accordingly.

In cases where sequence analysis or other prediction algorithms suggest the presence of an RNA motif or fold of known structure, one strategy to save computational time is to use an instance of the known structure as a template for modeling the sequence of interest—this method is called "threading." Threaded fragments of structures, such as kink-turn motifs or loops, can in turn be used as input PDBs for global modeling or remodeling of RNAs and can help to focus sampling on regions of entirely unknown structure. See command line [34] in the Appendix; further documentation is also available at RosettaCommons (https://www.rosettacommons.org/docs/latest/rna-thread.html).

4. EVALUATION

The pipeline we have described in the preceding text achieves *de novo* models of RNAs with subhelical (~10 Å) resolution, based on benchmark and blind prediction studies. Independent validation or falsification of models at this resolution can be challenging, because available chemical mapping and MOHCA-seq constraints are usually included in the modeling. We recommend two strategies to test the final models. First, check whether tertiary features of the RNA can be reconstituted without some of the available constraints; e.g., if mutate-and-map experiments identify tertiary contacts in the RNA, then exclude MOHCA-seq proximity constraints from the modeling and check for agreement of the final models with MOHCA-seq data. Recovery of proximities indicated by MOHCA-seq independent of modeling with those constraints lends support to those tertiary features. Second, one can perform mutational analysis to verify new tertiary contacts suggested by the modeling by using chemical mapping or MOHCA-seq experiments to assess the effects of mutations predicted to disrupt those new contacts or mutations that may rescue the structure through formation of compensatory base pairs (Tian et al., 2014; Xue et al., 2014).

5. CONCLUSION

Three-dimensional modeling of RNAs has improved greatly in recent years, aided by advances in both experimental methods and computational strategies for predicting secondary and tertiary structures. In this chapter, we have described a general workflow for modeling RNA 3D folds using the Rosetta framework for macromolecular modeling, guided by data from solution-state chemical mapping experiments. These experiments, particularly the two-dimensional M^2 and MOHCA-seq measurements, provide constraints for modeling by defining an RNA's secondary structure elements and identifying tertiary proximities within its fold. This experimental/ computational pipeline has allowed us to recover the tertiary folds of RNA-Puzzles challenges and continues to reveal avenues for exploring biological questions through *in vitro* and *in vivo* experiments.

The ultimate goal of prediction and design of RNA structures at consistent atomic accuracy has not yet been achieved, but continuing developments in computational and hybrid methods hold promise for making strides toward this goal. In particular, interfacing current methods for recovering RNA folds at medium resolution with new strategies for modeling small RNA motifs at near-atomic-accuracy; incorporating insights about local RNA tertiary conformations from NMR constraints or chemical mapping reagents into global modeling; and improving methods for classifying, sampling, and constructing RNA motifs are likely to have strong impacts in RNA structure modeling in the coming years.

ACKNOWLEDGMENTS

The writing of this chapter was supported by National Institutes of Health (5T32 GM007276 to C. Y. C.; R01 GM102519 to R. D.), the Burroughs-Wellcome Foundation (CASI to R. D.), and Stanford Bio-X and HHMI international fellowships (F. C. C.). Computational resources were provided by the Stanford BioX^3 computing cluster. We thank Caleb Geniesse and RosettaCommons for testing and helpful comments.

APPENDIX. EXAMPLE COMMAND LINES AND FILES FOR RNA MODELING IN ROSETTA

Command lines, input files, and example output files can be found in the Rosetta/demos/public/mohca_seq folder, which is included in the released Rosetta software package.

Documentation for setting up Rosetta and RNA tools:

https://www.rosettacommons.org/docs/latest/Build-Documentation.html

https://www.rosettacommons.org/docs/latest/RNA-tools.html

[F1] Example FASTA file:

```
>3P49_RNA.pdb
ggauaugaggagagauuucauuuuaaugaaacaccgaagaaguaaaucuuucagguaa
aaaggacucauauuggacgaaccucuggagagcuuaucuaagagauaacaccgaagga
gcaaagcuaauuuuagccuaaacucucagguaaaaggacggag
```

The RNA sequence must be lowercase.

[F2] Example secondary structure file:

```
.(((((((......((((((....)))))).(((...((((.....))))..)))
........))))))).......(((((......((((((...)))))).(((...
((((....((((....)))).....))))..))).......)))))
```

[1] Generate command lines for helix preassembly:

```
helix_preassemble_setup.py -secstruct [secondary structure file] -fasta [FASTA file]
```

[2] Example command line for helix preassembly:

```
rna_denovo -nstruct 100 -params_file helix0.params -fasta helix0.fasta -out:file:silent helix0.out -include_neighbor_base_stacks -minimize_rna true -rna::corrected_geo -score:rna_torsion_potential RNA11_based_new -chemical::enlarge_H_lj -score:weights stepwise/rna/rna_helix -cycles 1000 -output_res_num 2-9 65-72
```

[3] Run command lines for helix preassembly (local):

```
source CMDLINES
```

[4] Prepare native/reference structure for Rosetta, if available:

```
make_rna_rosetta_ready.py 3P49.pdb
```

Outputs reformatted native model as "3p49_RNA.pdb," to be input to README_SETUP. In the glycine riboswitch example presented here, the 3P49 crystal structure includes a protein-binding loop that is not part of the construct used for *de novo* modeling. Command lines [5] through [14] show how to replace the extraneous residues with a tetraloop matching the experimentally probed construct using a short FARFAR modeling run.

[5] Cut out a segment of a model:

```
pdbslice.py [3p49_RNA.pdb] -subset [1-21 36-169] [slice_]
```

The first input is the model from which you want to excise regions of interest. The second input is the range of nucleotides that you want to keep in your model. The third input is the prefix that will be added to the beginning of the input model's filename. Here, the protein-binding loop is excised by specifying the range of residues given in the command line.

[6] Renumber a PDB:

```
renumber_pdb_in_place.py [slice_3p49_RNA.pdb] [1-21 26-159]
```

The first input is the PDB file to be renumbered and the second input is the desired final ranges of sequence positions. Gaps may be intentionally left in the final sequence range to allow for remodeling in the middle of the RNA. Here, a UUUA tetraloop will be built in place of the excised protein-binding loop.

[7] Example README_SETUP for *de novo* remodeling with a sliced input PDB:

```
rna_denovo_setup.py -fasta fasta -secstruct_file secstruct \
 -tag native \
 -working_res 1-159 \
 -s slice_3p49_RNA.pdb \
 -cycles 20000 \
 -ignore_zero_occupancy false \
```

Options:

-fasta [fasta]	Input FASTA file
-secstruct_file [secstruct]	Input secondary structure file
-tag	Name for output files
-working_res	Specify range of residues to model
-s slice_3p49_RNA.pdb	See below
-ignore_zero_occupancy false	

The "-s" flag allows users to input a list of PDB files to use in the modeling; the residues that are part of the input PDB files will not be moved relative to each other, though if multiple PDB files are input, the orientations of the residues in the separate files may change. In this example, the full-atom refinement algorithm will be applied in the same run as fragment assembly.

[8] Generate command line for FARFAR modeling:

```
source README_SETUP
```

[9] Example README_FARFAR:

```
rna_denovo -nstruct 500 -params_file native.params -fasta
native.fasta -out:file:silent native.out -include_neighbor_
base_stacks -minimize_rna true -s slice_3p49_RNA.pdb -input_res
1-21 26-159 -cycles 20000 -ignore_zero_occupancy false -output_
res_num 1-159
```

[10] Test command line for FARFAR modeling:

```
source README_FARFAR
```

This command runs a single local job on your computer. Wait until sampling begins successfully (command line output similar to "`Picked Fragment Library for sequence u and sec. struct H ... found 2308 potential fragments`"), then cancel the run and submit the job to the cluster.

[11] Submit jobs to the cluster:

```
rosetta_submit.py README_FARFAR out [16] [1]
```

The first number states how many processors to use for the run, while the second number states the maximum time each job will be allowed to run (walltime, in hours). Note that certain supercomputers only allow requests specific multiples of processors (e.g., the Stampede cluster requires a multiple of 16). Start the run with the appropriate command listed by the output above (e.g., source qsubMPI for the Stampede cluster).

[12] Concatenate all models from the `out` folder:

```
easy_cat.py out
```

Also outputs the number of models in the final silent file to the screen.

[13] Extract lowest-energy models to .pdb files for viewing in PyMOL:

```
extract_lowscore_decoys.py native.out [1]
```

Input the number of lowest-scoring models to extract from the silent file. Here, extract the single lowest-scoring model to use as the native model input for comparison to the *de novo* models.

[14] Rename lowest-score model for use as reference model:

```
mv native.out.1.pdb 3p49_native_RNA.pdb
```

[F3] Example pseudoenergy constraint file:

```
[atompairs]
02′ 2 C4′ 38 FADE 0 30 15 −4.00 4.00
02′ 2 C4′ 38 FADE −99 60 30 −36.00 36.00
02′ 1 C4′ 44 FADE 0 30 15 −4.00 4.00
02′ 1 C4′ 44 FADE −99 60 30 −36.00 36.00
02′ 5 C4′ 60 FADE 0 30 15 −4.00 4.00
02′ 5 C4′ 60 FADE −99 60 30 −36.00 36.00
02′ 2 C4′ 64 FADE 0 30 15 −4.00 4.00
02′ 2 C4′ 64 FADE −99 60 30 −36.00 36.00
02′ 25 C4′ 54 FADE 0 30 15 −4.00 4.00
02′ 25 C4′ 54 FADE −99 60 30 −36.00 36.00
02′ 45 C4′ 64 FADE 0 30 15 −4.00 4.00
02′ 45 C4′ 64 FADE −99 60 30 −36.00 36.00
02′ 45 C4′ 75 FADE 0 30 15 −4.00 4.00
02′ 45 C4′ 75 FADE −99 60 30 −36.00 36.00
02′ 32 C4′ 88 FADE 0 30 15 −4.00 4.00
02′ 32 C4′ 88 FADE -99 60 30 −36.00 36.00
02′ 42 C4′ 84 FADE 0 30 15 −4.00 4.00
02′ 42 C4′ 84 FADE −99 60 30 −36.00 36.00
```

```
O2′ 48 C4′ 84 FADE 0 30 15 −4.00 4.00
O2′ 48 C4′ 84 FADE −99 60 30 −36.00 36.00
O2′ 55 C4′ 88 FADE 0 30 15 −4.00 4.00
O2′ 55 C4′ 88 FADE −99 60 30 −36.00 36.00
O2′ 55 C4′ 108 FADE 0 30 15 −4.00 4.00
O2′ 55 C4′ 108 FADE −99 60 30 −36.00 36.00
O2′ 58 C4′ 118 FADE 0 30 15 −4.00 4.00
O2′ 58 C4′ 118 FADE −99 60 30 −36.00 36.00
O2′ 67 C4′ 119 FADE 0 30 15 −4.00 4.00
O2′ 67 C4′ 119 FADE −99 60 30 −36.00 36.00
O2′ 67 C4′ 121 FADE 0 30 15 −4.00 4.00
O2′ 67 C4′ 121 FADE −99 60 30 −36.00 36.00
O2′ 78 C4′ 113 FADE 0 30 15 −4.00 4.00
O2′ 78 C4′ 113 FADE −99 60 30 −36.00 36.00
O2′ 78 C4′ 135 FADE 0 30 15 −4.00 4.00
O2′ 78 C4′ 135 FADE −99 60 30 −36.00 36.00
O2′ 42 C4′ 157 FADE 0 30 15 −4.00 4.00
O2′ 42 C4′ 157 FADE −99 60 30 −36.00 36.00
O2′ 74 C4′ 156 FADE 0 30 15 −4.00 4.00
O2′ 74 C4′ 156 FADE −99 60 30 −36.00 36.00
O2′ 100 C4′ 148 FADE 0 30 15 −4.00 4.00
O2′ 100 C4′ 148 FADE −99 60 30 −36.00 36.00
O2′ 100 C4′ 145 FADE 0 30 15 −4.00 4.00
O2′ 100 C4′ 145 FADE −99 60 30 −36.00 36.00
O2′ 113 C4′ 153 FADE 0 30 15 −4.00 4.00
O2′ 113 C4′ 153 FADE −99 60 30 −36.00 36.00
O2′ 135 C4′ 154 FADE 0 30 15 −4.00 4.00
O2′ 135 C4′ 154 FADE −99 60 30 −36.00 36.00
O2′ 5 C4′ 119 FADE 0 30 15 −4.00 4.00
O2′ 5 C4′ 119 FADE −99 60 30 −36.00 36.00
O2′ 25 C4′ 88 FADE 0 30 15 −0.80 0.80
O2′ 25 C4′ 88 FADE −99 60 30 −7.20 7.20
O2′ 37 C4′ 62 FADE 0 30 15 −0.80 0.80
O2′ 37 C4′ 62 FADE −99 60 30 −7.20 7.20
O2′ 79 C4′ 103 FADE 0 30 15 −0.80 0.80
O2′ 79 C4′ 103 FADE −99 60 30 −7.20 7.20
O2′ 15 C4′ 88 FADE 0 30 15 −0.80 0.80
O2′ 15 C4′ 88 FADE −99 60 30 −7.20 7.20
O2′ 32 C4′ 108 FADE 0 30 15 −0.80 0.80
```

```
O2′ 32 C4′ 108 FADE −99 60 30 −7.20 7.20
O2′ 9 C4′ 138 FADE 0 30 15 −0.80 0.80
O2′ 9 C4′ 138 FADE −99 60 30 −7.20 7.20
O2′ 25 C4′ 118 FADE 0 30 15 −0.80 0.80
O2′ 25 C4′ 118 FADE −99 60 30 −7.20 7.20
```

[15] Example README_SETUP:

```
rna_denovo_setup.py -fasta fasta -secstruct_file secstruct \
    -fixed_stems \
    -no_minimize \
    -tag glycine_riboswitch \
    -working_res 1-159 \
    -native 3p49_native_RNA.pdb \
    -cst_file constraints \
    -staged_constraints \
    -cycles 20000 \
    -ignore_zero_occupancy false \
    -silent helix0.out helix1.out helix2.out helix3.out helix4.out helix5.out helix6.out helix7.out \
    -input_silent_res 2-9 65-72 16-21 26-31 33-35 54-56 39-42 48-51 81-85 155-159 92-97 101-106 108-110 145-147 114-117 139-142 \
```

Options:

-fasta [fasta]	Input FASTA file
-secstruct_file [secstruct]	Input secondary structure file
-fixed_stems	Specify whether helices should be fixed
-no_minimize	Specify not to perform full-atom refinement; minimization will be performed in the next stage of modeling
-tag	Name for output files
-working_res	Specify range of residues to model
-native [native.pdb]	Input reference or native model; used for benchmarking cases and will return rms calculations for all models (see command line [5])
-cst_file [constraints]	Input file with pseudoenergy constraints
-staged_constraints	Apply constraints

-ignore_zero_occupancy false	
-silent [helix0.out helix1.out . . .]	Input silent files with preassembled helices
-input_silent_res [2–9 65–72 16–21 26–31 . . .]	Specify position ranges of helices in silent files

[16] Generate command line for FARFAR modeling:

```
source README_SETUP
```

[17] Example README_FARFAR:

```
rna_denovo -nstruct 500 -params_file glycine_riboswitch.params -fasta glycine_riboswitch.fasta -out:file:silent glycine_riboswitch.out -include_neighbor_base_stacks -minimize_rna false -native glycine_riboswitch_3p49_native_RNA.pdb -in:file:silent helix0.out helix1.out helix2.out helix3.out helix4.out helix5.out helix6.out helix7.out -input_res 2-9 65-72 16-21 26-31 33-35 54-56 39-42 48-51 81-85 155-159 92-97 101-106 108-110 145-147 114-117 139-142 -cst_file glycine_riboswitch_constraints -staged_constraints -cycles 20000 -ignore_zero_occupancy false -output_res_num 1-159
```

[18] Test command line for FARFAR modeling:

```
source README_FARFAR
```

[19] Submit jobs to the cluster:

```
rosetta_submit.py README_FARFAR out [96] [16]
```

[20] Concatenate all models from the `out` folder:

```
easy_cat.py out
```

[21] Extract lowest-energy models to .pdb files for viewing in PyMOL:

```
extract_lowscore_decoys.py glycine_riboswitch.out [15]
```

[22] Example MINIMIZE:

```
parallel_min_setup.py -silent glycine_riboswitch.out -tag glycine_riboswitch_min -proc [96] -nstruct [2000] -out_folder min_out -out_script min_cmdline "-native glycine_riboswitch_3p49_native_RNA.pdb -cst_fa_file
```

```
glycine_riboswitch_constraints -params_file glycine_
riboswitch.params-ignore_zero_occupancy false -skip_
coord_constraints"
```

The first number states how many processors to use for the run, while the second number is 1/6 the total number of previously generated FARNA models. If you are running on a supercomputer that only allows specific multiples of processors, use an appropriate number for the first input.

[23] Generate command lines for full-atom refinement:

```
source MINIMIZE
```

[24] Example command line from min_cmdline to run as test:

```
rna_minimize -native glycine_riboswitch_3p49_native_RNA.pdb
-cst_fa_fileglycine_riboswitch_constraint-params_fileglycine_
riboswitch.params -ignore_zero_occupancy false -skip_coord_
constraints -in:file:silentmin_out/0/0.silent -out:file:silent
min_out/0/glycine_riboswitch_min.out
```

[25] Submit jobs to the cluster:

```
rosetta_submit.py min_cmdline min_out [1] [16]
```

The first number states how many processors to use for each line in min_cmdline. Here, enter 1 for the first input so that the total number of processors used will be equal to the number of processors entered with the "-proc" flag in command line [12], above. The second number states the maximum time each job will be allowed to run (walltime). Start the run with the appropriate command listed by the output above (e.g., source qsubMPI for the Stampede cluster).

[26] Concatenate all models from the `min_out` folder:

```
easy_cat.py min_out
```

[27] Sort models by Rosetta energy and select a subset for clustering:

```
silent_file_sort_and_select.py [glycine_riboswitch_
min.out]-select [1-60]-o [glycine_riboswitch_min_sort.
out]
```

The range of models under the `-select` tag includes 0.5% of the total number of FARNA models generated previously. Outputs a new silent file containing selected number of lowest-energy models.

[28] Cluster models:

```
cluster -in:file:silent glycine_riboswitch_min_sort.
out -in:file:fullatom -out:file:silent_struct_type
binary -export_only_low false -out:file:silent clus-
ter.out -cluster:radius [radius]
```

Select a radius so that 1/6 of the models in the input sorted silent file are in the largest cluster (cluster0) of models.

[29] Copy clustered .out file to a new file to isolate cluster0:

```
cp cluster.out cluster0.out
```

[30] Extract lowest-energy models to .pdb files for viewing in PyMOL:

```
extract_lowscore_decoys.py cluster0.out [15]-no_
replace_names
```

Input the number of models in cluster0. The -no_replace_names tag preserves the filenames of the cluster members to reflect their order in the cluster, rather than renaming them in order of Rosetta energy score.

[31] Cut out a segment of a model:

```
pdbslice.py [3p49_native_RNA.pdb] -subset [2-72
81-159] [slice_kinkturn_]
```

Here, the 3P49 crystal structure includes an additional G at position 0, which must be excised to allow the leader sequence to be added to the 5′-end, and the internal linker that forms the kink-turn motif with the leader sequence is also excised to allow remodeling.

[32] Renumber a PDB:

```
renumber_pdb_in_place.py [slice_kinkturn_3P49_native_
RNA.pdb] [10-80 89-167]
```

Here, the PDB is renumbered to allow the leader sequence to be added at the 5′-end.

[F4] Example revised FASTA file:

```
>3P49_RNA_kinkturn.pdb
ucggaugaagauaugaggagagauuucauuuuaaugaaacaccgaagaaguaaaucuu
ucagguaaaaaggacucauauuggacgaaccucuggagagcuuaucuaagagauaaca
ccgaaggagcaaagcuaauuuuagccuaaacucucagguaaaaggacggag
```

[F5] Example revised secondary structure file:

```
(((......(((((((......((((((....)))))).(((...(((.....))
))..)))........))))))))...)))..(((((......((((((...)))))).
(((...((((....((((....)))).....)))).))).......)))))
```

[33] Example README_SETUP for *de novo* remodeling with a sliced input PDB:

```
rna_denovo_setup.py -fasta fasta2 -secstruct_file sec-
struct2 \
  -fixed_stems \
  -tag glycine_rbsw_kinkturn \
  -working_res 1-167 \
  -s slice_kinkturn_3P49_native_RNA.pdb \
  -cycles 20000 \
  -ignore_zero_occupancy false \
```

[34] Thread an RNA sequence into a template structure:

```
rna_thread -in:file:fasta [fasta] -in:file:s [template
PDB] -o [output PDB]
```

The first input is a FASTA file containing two RNA sequences: (1) the sequence of interest, onto which the structure of the template sequence will be threaded and (2) the template sequence. The template sequence should be truncated to the regions into which the sequence of interest will be threaded; use hyphens ("-") to align the template sequence with the target sequence in the FASTA file. The second input, the template structure in PDB format, should be similarly truncated, using `pdbslice.py` if necessary. If the template PDB is not correctly formatted for Rosetta modeling, use `make_rna_rosetta_ready.py` to reformat it. The last input is the name of the output PDB.

Further documentation for RNA threading in Rosetta can be found at the RosettaCommons (https://www.rosettacommons.org/docs/latest/rna-thread.html).

REFERENCES

Butler, E. B., Xiong, Y., Wang, J., & Strobel, S. A. (2011). Structural basis of cooperative ligand binding by the glycine riboswitch. *Chemistry & Biology*, *18*(3), 293–298. http://dx.doi.org/10.1016/j.chembiol.2011.01.013.

Cheng, C., Chou, F.-C., Kladwang, W., Tian, S., Cordero, P., & Das, R. (2014). MOHCA-seq: RNA 3D models from single multiplexed proximity-mapping experiments. *bioRxiv*. http://dx.doi.org/10.1101/004556.

Chou, F. C., Lipfert, J., & Das, R. (2014). Blind predictions of DNA and RNA tweezers experiments with force and torque. *PLoS Computational Biology*, *10*(8), e1003756. http://dx.doi.org/10.1371/journal.pcbi.1003756.

Chou, F. C., Sripakdeevong, P., Dibrov, S. M., Hermann, T., & Das, R. (2013). Correcting pervasive errors in RNA crystallography through enumerative structure prediction. *Nature Methods*, *10*(1), 74–76. http://dx.doi.org/10.1038/nmeth.2262.

Cordero, P., Kladwang, W., VanLang, C. C., & Das, R. (2012). Quantitative dimethyl sulfate mapping for automated RNA secondary structure inference. *Biochemistry*, *51*(36), 7037–7039. http://dx.doi.org/10.1021/bi3008802.

Cordero, P., Kladwang, W., VanLang, C. C., & Das, R. (2014). The mutate-and-map protocol for inferring base pairs in structured RNA. *Methods in Molecular Biology*, *1086*, 53–77. http://dx.doi.org/10.1007/978-1-62703-667-2_4.

Cordero, P., Lucks, J. B., & Das, R. (2012). An RNA mapping database for curating RNA structure mapping experiments. *Bioinformatics*, *28*(22), 3006–3008. http://dx.doi.org/10.1093/bioinformatics/bts554.

Cruz, J. A., Blanchet, M. F., Boniecki, M., Bujnicki, J. M., Chen, S. J., Cao, S., et al. (2012). RNA-Puzzles: A CASP-like evaluation of RNA three-dimensional structure prediction. *RNA*, *18*(4), 610–625. http://dx.doi.org/10.1261/rna.031054.111.

Das, R., & Baker, D. (2007). Automated de novo prediction of native-like RNA tertiary structures. *Proceedings of the National Academy of Sciences of the United States of America*, *104*(37), 14664–14669. http://dx.doi.org/10.1073/pnas.0703836104.

Das, R., Karanicolas, J., & Baker, D. (2010). Atomic accuracy in predicting and designing noncanonical RNA structure. *Nature Methods*, *7*(4), 291–294. http://dx.doi.org/10.1038/nmeth.1433.

Das, R., Kudaravalli, M., Jonikas, M., Laederach, A., Fong, R., Schwans, J. P., et al. (2008). Structural inference of native and partially folded RNA by high-throughput contact mapping. *Proceedings of the National Academy of Sciences of the United States of America*, *105*(11), 4144–4149. http://dx.doi.org/10.1073/pnas.0709032105.

Ditzler, M. A., Otyepka, M., Sponer, J., & Walter, N. G. (2010). Molecular dynamics and quantum mechanics of RNA: Conformational and chemical change we can believe in. *Accounts of Chemical Research*, *43*(1), 40–47. http://dx.doi.org/10.1021/ar900093g.

Erion, T. V., & Strobel, S. A. (2011). Identification of a tertiary interaction important for cooperative ligand binding by the glycine riboswitch. *RNA*, *17*(1), 74–84. http://dx.doi.org/10.1261/rna.2271511.

Hajdin, C. E., Bellaousov, S., Huggins, W., Leonard, C. W., Mathews, D. H., & Weeks, K. M. (2013). Accurate SHAPE-directed RNA secondary structure modeling, including pseudoknots. *Proceedings of the National Academy of Sciences of the United States of America*, *110*(14), 5498–5503. http://dx.doi.org/10.1073/pnas.1219988110.

Kim, H., Cordero, P., Das, R., & Yoon, S. (2013). HiTRACE-Web: An online tool for robust analysis of high-throughput capillary electrophoresis. *Nucleic Acids Research*, *41*(Web Server issue), W492–W498. http://dx.doi.org/10.1093/nar/gkt501.

Kladwang, W., Chou, F. C., & Das, R. (2012). Automated RNA structure prediction uncovers a kink-turn linker in double glycine riboswitches. *Journal of the American Chemical Society*, *134*(3), 1404–1407. http://dx.doi.org/10.1021/ja2093508.

Kladwang, W., Mann, T. H., Becka, A., Tian, S., Kim, H., Yoon, S., et al. (2014). Standardization of RNA chemical mapping experiments. *Biochemistry*, *53*(19), 3063–3065. http://dx.doi.org/10.1021/bi5003426.

Kladwang, W., VanLang, C. C., Cordero, P., & Das, R. (2011). A two-dimensional mutate-and-map strategy for non-coding RNA structure. *Nature Chemistry*, *3*(12), 954–962. http://dx.doi.org/10.1038/nchem.1176.

Leaver-Fay, A., Tyka, M., Lewis, S. M., Lange, O. F., Thompson, J., Jacak, R., et al. (2011). ROSETTA3: An object-oriented software suite for the simulation and design of macromolecules. *Methods in Enzymology*, *487*, 545–574. http://dx.doi.org/10.1016/B978-0-12-381270-4.00019-6.

Lee, J., Kladwang, W., Lee, M., Cantu, D., Azizyan, M., Kim, H., et al. (2014). RNA design rules from a massive open laboratory. *Proceedings of the National Academy of Sciences of the United States of America*, *111*(6), 2122–2127. http://dx.doi.org/10.1073/pnas.1313039111.

Lyskov, S., Chou, F. C., Conchuir, S. O., Der, B. S., Drew, K., Kuroda, D., et al. (2013). Serverification of molecular modeling applications: The Rosetta online server that includes everyone (ROSIE). *PLoS One*, *8*(5), e63906. http://dx.doi.org/10.1371/journal.pone.0063906.

Petrov, A. I., Zirbel, C. L., & Leontis, N. B. (2013). Automated classification of RNA 3D motifs and the RNA 3D Motif Atlas. *RNA*, *19*(10), 1327–1340. http://dx.doi.org/10.1261/rna.039438.113.

Rahrig, R. R., Petrov, A. I., Leontis, N. B., & Zirbel, C. L. (2013). R3D align web server for global nucleotide to nucleotide alignments of RNA 3D structures. *Nucleic Acids Research*, *41*(Web Server issue), W15–W21. http://dx.doi.org/10.1093/nar/gkt417.

Reuter, J. S., & Mathews, D. H. (2010). RNAstructure: Software for RNA secondary structure prediction and analysis. *BMC Bioinformatics*, *11*, 129. http://dx.doi.org/10.1186/1471-2105-11-129.

Sarver, M., Zirbel, C. L., Stombaugh, J., Mokdad, A., & Leontis, N. B. (2008). FR3D: Finding local and composite recurrent structural motifs in RNA 3D structures. *Journal of Mathematical Biology*, *56*(1–2), 215–252. http://dx.doi.org/10.1007/s00285-007-0110-x.

Schrodinger, LLC (2010). *The PyMOL Molecular Graphics System, version 1.3r1.*

Seetin, M. G., Kladwang, W., Bida, J. P., & Das, R. (2014). Massively parallel RNA chemical mapping with a reduced bias MAP-seq protocol. *Methods in Molecular Biology*, *1086*, 95–117. http://dx.doi.org/10.1007/978-1-62703-667-2_6.

Shortle, D., Simons, K. T., & Baker, D. (1998). Clustering of low-energy conformations near the native structures of small proteins. *Proceedings of the National Academy of Sciences of the United States of America*, *95*(19), 11158–11162.

Sripakdeevong, P., Beauchamp, K., & Das, R. (2012). Why can't we predict RNA structure at atomic resolution? In N. B. Leontis & E. Westhof (Eds.), *RNA 3D structure analysis and prediction*. Heidelberg, New York: Springer, 400 p.

Sripakdeevong, P., Cevec, M., Chang, A. T., Erat, M. C., Ziegeler, M., Zhao, Q., et al. (2014). Structure determination of noncanonical RNA motifs guided by (1)H NMR chemical shifts. *Nature Methods*, *11*(4), 413–416. http://dx.doi.org/10.1038/nmeth.2876.

Sripakdeevong, P., Kladwang, W., & Das, R. (2011). An enumerative stepwise ansatz enables atomic-accuracy RNA loop modeling. *Proceedings of the National Academy of Sciences of the United States of America*, *108*(51), 20573–20578. http://dx.doi.org/10.1073/pnas.1106516108.

Tian, S., Cordero, P., Kladwang, W., & Das, R. (2014). High-throughput mutate-map-rescue evaluates SHAPE-directed RNA structure and uncovers excited states. *RNA*, *20*(11), 1815–1826. http://dx.doi.org/10.1261/rna.044321.114.

Tinoco, I., Jr., Borer, P. N., Dengler, B., Levin, M. D., Uhlenbeck, O. C., Crothers, D. M., et al. (1973). Improved estimation of secondary structure in ribonucleic acids. *Nature: New Biology*, *246*(150), 40–41.

Xue, S., Tian, S., Fujii, K., Kladwang, W., Das, R., & Barna, M. (2014). RNA regulons in Hox 5' UTRs confer ribosome specificity to gene regulation. *Nature*. http://dx.doi.org/10.1038/nature14010.

Yoon, S., Kim, J., Hum, J., Kim, H., Park, S., Kladwang, W., et al. (2011). HiTRACE: High-throughput robust analysis for capillary electrophoresis. *Bioinformatics*, *27*(13), 1798–1805. http://dx.doi.org/10.1093/bioinformatics/btr277.

CHAPTER THREE

Computational Methods Toward Accurate RNA Structure Prediction Using Coarse-Grained and All-Atom Models

Andrey Krokhotin, Nikolay V. Dokholyan[1]

Department of Biochemistry and Biophysics, School of Medicine, University of North Carolina, Chapel Hill, North Carolina, USA

[1]Corresponding author: e-mail address: dokh@unc.edu

Contents

Abstract

Computational methods can provide significant insights into RNA structure and dynamics, bridging the gap in our understanding of the relationship between structure and biological function. Simulations enrich and enhance our understanding of data derived on the bench, as well as provide feasible alternatives to costly or technically challenging experiments. Coarse-grained computational models of RNA are especially important in this regard, as they allow analysis of events occurring in timescales relevant to RNA biological function, which are inaccessible through experimental methods alone. We have developed a three-bead coarse-grained model of RNA for discrete molecular dynamics

Methods in Enzymology, Volume 553
ISSN 0076-6879
http://dx.doi.org/10.1016/bs.mie.2014.10.052

simulations. This model is efficient in *de novo* prediction of short RNA tertiary structure, starting from RNA primary sequences of less than 50 nucleotides. To complement this model, we have incorporated additional base-pairing constraints and have developed a bias potential reliant on data obtained from hydroxyl probing experiments that guide RNA folding to its correct state. By introducing experimentally derived constraints to our computer simulations, we are able to make reliable predictions of RNA tertiary structures up to a few hundred nucleotides. Our refined model exemplifies a valuable benefit achieved through integration of computation and experimental methods.

1. INTRODUCTION

The central dogma of molecular biology casts ribonucleic acid (RNA) as the biological intermediary of genetic information transmission between deoxyribonucleic acid (DNA) and proteins. Over the past few decades, these essential biopolymers have been implicated in tasks reaching far beyond their canonical designation. Indeed, many noncoding RNAs bear important and diverse physiological functions: catalysis, regulation of cell activity, and cellular response to disease states. This diversity in function is attributed to the variety of three-dimensional (3D) structures accommodated by RNAs. Riboswitches, which are at the focus of this book, are a fundamental example of RNAs involved in cellular regulation through vast structural rearrangement. They are segments of mRNA located primarily in the 5′ UTR that form unique 3D structures permissive to conformational change upon metabolite binding. Riboswitch conformational shifts regulate the downstream RNA coding sequence (Serganov & Nudler, 2013). This example emphasizes the importance of 3D structural knowledge for elucidation of RNA functioning mechanisms.

Traditional analytical methods to obtain 3D structure, such as X-ray crystallography or NMR, are limited in applicability to RNA, due to the high conformational flexibility of RNA polymers. X-ray crystallography is particularly sensitive in this regard, and special care must be taken to avoid RNA aggregation and/or misfolding prior to crystal formation (Reyes, Garst, & Batey, 2009). NMR experiments with RNA suffer from poor long-range correlations, due to the low proton density of RNA molecules (Addess & Feigon, 1996). As a result, the Protein Data Bank (PDB; Berman et al., 2000) currently contains 3032 solved RNA structures, out of approximately 6.6 million known noncoding sequences (RNA central, http://rnacentral.org). Hence, developing accurate computational methods for

RNA structure prediction is crucial to circumventing the bottleneck imposed by the constraints of current experimental methods.

Computational methods traditionally used for structure prediction include homology modeling, Monte Carlo-based methods and molecular dynamics (MD) simulations. Among these approaches, MD is the only approach that can provide a link between structure and dynamics of macromolecules. MD simulations rely on force fields to describe atomic interactions as a sum of pairwise interactions. Some of the most successful force fields include AMBER, CHARMM, GROMOS, and OPLS (Ponder & Case, 2003). Historically, greater emphasis has been placed on protein modeling, such that these force fields were extensively optimized to describe proteins. However, these force fields are not easily translatable to RNA modeling, due to the complex electrostatics introduced by the sugar and phosphate moieties of RNAs. The flexible conformations distinct to RNAs are not fully represented by the point charges adopted in common force fields for protein modeling. Additionally, the phosphodiester represents a highly polarizable anion, which is affected by solvation (McDowell, Špačková, Šponer, & Walter, 2007). The latter effect is typically neglected in most of the classical force fields, presenting further inaccuracy when extrapolating protein modeling to RNA. During the last two decades, an increasing number of known native RNA structures has improved the quality of current force fields to model RNA, such as variations of the Cornell force field implemented in the AMBER suite: parm94, parm98, parm99, or parmbsc0, and parmOL (Cornell et al., 1995; Pérez et al., 2007).

An extreme complexity of force fields and large number of atoms in all-atom simulations dictates the need to use extensive computational resources, to produce trajectories in the time frame relevant for biological functions. Current state of the art simulations, running in parallel on many processors, allow one to access millisecond range for relatively short proteins (Freddolino, Harrison, Liu, & Schulten, 2010). RNA simulations are even more resource consuming due to the higher number of degrees of freedom in RNA molecules versus proteins: in order to describe the RNA backbone one needs to employ six torsional angles as compared to only two torsional angles for proteins.

Coarse-grained models with a less exhaustive representation of RNA molecules can be computationally cheap and efficient alternatives to detailed all-atom simulations. The energy landscape of coarse-grained models is less rugged and can be sampled more efficiently, while retaining fundamental physicochemical properties of the system (Sim, Minary, & Levitt, 2012).

We have developed a coarse-grained three-bead model of RNA for an efficient implementation of MD, known as discrete molecular dynamics (DMD). Using this model, we are able to correctly predict 3D structures of short RNAs (less than 50 nucleotides) (Ding et al., 2008). With additional experimental constraints our model can be expanded to predict the structure of larger RNAs (up to a few hundred nucleotides) (Ding, Lavender, Weeks, & Dokholyan, 2012), entering the realm of functional RNAs, such as riboswitches, that are ~35–200 nucleotides in length (Serganov & Patel, 2007). The constraints we use include information on base-pairing and solvent accessibilities of different nucleotides. We have also established the iFoldRNA Web server for prediction of RNA tertiary structure, which is freely available for researchers worldwide (Sharma, Ding, & Dokholyan, 2008).

2. DISCRETE MOLECULAR DYNAMICS

MD is routinely used for modeling the time-dependent motions (trajectories) of biological macromolecules (proteins and RNAs). In conventional implementations of MD, forces acting on the atoms (particles) are calculated as derivatives of potentials. These forces are substituted into Newton's equations of motion, which are iteratively solved for every particle in the system through femtosecond time steps. While precise, MD is also computationally costly. Through modern parallelization one is capable of modeling trajectories on the order of milliseconds. Unfortunately, these exhaustive simulations are still unable to address the majority of relevant biological processes, including folding of large macromolecules (i.e., proteins and RNAs) that occur on much longer timescales. To increase the time range available for simulations, we use a simplified description of forces. In DMD, the simplification is realized by approximating the potential force fields describing interactions between particles as stepwise potentials (Fig. 1). In this approximation, a particle changes its velocity only upon collision, while passing the region of a potential barrier. Between collisions, all velocities remain the same. The change of momentum as well as kinetic and potential energies of a particle is determined solely by conservation laws. If the height of a potential barrier is larger than the particle's kinetic energy, the particle bounces back retaining kinetic energy. The advantage of this tactic is that we do not explicitly calculate physical forces nor integrate equations of motion. DMD is a generalized approach, which has been successfully applied to biomolecules (Dokholyan, Buldyrev, Stanley, & Shakhnovich, 1998; Zhou, Karplus, Wichert, & Hall, 1997).

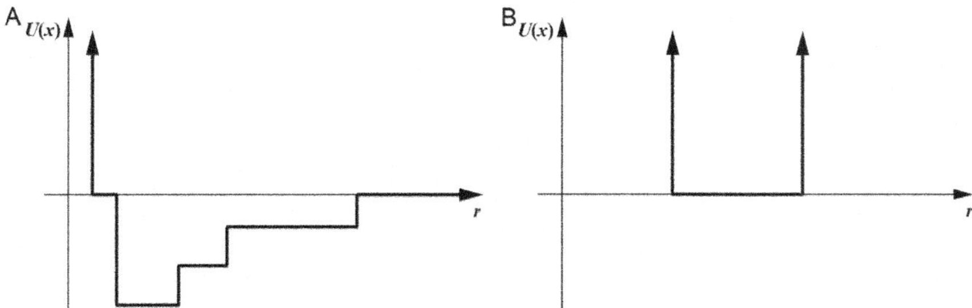

Figure 1 Example potentials used by discrete molecular dynamics (DMD). (A) Discretized van der Waals potential. (B) An infinite square-well potential used to describe covalent bonding.

Unlike MD, which is driven by physical forces, DMD is driven by collision events. This is a consequence of the discretization of interaction potentials characteristic of DMD. In the limit of very fine discretization of potentials, with step size approaching zero, DMD becomes identical to MD. Because DMD is event driven, the efficiency of the algorithm used to search for the next collision event is important for fast performance of DMD. We utilize a collision list approach, which restricts the search for the next collision to atoms located only in the same local area. We developed an efficient implementation of DMD (Proctor, Ding, & Dokholyan, 2011), which can be equally well applied for simulation of different systems starting from uncoupled molecules in an ideal gas system to large macromolecules, such as proteins and RNA. Recently, we also developed a parallel version of DMD (Shirvanyants, Ding, Tsao, Ramachandran, & Dokholyan, 2012). The parallelization was achieved using of a special algorithm, which predicts the possible collisions between atoms that are later accepted or rejected in accordance to the whole dynamics of a simulated system.

3. THREE-BEAD MODEL

3.1. Model geometry and interaction potential

We have developed a coarse-grained representation of RNA structure (Ding et al., 2008). Each nucleotide in our model consists of three beads, representing phosphate (P_i), sugar (S_i), and the nucleo-base (B_i). The beads are placed in the center of mass of corresponding chemical moieties (Fig. 2A). Two types of interactions are considered: bonded and nonbonded. Bonded interactions are used to describe chain connectivity and local geometry and include constraints on the lengths of the bonds connecting adjacent beads

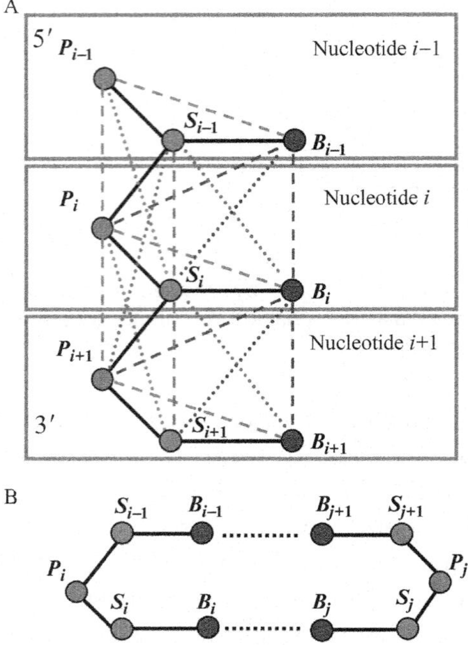

Figure 2 (A) RNA representation in three-bead model. P_i, S_i, and B_i represent phosphate, sugar, and nucleo-base correspondingly. Bonded interactions are shown with constraints on the bond length between nearest beads (solid lines), and on bond angles (dashed lines). (B) Base-pairing model. A hydrogen bond is formed if B_i and B_j come within interaction range, and distances between B_i and S_j, as well as between B_j and S_i provide correct orientation of the bond. *Adapted from Ding et al. (2008) with permission.*

(solid lines), on the bond angles (dashed lines), and on the dihedral angles (dot-dashed lines). The width of infinite square-well potentials for bonded interactions is determined empirically using structures deposited in the PDB. The assigned width is slightly larger than the standard deviation observed in PDB structures, which is a compromise between the desire to decrease the number of collisions (by increasing the well width), and to avoid large distortions in the structure (by minimizing the well width). Physical terms describing nonbonded interactions inside RNA include base-pairing (A–U, G–C, U–G), base-stacking, phosphate–phosphate repulsion, and hydrophobic interaction.

3.1.1 Base-pairing

Base-pairing is formed through hydrogen bonds between nucleo-bases of the corresponding nucleotides. Hydrogen bonds can be formed if B_i and

B_j fall within interaction range. To ensure that hydrogen binding is orientation dependent, we also check the distances between B_i and S_j as well as between B_j and S_i (Ding, Borreguero, Buldyrey, Stanley, & Dokholyan, 2003) (Fig. 2B).

3.1.2 Phosphate repulsion

The negative charges on phosphate atoms repel them from each other. However, in solution, the interaction potential between two phosphates departs from the ideal Coulomb law due to the screening effects of water on ions. To account for these effects, we use a Debye–Hückel model for phosphate–phosphate interactions where the potential is discretized with a step of 1 Å, and a cutoff distance of 10 Å.

3.1.3 Hydrophobic interactions

We account for the hydrophobicity of base pairs by including attractive interactions between all the bases. To avoid artificial overpacking, we introduce an additional energy term $E = \mathrm{d}E \cdot \Theta(n - n_{\mathrm{max}})$, where d$E$ is a repulsion coefficient, $\Theta(n)$ equals 0 when $n \leq 0$ and equals n otherwise; n is the number of nucleotides found in a sphere of 6.5 Å radius. Upon analyzing available RNA structures in the PDB, we found n_{max} to be 4.2.

3.1.4 Base stacking

Base stacking is modeled by assuming every base can form no more than two stacking contacts, with stacked bases aligned linearly. After completing analysis of RNA structures from the PDB, we defined cutoff base–base distances for different base types: 4.65 Å between purines, 4.60 Å between pyrimidines, and 3.80 Å between purine and pyrimidine. To enforce the linearity of stacked bases, we additionally require that two bases forming stacking interactions to the third one should maintain a minimum distance of 6.5 Å.

3.1.5 Parameterization of the hydrogen-bond, base-stacking, and hydrophobic interactions

Stacking and hydrophobic interaction parameters for all pairs of bases were determined by decomposition of the sequence-dependent free-energy parameters for individual nearest-neighbor hydrogen-bond model (INN-HB) (Mathews, Sabina, Zuker, & Turner, 1999). We assume that neighboring base pairs in INN-HB interact through hydrogen-bond, base-stacking, and hydrophobic interactions. Usually bases B_i and B_{i+1} stack on top of each other, except when B_{i+1} and B_j are purines. In the latter case, B_{i+1} stacks on top of B_j, as opposed to B_{i+1} on B_i. The distance between B_i and B_{j-1} usually exceeds the cutoff value 6.5 Å. Therefore, we use the following equation to evaluate the pairwise interactions:

$$E\begin{pmatrix} 5' & B_i & B_{i+1} & 3' \\ 3' & B_j & B_{j-1} & 5' \end{pmatrix} = \begin{cases} \left(E^{\mathrm{HB}}_{B_iB_j} + E^{\mathrm{HB}}_{B_{i+1}B_{j-1}}\right)/2 + E^{\mathrm{Stack}}_{B_jB_{i+1}} \\ \quad + E^{\mathrm{Hydrophobic}}_{B_iB_{i+1}} + E^{\mathrm{Hydrophobic}}_{B_jB_{j-1}}, & B_{i+1}, B_j = \text{purines} \\ \left(E^{\mathrm{HB}}_{B_iB_j} + E^{\mathrm{HB}}_{B_{i+1}B_{j-1}}\right)/2 + E^{\mathrm{Stack}}_{B_iB_{i+1}} + E^{\mathrm{Stack}}_{B_jB_{j-1}} \\ \quad + E^{\mathrm{Hydrophobic}}_{B_{i+1}B_j}, & \text{otherwise} \end{cases}$$

Here, E^{Stack}, E^{HB}, and $E^{\mathrm{Hydrophobic}}$ are interaction strengths of stacking, base-pair, and hydrophobic interactions. Their values were determined from experimentally tabulated values of interaction energies between neighboring base pairs (Mathews et al., 1999) using singular value decomposition.

3.1.6 Loop entropy

Correctly accounting for loop entropy is essential for predicting RNA folding kinetics (Tinoco & Bustamante, 1999). Due to the coarse-grained nature of our model, the entropy is often underestimated in DMD simulations, causing formation of long unnatural loops. To deal with this problem, we explicitly take entropy into account, using the experimentally determined values of free energy measured for different types of RNA loops (Mathews et al., 1999). We compute the free energy of loops in DMD simulations based on formation or breakage of base-pair contacts. The algorithm used to form base pairs in DMD includes the evaluation of the loop free-energy difference ΔG^{loop} due to base-pair formation followed by the formation of a base pair with probability calculated according to the Boltzmann distribution ($p = \exp(-\beta \Delta G^{\mathrm{loop}})$. Kinetic energy of the nucleotides should be enough to overcome the potential barrier for base-pair formation. The breakage of a base pair is only governed by energy, momentum, and angular momentum conservation. We do not use stochastic procedures for breaking base pairs because this process is always entropically favorable.

The total potential energy E used for DMD simulation of RNA molecules is calculated as a sum of all interaction terms:

$$E = E_{\mathrm{Bonded}} + E_{\mathrm{Hbond}} + E_{\mathrm{Stack}} + E_{\mathrm{Hydrophobic}} + E_{\mathrm{Overpacking}} + G_{\mathrm{loop}}$$

3.2. Prediction of small RNA structure

The described method was tested on a set of 153 short RNAs, ranging in length from 10 to 100 nucleotides (Ding et al., 2008). All of these RNAs have experimentally obtained 3D structures deposited in the PDB. We used replica exchange DMD simulations in order to increase sampling efficiency (Sugita & Okamoto, 1999). In each simulation, we ran eight replicas at temperatures: 0.200, 0.208, 0.214, 0.220, 0.225, 0.230, 0.235, and 0.240. The temperature is given in abstract units of kcal/(mol k_B). The replicas were allowed to exchange every 2000 DMD time steps, with 2×10^6 total time steps. The quality of reconstructed structures was monitored using two observables: (1) root-mean-square deviation (RMSD), calculated between backbone phosphate atoms of the model and experimental structure, (2) the fraction of correctly predicted native base pairs (Q-value). We observed that for RNAs with length less than 50 nucleotides, the predicted RMSD fell below 6 Å (Fig. 3A). Models of longer RNA molecules demonstrated larger RMSD; however, 84% of all predicted structures have RMSD less than 4 Å. We also observed that predicted structures corresponding to the lowest free energy do not generally have the lowest RMSD. It can be attributed to the inaccuracy of the force field and coarse-graining nature of the RNA model. For the majority of the lowest free-energy RNA structures, the observed Q-values were close to unity (Fig. 3B), suggesting the correct formation of native base pairs. The average Q-value for all 153 RNA molecules was 94%. For comparison with existing secondary structure prediction methods, we calculated the Q-values using Mfold (Zuker, 2003). Its average Q-value was 91%.

Pseudoknot structure (PDB code: 1A60) demonstrates an example of the performance of our method. This pseudoknot represents the T-arm and acceptor stem of the turnip yellow mosaic virus (TYMV) and shares structural similarity with TYMV genomic tRNA (Kolk et al., 1998). We found the RMSD of the lowest free-energy structure to be 4.58 Å, while the RMSD of the structure with minimal deviation from the crystal structure was 2.03 Å (Fig. 4A). We studied the folding thermodynamics of this RNA using weighted histogram analysis method (Kumar, Rosenberg, Bouzida, Swendsen, & Kollman, 1992) as implemented in the MMTSB toolset (Feig, Karanicolas, & Brooks, 2004). The specific heat (Fig. 4B) has two peaks: one peak around $T^* = 0.245$ and another one around $T^* = 0.21$. This result suggests the presence of intermediate states in the folding pathway. To characterize folding intermediates, we computed the two-dimensional potential of mean force (2D-PMF) as a function of

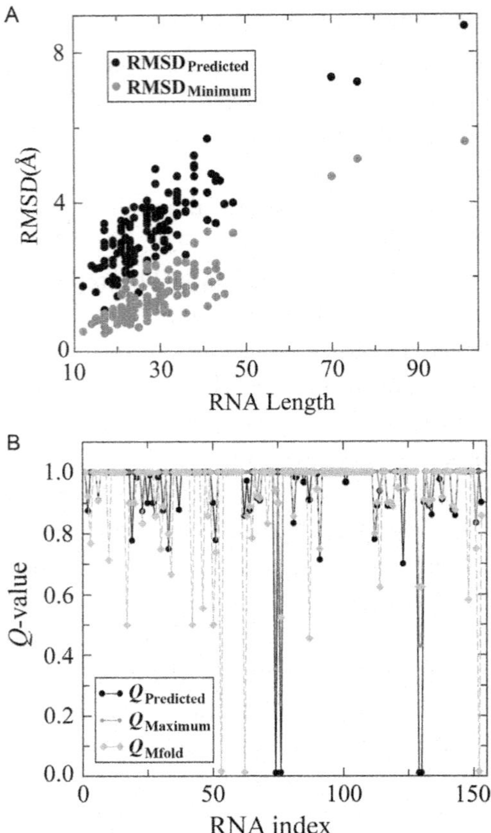

Figure 3 (A) RMSD calculated between RNA structure obtained in the course of DMD simulations and experimental structure from the PDB. Black points correspond to the predicted structures with the lowest free energy and red points correspond to structures with minimum RMSD. (B) *Q*-values for the predicted structures with the lowest free energy (black dots), the maximum *Q*-values (red dots) and structures obtained using Mfold (green dots). *Adapted from Ding et al. (2008) with permission.* (See the color plate.)

the total number of base pairs and the number of native base pairs at two peak temperatures (Fig. 4C and D). The 2D-PMF plots show two intermediate states with distinct free-energy basins. The first state corresponds to the folded 5′ hairpin, while the second state corresponds to the formation of one of the helix stems in the 3′ pseudoknot. Progression of the RNA folding toward the native state can be demonstrated at the contact frequencies of the folding intermediates (Fig. 4E and F) and the contact map of the native state (Fig. 4G).

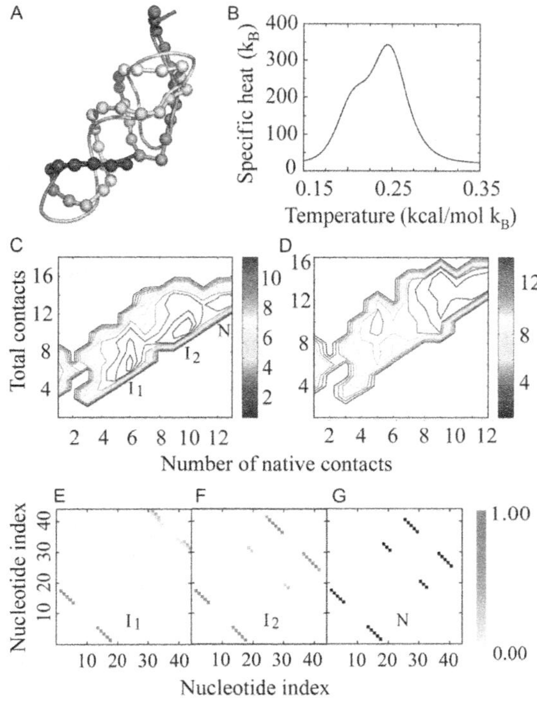

Figure 4 Folding kinetics of pseudoknot RNA. (A) Structure obtained as a result of DMD simulation (ribbon backbone trace with backbone spheres) superimposed over experimentally determined RNA structure (PDB code: 1A60). Backbone ribbons are colored from blue (N terminus) to red (C-terminus). (B) Specific heat as a function of temperature. (C) Two-dimensional potential of mean force (2D-PMF) for pseudoknot folding at $T^* = 0.245$ (the major peak in the specific heat). I_1, I_2, and N denote two intermediate states and native state correspondingly. (D) 2D-PMF at $T^* = 0.21$. (E) Contact frequencies at the I_1 intermediate state, corresponding to the formation of 5′ hairpin. (F) Contact frequencies at the I_2 intermediate state, corresponding to the formation of the helix stem in the 3′ pseudoknot. (G) Contact map of the native state as observed in the experimental NMR structure (PDB code: 1A60). *Adapted from Ding et al. (2008) with permission.* (See the color plate.)

4. USE OF HYDROXYL-RADICAL PROBING TO REFINE RNA THREE-DIMENSIONAL STRUCTURE

4.1. Folding RNA with experimental constraints

The quality of RNA tertiary structure prediction worsens with the increase of RNA length. There are two major reasons for this: (1) the inaccuracy of existing force fields, in particular, describing electrostatics by point charges and ignoring polarizability effects, and (2) the need to sample larger conformational space in order to find the correct RNA fold. However, if any structural experimental data is available, it can be used to constrain RNA simulations and enhance sampling efficiency. For example, *a priori* knowledge of RNA secondary structure greatly restrains possible tertiary folds (Bailor, Mustoe, Brooks, & Al-Hashimi, 2011; Hajdin, Ding, Dokholyan, & Weeks, 2010). The secondary structure can be determined with high fidelity using comparative sequence analysis (Cannone et al., 2002; Gutell, Power, Hertz, Putz, & Stormo, 1992; Michel & Westhof, 1990) or SHAPE experiments (Deigan, Li, Mathews, & Weeks, 2008; Weeks, 2010). The conformational space can be constrained even further if the experimental information on the tertiary contacts, not involved in base-pairing, is available (Gherghe, Leonard, Ding, Dokholyan, & Weeks, 2009; Lavender, Ding, Dokholyan & Weeks, 2010). Here, we describe a method that relies on the information extracted from hydroxyl-radical probing (HRP) experiments to constrain RNA folds (Ding et al., 2012). This method shows high efficiency in predicting the structure of long RNAs. HRP experiments do not provide direct information about tertiary contacts between any particular pair of nucleotides. Instead they probe the solvent accessibility of nucleotides that can be related to the expected number of contacts that each particular nucleotide makes with its through-space neighbors. In general, nucleotides showing lower HRP reactivity are more buried and have more through-space contacts. This information can be translated into statistics-based potential biasing RNA toward correct 3D structure.

4.2. Definition of through-space contacts

The bias potential introduced here is based on the number of contacts that every nucleotide makes with its through-space neighbors. We define such contacts based on the distance between sugar pseudoatoms in our three-bead model. We exclude nucleotides located within four nucleotides in the sequence, and within four nucleotides from the nucleotide forming a base pair with the nucleotide in question, as these contacts reflect primary and secondary structure rather than higher order tertiary contacts. The number of through-space neighbor contacts depends on the cutoff distance d_{cutoff} used to define the contact. If this distance is too small, the number of contacts will be underestimated. On the other hand if the cutoff distance is too large, the number of contacts will be overestimated, and will not correlate with the

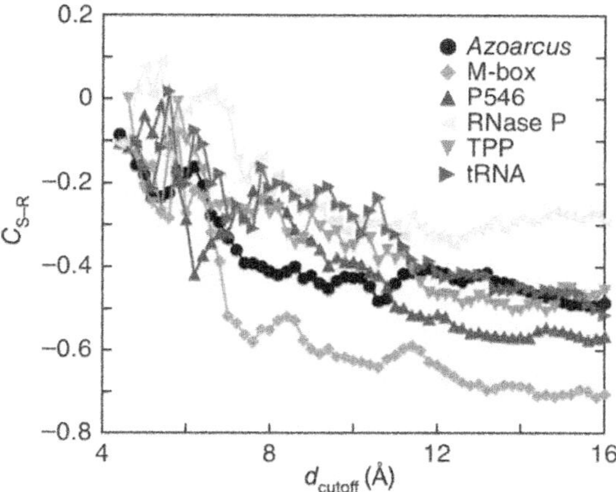

Figure 5 The structure–reactivity correlation, C_{S-R}, as a function of the cutoff distance, used to define a contact. HRP reactivities were smoothed over a sliding window of three nucleotides to reduce noise (as described in the text). The data points are shown for six RNAs from the training set (Table 1). *Adapted from Ding et al. (2012) with permission.*

values of HRP reactivity. In order to optimize d_{cutoff}, we calculated how the structure–reactivity correlation C_{S-R} depends on d_{cutoff}, where C_{S-R} was defined as a Pearson correlation coefficient between nucleotide HRP reactivity and the number of contacts it makes within cutoff distance (Fig. 5). We used data collected from six RNAs as a training set to optimize model parameters (Table 1). The calculated C_{S-R} is negative because a lower HRP reactivity corresponds to a more buried nucleotide with a larger number of through-space contacts. We noticed that at $d_{cutoff} = 14$ Å, C_{S-R} reaches its minimum value. This optimized value of d_{cutoff} is further used to calculate the number of contacts.

4.3. HRP bias potential

Our HRP bias potential consists of two terms. The first term is the sum of pairwise attractive potentials assigned to every pair of nucleotides. The second term is an overburial repulsion potential that penalizes the energy function when a given nucleotide exceeds its maximum number of allowed contacts (N_{max}) (Fig. 6A). The strength of attraction and repulsion potentials as well as N_{max} is assigned to every nucleotide depending on its HRP reactivity. In general, the bias interaction potential can be written as:

$$E_{bias} = \sum_{i<j} E_{ij}^{attr} + \sum_{i} E_{i}^{overbury}$$

Table 1 The first six RNAs comprise the training set used for algorithm optimization and determination of its applicable range and the last four RNAs comprise the verification set used to rate the performance of the algorithm

				Large clusters			
RNA	**Length (nt)**	***f(0.25)***	C_{S-R}	**Number of clusters**	***n* (out of 100)**	**RMSD (Å)**	***P*-value**
Azoarcus group I intron	214	**0.43**	−0.45	1	100	16.8±2.1	$<10^{-6}$
M-box riboswitch	161	**0.35**	−0.68	1	100	11.6±2.3	$<10^{-6}$
P546 domain	158	**0.37**	−0.57	3	66	19.8±1.4	3.0×10^{-3}
					32	15.1±1.9	$<10^{-6}$
RNase P specificity domain	152	0.25	−0.30	3	93	24.9±1.2	0.67
					4	24.1±3.2	0.50
					3	22.7±0.5	0.22
TPP riboswitch	80	0.21	−0.50	4	44	12.6±1.3	0.10
					42	14.6±1.9	0.44
					12	9.9±1.8	2.9×10^{-3}
$tRNA^{Asp}$	75	0.25	−0.45	8	29	14.1±1.3	0.50
					23	17.5±1.5	0.97
					18	16.4±0.8	0.90
					8	18.9±0.8	0.99
					6	12.3±1.5	0.17
					6	15.7±0.4	0.82
					5	18.9±0.8	0.99
O. iheyensis group II intron	412	0.21	−0.30	–	–	–	–
RNase P catalytic domain	231	**0.28**	−0.50	4	46	19.2±1.6	$<10^{-6}$
					41	21.6±2.1	$<10^{-6}$
					8	25.0±0.7	1.3×10^{-4}
					5	24.4±2.0	3.5×10^{-5}

Table 1 The first six RNAs comprise the training set used for algorithm optimization and determination of its applicable range and the last four RNAs comprise the verification set used to rate the performance of the algorithm—cont'd

				Large clusters			
RNA	**Length (nt)**	***f(0.25)***	C_{S-R}	**Number of clusters**	***n* (out of 100)**	**RMSD (Å)**	***P*-value**
Lysine riboswitch	174	**0.36**	−0.57	3	57	12.0 ± 1.6	$<10^{-6}$
					42	18.1 ± 1.0	1.6×10^{-6}
glmS ribozyme	152	**0.35**	−0.55	2	74	16.6 ± 1.7	1.5×10^{-5}
					26	8.5 ± 1.3	$<10^{-6}$

The structure–reactivity correlation, C_{S-R}, was calculated with reference to the experimental structure. For each RNA, one hundred of the lowest free-energy structures were selected and clustered with the cluster centroid chosen as a representative structure. Low populated clusters with less than two structures were excluded. RMSD was calculated between the centroid of a cluster and the corresponding experimental structure. *P*-values were calculated based on the average RMSD with respect to the accepted experimental structure. Structure refinement was not performed for *O. iheyensis* group II intron RNA in the test set because of the low *f*(0.25) value.
Adapted from Ding et al. (2012) with permission.

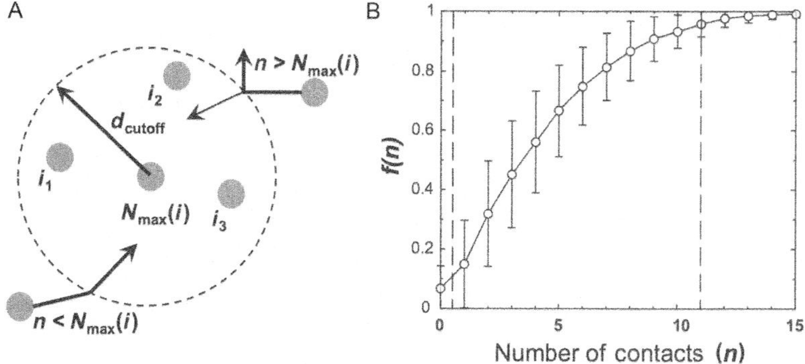

Figure 6 (A) Two nucleotides make a contact if they fall within cutoff distance, d_{cutoff}. Every nucleotide is assigned a threshold number of contacts N_{max}. If the number of already existing contacts is less than N_{max}, a new contact is energetically favorable (shown by an inward arrow). However, if the number of contacts exceeds N_{max}, every new approaching nucleotide can form a contact only if the total DMD kinetic energy is sufficient to overcome the energy penalty for overpacking; otherwise it bounces back (shown by outward arrow). (B) The cumulative distribution of the fraction of nucleotides, *f*(*n*), as a function of the number of contacts, *n*, as follows from statistical analysis of all single chain RNAs in the PDB. The values of NC_{min} and NC_{max} are shown by vertical dashed lines. *Adapted from Ding et al. (2012) with permission.*

Here, $E_{ij}^{\text{attr}} = \min\{E_{\text{attr}}(i), E_{\text{attr}}(j)\}F(r_{ij})$. $E_{\text{attr}}(i)$ defines the strength of attractive interaction for every particular nucleotide based on its reactivity value, and

$$F(x) = \begin{cases} 0 & d_{\text{cutoff}} < r \\ 1 & R_{\text{hc}} < r < d_{\text{cutoff}} \\ \infty & r \leq R_{\text{hc}} \end{cases}$$

Here, $d_{\text{cutoff}} = 14$ Å (as defined above), and R_{hc} is a hardcore diameter, 3.0 Å.

$$E^{\text{overbury}}(i) = \mathrm{d}E_{\text{rep}}(i)(n_{\text{C}}(i) - N_{\max}(i))\Theta(n_{\text{C}}(i) - N_{\max}(i))$$

Here, $n_{\text{C}}(i)$ is the number of contacts formed by nucleotide i, $\mathrm{d}E_{\text{rep}}(i)$ is the penalty energy for overburying, which depends on HRP-reactivity value of nucleotide i, and Θ is a unit step function:

$$\Theta(x) = \begin{cases} 0 & x < 0 \\ 1 & x \geq 0 \end{cases}$$

4.4. Assignment of interaction parameters

The maximally and minimally buried nucleotides make the largest ($NC_{\max}$) and smallest ($NC_{\min}$) possible number of through-space contacts, respectively. To determine these numbers, we performed a statistical analysis of contacts between nucleotides in all single chain RNAs from the PDB (Berman et al., 2000). Based on the cumulative distribution of the fraction of nucleotides $f(n)$ as a function of the number of contacts n (Fig. 6B), we derived that $NC_{\min} = 0.5$ and $NC_{\max} = 11$. By choosing these values, we avoided the tails of the $f(n)$ distribution. We want to assign $NC_{\min}$ and $NC_{\max}$ to nucleotides exhibiting maximal and minimal reactivity values, correspondingly. HRP experiments are intrinsically noisy, so to account for the noise we average reactivities over a sliding window of three nucleotides. We also define threshold values of HRP reactivity $R_{\min}$ and $R_{\max}$. Nucleotides with reactivity R, lower than $R_{\min}$ are maximally buried and assigned $N_{\max} = NC_{\max}$, while nucleotides exceeding $R_{\max}$ are maximally exposed and assigned $N_{\max} = NC_{\min}$. The $R_{\min}$ and $R_{\max}$ values are calculated as averages of subsets of rank-ordered R values from 2% to 20% and from 80% to 98% correspondingly. The top and bottom 2% of R values are not taken into account to reduce the effect of extreme R values observed

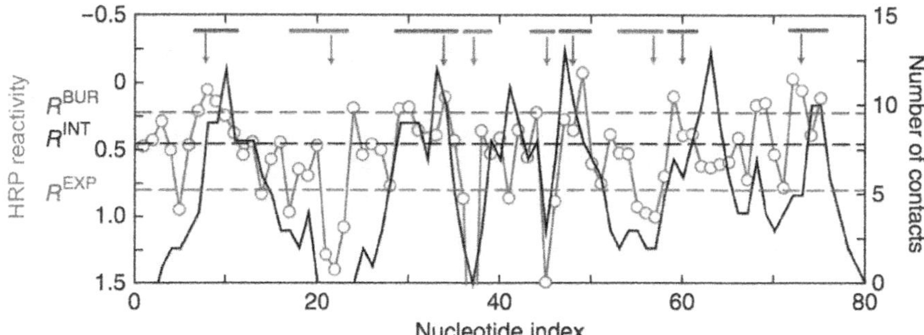

Figure 7 Experimentally measured HRP reactivities (red (light gray in the print version)) and the number of through-space contacts (black) for TPP riboswitch. Buried/exposed nucleotide segments are highlighted in blue (dark gray in the print version)/red (light gray in the print version) bars at the top of the figure. Arrows pointing downward show nucleotides, which can be identified as buried or exposed with high confidence. The dashed horizontal lines represent threshold values used to define exposed (R^{EXP}), intermediate (R^{INT}), or buried (R^{BUR}) nucleotides. *Adapted from Ding et al. (2012) with permission.*

in a typical HRP experiment. For nucleotides with intermediate reactivities ($R_{min} < R < R_{max}$), we define N_{max} as

$$N_{max}(i) = NC_{max} + (NC_{min} - NC_{max})(R(i) - \langle R_{min} \rangle)/(\langle R_{max} \rangle - \langle R_{min} \rangle)$$

The intrinsic noise inherent to HRP experiments (Fig. 7) render estimations of nucleotide solvent exposure less reliable. However, we can identify nucleotides as buried or exposed with high confidence if those nucleotides are found in buried or exposed RNA segments. To define buried or exposed segments, we first define three values, R^{EXP}, R^{INT}, and R^{BUR}, corresponding to the threshold values of exposed, intermediate, and buried residues. Intermediate R values can be defined in two different ways: as an average value of the reactivities, $\langle R \rangle$, or as the median value of the rank-ordered reactivities. To avoid ambiguity we use the mean of these two values, denoted as R^{INT}. We define buried segments as segments with more than three consecutive nucleotides having $R < R^{INT}$ and at least one nucleotide having $R < R^{BUR}$. Inside each buried segment, a nucleotide with the lowest R (excluding the first and the last nucleotide in the segment) is considered as buried with high confidence. Exposed segments are defined as segments with more than three consecutive nucleotides having $R > R^{\mathrm{INT}}$, and at least one nucleotide having $R > R^{\mathrm{EXP}}$, and for two-nucleotide segments with both nucleotides having $R > R^{\mathrm{EXP}}$. Inside each exposed segment, a nucleotide having the largest reactivity is considered as exposed with high confidence.

Two attractive energy scales were used $E_{\mathrm{attr}}^{\mathrm{low}} = -10\,\mathrm{kcal/mol}$ and $E_{\mathrm{attr}}^{\mathrm{high}} = -0.05\,\mathrm{kcal/mol}$. The strong attractive energy $E_{\mathrm{attr}}^{\mathrm{low}}$ was assigned to the buried nucleotides identified with high confidence, and intermediate values $(E_{\mathrm{attr}}^{\mathrm{low}} + E_{\mathrm{attr}}^{\mathrm{high}})/2$ were assigned to their immediate neighbors. For all remaining nucleotides $E_{\mathrm{attr}}^{\mathrm{high}}$ was assigned. The strong repulsive overburial energy was assigned to all high-confidence buried and exposed nucleotides. The repulsive energy $\mathrm{d}E_{\mathrm{rep}}(i) = -E_{\mathrm{attr}}(i)$ was assigned to all other nucleotides, where $E_{\mathrm{attr}}(i)$ is equal to $E_{\mathrm{attr}}^{\mathrm{low}}$ or $E_{\mathrm{attr}}^{\mathrm{high}}$. By making overburial repulsive potential equal to attractive potential, we allow nucleotides to make additional contacts without a net energy penalty.

4.5. Simulation protocol

The simulation protocol, which we apply along with HRP bias potential, is an essential part of this method. The aim of this protocol is to increase the sampling efficiency of conformational space and hence increase the precision of our method. The protocol consists of three steps (Fig. 8). At the first step, we perform DMD simulations with the input RNA sequence and base-pairing constraints. This allows us to form an initial 3D structure of modeled RNA. For all the RNAs shown in Table 1, high-resolution structures are available in PDB (Berman et al., 2000). However in the course of *de novo* prediction, reliable base-pairing information can be obtained from sequence comparative analysis (Cannone et al., 2002; Gutell et al., 1992; Michel & Westhof, 1990) or from SHAPE data (Deigan et al., 2008; Weeks, 2010). At the second step, after the formation of native secondary structure, we perform a second round of simulations with HRP bias potential (Fig. 2C). After each round of simulations, we check if the experimental constraints are satisfied and if they are not, we repeat the simulation. In order to sample conformational space of RNAs more efficiently, we use DMD with replica exchange. At the third step, we select one hundred structures based on low energy and high $C_{\mathrm{S-R}}$ values. We cluster these structures and take centroids of every cluster as a representative structure.

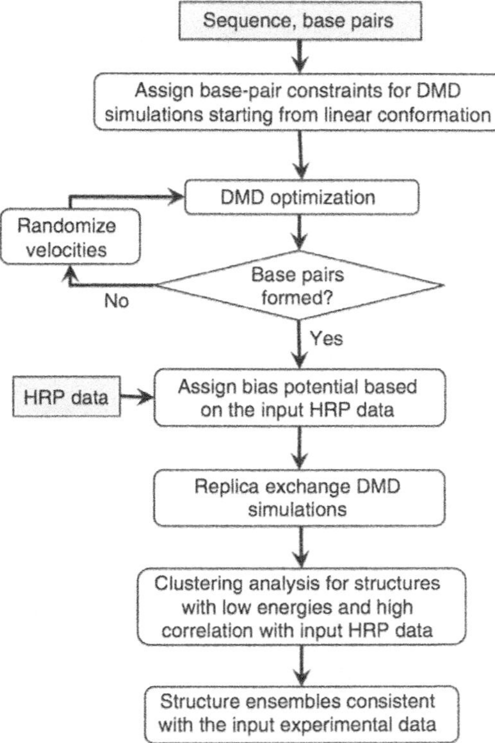

Figure 8 RNA simulation algorithm with the use of base-pairing constraints and constraints from hydroxyl-radical probing experiments. *Adapted from Ding et al. (2012) with permission.*

4.6. Results, significance of predictions, and the scope of applicability of the method

We optimized the modeling procedure on a training set of six RNA: the *Azoarcus* group I intron, the M-box riboswitch, the P546 domain of the *Tetrahymena thermophila* group I intron, the RNase P specificity domain, the thiamine pyrophosphate (TPP) riboswitch, and the yeast aspartic acid transfer ($tRNA^{Asp}$) RNA (Table 1). These are structurally diverse RNAs, with lengths ranging from 75 to 214 nucleotides. We used replica exchange DMD with eight replicas at the temperatures: 0.200, 0.225, 0.250, 0.270, 0.300, 0.333, 0.367, and 0.400. Replicas were exchanged every 1000

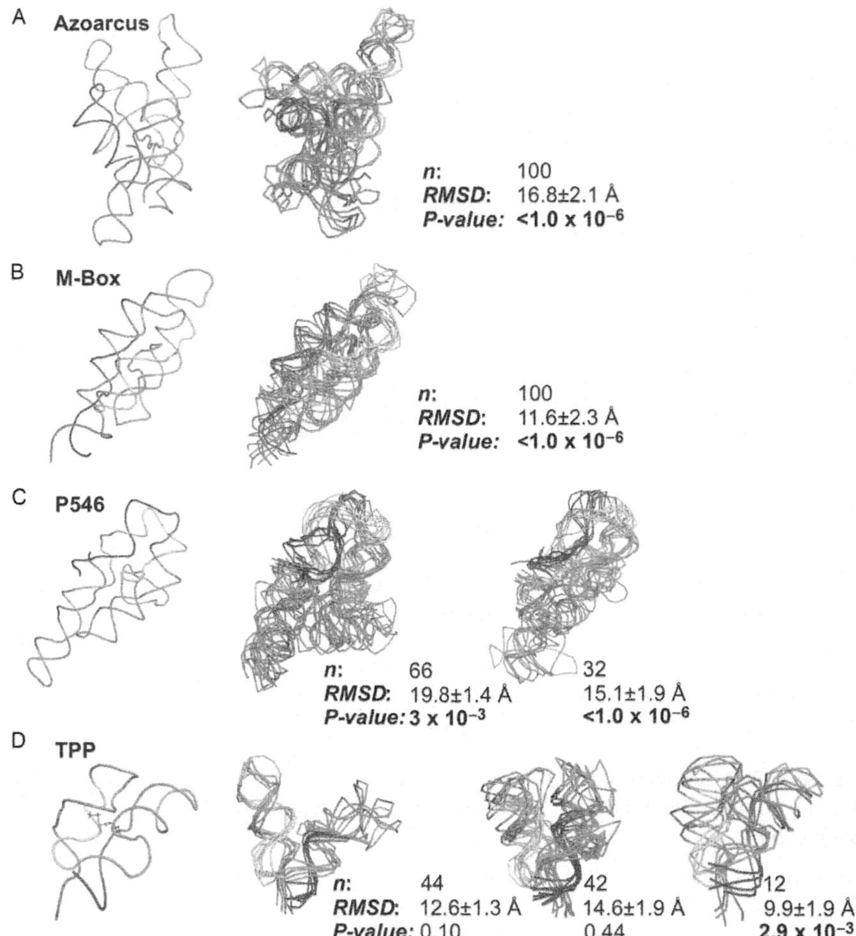

Figure 9 Predicted structures of RNAs from the training set. RNAs are shown with backbone traces. The accepted structures are shown on the left. Representative structures for each of the highly populated cluster are shown on the right. Backbones are colored from blue to red in the direction from 5′ to 3′. The mean RMSD and *P*-values are shown, with significant *P*-values emphasized in bold. *Adapted from Ding et al. (2012) with permission.* (See the color plate.)

DMD time steps. The simulations were run for 5×10^5 DMD time steps. The temperature is measured in abstract units of kcal/(mol k_B). To evaluate the accuracy of our method we compared predicted structures with available structures in the PDB (Table 1; Fig. 9). In particular, for each representative predicted structure we calculated RMSD relative to the structure from the PDB, and a *P*-value showing the significance of the prediction. *P*-values lower than 0.01 correspond to highly statistically significant predictions (Hajdin et al., 2010). It is interesting to note that observed RMSD values corresponding to significant predictions are different for RNAs of different length. For example, a prediction with an RMSD of 7–10 Å is not significant for small RNAs, but is highly significant for RNAs beyond 100 nucleotides. The structures of the three largest RNAs in the training set were predicted with high significance values (Table 1). In addition, all of the predicted structures for the *Azoarcus* group I intron and for M-box riboswitch fell into a single cluster and the P546 domain showed two resulting clusters. The structures for TPP riboswitch fell into three clusters, and only one of them was highly significant. For $tRNA^{Asp}$ and RNase P specificity domain no predictions were statistically significant (Table 1). We noticed that the quality of prediction is related to the packing density of RNA. Those RNAs with higher order RNA packing returned better predictions. The extent of RNA packing can be estimated *a priori* as a fraction of nucleotides $f(r)$ falling below a given reactivity r. At $r=0.25$, the RNAs with extensive packing (the *Azoarcus* group I intron, M-box riboswitch, and P546 domain) have significantly larger $f(r)$ values than other RNAs (Table 1). We determined that our HRP directed refinement method generally predicts correct structure for RNAs with $f(0.25)>0.25$. After the optimization, we tested our methods on a set of four RNAs: *Oceanobacillus iheyensis* group II intron, RNase P catalytic domain, lysine riboswitch, and the glmS ribozyme (ranged from 152 to 412 nucleotides in length). We observed that three RNAs from the training set yielded highly significant predicted structures grouped in a few clusters. For the last RNA (*O. iheyensis* group II intron), we did not find any reasonable structures at all, which confirms the empirically derived rule of $f(0.25)$.

5. ALL-ATOM STRUCTURE RECONSTRUCTION

We developed an algorithm to translate RNA from a three-bead representation to an all-atom structure.

The algorithm consists of two steps. In the first step, we build an initial structure from template A, G, C, and U nucleotides, which are randomly chosen from the PDB (Berman et al., 2000). To accomplish this step, we define a three-bead representation to every template nucleotide and determine an optimal translation and rotation that superimposes the three-bead template nucleotide over the three-bead nucleotide from the RNA. The resulting transformation of coordinate system applied to all atoms of template nucleotides provides an initial all-atom RNA structure, with many nonideal bonds, angles, and dihedral angles between adjacent nucleotides. In the second step, we further refine the all-atom structure through all-atom DMD simulations to remove clashes and connect bonds. We also impose base-pairing constraints derived from the coarse-grained simulations (or experimental input). Our all-atom DMD relies on the Medusa force field (Yin, Biedermannova, Vondrasek, & Dokholyan, 2008). Severe clashes in the initial all-atom structure can lead to instantaneous heating of the RNA and cause large distortions compared to the initial three-bead model before the system cools down to the thermostat temperature. We overcome this obstacle using a high heat-exchange coefficient, which allows the system to cool quickly thus preventing it from destruction.

6. iFoldRNA

We created and support the iFoldRNA Web server (Sharma et al., 2008) for the prediction of RNA structure and the analysis of various thermodynamic properties of RNA folding. iFoldRNA operates in two different prediction modes. In the first mode, RNA structures can be predicted *de novo* without any additional constraints. This approach works well for short RNAs with a length of RNA limited to 50 nucleotides at most (Ding et al., 2008). In the second mode, a user can provide base-pairing constraints and an optional HRP-reactivity profile. If an HRP-reactivity profile is provided, the additional biasing potential is used to guide RNA folding (Ding et al., 2012). To sample RNA conformational space more efficiently, the server runs eight replicas at different temperatures. The default values of replica temperatures, replica exchange time and number of steps are the same as described in the text above for the unconstraint and constraint simulations correspondingly. Constant temperature throughout the course of the simulation is supported by the Andersen thermostat (Andersen, 1980). To perform simulations, we use a Linux-based computing cluster with more than 1840 computing cores from the University of North Carolina. During the simulation, the state of the system in each of the eight trajectories is saved every 100 DMD steps. After the DMD simulation is finished, saved states are subjected to automated clustering. The centroids of these clusters are selected as representative structures. The default value for the number of output clusters is 1. The Web interface of iFoldRNA server allows one to rewrite the default values for the maximum number of simulation DMD time steps, replica temperatures, and the number of output clusters.

For each of the representative structures, we reconstruct an all-atom geometry based on the three-bead model obtained from the previous steps of the simulation. Simulation time depends on the size of RNA molecule and on the constraints applied during the simulation. We observed that simulation timescales linearly with RNA size. As an example, the simulation of M-box riboswitch (160 nucleotides) takes 170 CPU hours. The wall-clock time is one-eighth of the total CPU time.

7. CONCLUSIONS

Computational modeling can be used to obtain information about RNA 3D structure and functions that are otherwise difficult and expensive to obtain experimentally. Methods like DMD and geometry coarse graining are especially important in this respect because they allow access to biologically relevant timescales, which is beyond the reach of ordinary MD in the extreme complexity of RNA modeling. We have developed an RNA simulation package based on our coarse-grained three-bead RNA model coupled to an efficient implementation of MD, known as DMD. Our approach efficiently predicts the structure of short RNAs (less than 50 nucleotides) *de novo*, from only RNA sequence. Additionally, experimental constraints, such as base-pairing information or reactivities derived from HRP experiments, allow us to get precise predictions for RNAs in the range of a few hundred nucleotides. Our method consistently shows good performance at the RNA-Puzzle competition (Cruz et al., 2012). RNA modeling is a rapidly developing field and in order to keep pace we continue to develop our method. In particular, we are currently optimizing the Medusa force field used to run all-atom RNA simulations. We are also working to develop better ways to incorporate experimental constraints into DMD simulations.

REFERENCES

Addess, K. J., & Feigon, J. (1996). Introduction to 1H NMR spectroscopy of DNA. In S. M. Hecht (Ed.), *Bioorganic chemistry: Nucleic acids* (pp. 163–185). New York: Oxford University Press.

Andersen, H. C. (1980). Molecular dynamics simulations at constant pressure and/or temperature. *The Journal of Chemical Physics*, *72*(4), 2384–2393.

Bailor, M. H., Mustoe, A. M., Brooks, C. L., III, & Al-Hashimi, H. M. (2011). Topological constraints: Using RNA secondary structure to model 3D conformation, folding pathways, and dynamic adaptation. *Current Opinion in Structural Biology*, *21*(3), 296–305.

Berman, H. M., Westbrook, J., Feng, Z., Gilliland, G., Bhat, T. N., Weissig, H., et al. (2000). The protein data bank. *Nucleic Acids Research*, *28*(1), 235–242.

Cannone, J. J., Subramanian, S., Schnare, M. N., Collett, J. R., D'Souza, L. M., Du, Y., et al. (2002). The comparative RNA web (CRW) site: An online database of comparative sequence and structure information for ribosomal, intron, and other RNAs. *BMC Bioinformatics, 3*(1), 2.

Cornell, W. D., Cieplak, P., Bayly, C. I., Gould, I. R., Merz, K. M., Ferguson, D. M., et al. (1995). A second generation force field for the simulation of proteins, nucleic acids, and organic molecules. *Journal of the American Chemical Society, 117*(19), 5179–5197.

Cruz, J. A., Blanchet, M. F., Boniecki, M., Bujnicki, J. M., Chen, S. J., Cao, S., et al. (2012). RNA-Puzzles: A CASP-like evaluation of RNA three-dimensional structure prediction. *RNA, 18*(4), 610–625.

Deigan, K. E., Li, T. W., Mathews, D. H., & Weeks, K. M. (2008). Accurate SHAPE-directed RNA structure determination. *Proceedings of the National Academy of Sciences, 106*(1), 97–102, pnas-0806929106.

Ding, F., Borreguero, J. M., Buldyrey, S. V., Stanley, H. E., & Dokholyan, N. V. (2003). Mechanism for the a-helix to b-hairpin transition. *Proteins: Structure, Function, and Genetics, 53*, 220–228.

Ding, F., Lavender, C. A., Weeks, K. M., & Dokholyan, N. V. (2012). Three-dimensional RNA structure refinement by hydroxyl radical probing. *Nature Methods, 9*(6), 603–608.

Ding, F., Sharma, S., Chalasani, P., Demidov, V. V., Broude, N. E., & Dokholyan, N. V. (2008). Ab initio RNA folding by discrete molecular dynamics: From structure prediction to folding mechanisms. *RNA, 14*(6), 1164–1173.

Dokholyan, N. V., Buldyrev, S. V., Stanley, H. E., & Shakhnovich, E. I. (1998). Discrete molecular dynamics studies of the folding of a protein-like model. *Folding and Design, 3*(6), 577–587.

Feig, M., Karanicolas, J., & Brooks, C. L., III. (2004). MMTSB Tool Set: Enhanced sampling and multiscale modeling methods for applications in structural biology. *Journal of Molecular Graphics and Modelling, 22*(5), 377–395.

Freddolino, P. L., Harrison, C. B., Liu, Y., & Schulten, K. (2010). Challenges in protein-folding simulations. *Nature Physics, 6*(10), 751–758.

Gherghe, C. M., Leonard, C. W., Ding, F., Dokholyan, N. V., & Weeks, K. M. (2009). Native-like RNA tertiary structures using a sequence-encoded cleavage agent and refinement by discrete molecular dynamics. *Journal of the American Chemical Society, 131*(7), 2541–2546.

Gutell, R. R., Power, A., Hertz, G. Z., Putz, E. J., & Stormo, G. D. (1992). Identifying constraints on the higher-order structure of RNA: Continued development and application of comparative sequence analysis methods. *Nucleic Acids Research, 20*(21), 5785–5795.

Hajdin, C. E., Ding, F., Dokholyan, N. V., & Weeks, K. M. (2010). On the significance of an RNA tertiary structure prediction. *RNA, 16*(7), 1340–1349.

Kolk, M. H., van der Graaf, M., Wijmenga, S. S., Pleij, C. W., Heus, H. A., & Hilbers, C. W. (1998). NMR structure of a classical pseudoknot: Interplay of single-and double-stranded RNA. *Science, 280*(5362), 434–438.

Kumar, S., Rosenberg, J. M., Bouzida, D., Swendsen, R. H., & Kollman, P. A. (1992). The weighted histogram analysis method for free-energy calculations on biomolecules. I. The method. *Journal of Computational Chemistry, 13*(8), 1011–1021.

Lavender, C. A., Ding, F., Dokholyan, N. V., & Weeks, K. M. (2010). Robust and generic RNA modeling using inferred constraints: A structure for the hepatitis C virus IRES pseudoknot domain. *Biochemistry, 49*(24), 4931–4933.

Mathews, D. H., Sabina, J., Zuker, M., & Turner, D. H. (1999). Expanded sequence dependence of thermodynamic parameters improves prediction of RNA secondary structure. *Journal of Molecular Biology, 288*(5), 911–940.

McDowell, S. E., Špačková, N., Šponer, J., & Walter, N. G. (2007). Molecular dynamics simulations of RNA: An in silico single molecule approach. *Biopolymers*, *85*(2), 169–184.
Michel, F., & Westhof, E. (1990). Modelling of the three-dimensional architecture of group I catalytic introns based on comparative sequence analysis. *Journal of Molecular Biology*, *216*(3), 585–610.
Pérez, A., Marchán, I., Svozil, D., Sponer, J., Cheatham, T. E., III, Laughton, C. A., et al. (2007). Refinement of the AMBER force field for nucleic acids: Improving the description of α/γ conformers. *Biophysical Journal*, *92*(11), 3817–3829.
Ponder, J. W., & Case, D. A. (2003). Force fields for protein simulations. *Advances in Protein Chemistry*, *66*, 27–86.
Proctor, E. A., Ding, F., & Dokholyan, N. V. (2011). Discrete molecular dynamics. *Wiley Interdisciplinary Reviews: Computational Molecular Science*, *1*(1), 80–92.
Reyes, F. E., Garst, A. D., & Batey, R. T. (2009). Strategies in RNA crystallography. *Methods in Enzymology*, *469*, 119–139.
Serganov, A., & Nudler, E. (2013). A decade of riboswitches. *Cell*, *152*(1), 17–24.
Serganov, A., & Patel, D. J. (2007). Ribozymes, riboswitches and beyond: Regulation of gene expression without proteins. *Nature Reviews Genetics*, *8*(10), 776–790.
Sharma, S., Ding, F., & Dokholyan, N. V. (2008). iFoldRNA: Three-dimensional RNA structure prediction and folding. *Bioinformatics*, *24*(17), 1951–1952. http://troll.med.unc.edu/ifoldrna.v2/index.php.
Shirvanyants, D., Ding, F., Tsao, D., Ramachandran, S., & Dokholyan, N. V. (2012). Discrete molecular dynamics: An efficient and versatile simulation method for fine protein characterization. *The Journal of Physical Chemistry B*, *116*(29), 8375–8382.
Sim, A. Y., Minary, P., & Levitt, M. (2012). Modeling nucleic acids. *Current Opinion in Structural Biology*, *22*(3), 273–278.
Sugita, Y., & Okamoto, Y. (1999). Replica-exchange molecular dynamics method for protein folding. *Chemical Physics Letters*, *314*(1), 141–151.
Tinoco, I., Jr., & Bustamante, C. (1999). How RNA folds. *Journal of Molecular Biology*, *293*(2), 271–281.
Weeks, K. M. (2010). Advances in RNA structure analysis by chemical probing. *Current Opinion in Structural Biology*, *20*(3), 295–304.
Yin, S., Biedermannova, L., Vondrasek, J., & Dokholyan, N. V. (2008). MedusaScore: An accurate force field-based scoring function for virtual drug screening. *Journal of Chemical Information and Modeling*, *48*(8), 1656–1662.
Zhou, Y., Karplus, M., Wichert, J. M., & Hall, C. K. (1997). Equilibrium thermodynamics of homopolymers and clusters: Molecular dynamics and Monte Carlo simulations of systems with square-well interactions. *The Journal of Chemical Physics*, *107*(24), 10691–10708.
Zuker, M. (2003). Mfold web server for nucleic acid folding and hybridization prediction. *Nucleic Acids Research*, *31*(13), 3406–3415.

CHAPTER FOUR

Improving RNA Secondary Structure Prediction with Structure Mapping Data

Michael F. Sloma*,†, David H. Mathews[1],*,†

*Department of Biochemistry & Biophysics, University of Rochester Medical Center, Box 712, Rochester, New York, USA

†Center for RNA Biology, University of Rochester Medical Center, Box 712, Rochester, New York, USA

[1]Corresponding author: e-mail address: David_Mathews@urmc.rochester.edu

Contents

Abstract

Methods to probe RNA secondary structure, such as small molecule modifying agents, secondary structure-specific nucleases, inline probing, and SHAPE chemistry, are widely used to study the structure of functional RNA. Computational secondary structure prediction programs can incorporate probing data to predict structure with high accuracy. In this chapter, an overview of current methods for probing RNA secondary structure is provided, including modern high-throughput methods. Methods for guiding secondary

Methods in Enzymology, Volume 553
ISSN 0076-6879
http://dx.doi.org/10.1016/bs.mie.2014.10.053

structure prediction algorithms using these data are explained, and best practices for using these data are provided. This chapter concludes by listing a number of open questions about how to best use probing data, and what these data can provide.

1. INTRODUCTION

RNA molecules actively participate in numerous cellular functions. In addition to acting as a template for translation to protein, RNA molecules catalyze key chemical reactions in the cell, such as protein chain elongation in translation (Ban, Nissen, Hansen, Moore, & Steitz, 2000; Noller, Hoffarth, & Zimniak, 1992) and the transesterification steps in pre-mRNA splicing (Fica et al., 2013; Kruger et al., 1982). RNA molecules are integral to the RNA interference pathway, which regulates gene expression (Fire et al., 1998), and RNA structures within mRNAs called riboswitches can regulate gene expression or splicing by modulating their structure in response to ligand binding (Nahvi et al., 2002; Serganov & Nudler, 2013; Winkler, Nahvi, & Breaker, 2002).

Because the structure of a biological molecule determines its function, an understanding of RNA structure is necessary for a complete understanding of all of these processes. A study of RNA secondary structure, i.e., the set of canonical base-pairings in the structure, is a useful first step in understanding the folding of an RNA molecule. The secondary structure generally forms faster and is more stable than the additional contacts that mediate the tertiary structure; therefore, secondary structure can largely be considered independently of the tertiary fold (Tinoco Jr. & Bustamante, 1999). Studying RNA at this level of abstraction has proved useful for identification of functional RNA in genomes (Gorodkin et al., 2010; Gruber, Findeiss, Washietl, Hofacker, & Stadler, 2010; Klein & Eddy, 2003; Macke et al., 2001; Nawrocki, Kolbe, & Eddy, 2009; Torarinsson, Sawera, Havgaard, Fredholm, & Gorodkin, 2006; Uzilov, Keegan, & Mathews, 2006; Yao, Weinberg, & Ruzzo, 2006), prediction of accessible sites for design of siRNA (Heale, Soifer, Bowers, & Rossi, 2005; Lu & Mathews, 2007; Tafer et al., 2008), and for design of new RNA folds (Garcia-Martin, Clote, & Dotu, 2013; Hofacker et al., 1994; Lee et al., 2014; Zadeh et al., 2010).

There are two general approaches to determine the secondary structure of an RNA molecule: comparative sequence analysis and computational secondary structure prediction. Comparative analysis infers the secondary structure using the principle that RNA structure is more conserved than sequence. Using a large number of homologous RNA sequences, called a family, the structure common to the family can be determined with high accuracy. Mutations at two nucleotides that preserve their ability to base pair, called compensating mutations, provide strong evidence that those nucleotides pair in the structure (Pace, Thomas, & Woese, 1999). From the pattern of compensating mutations, the secondary structure can be deduced with high accuracy. For example, the secondary structures of the small and large subunit ribosomal RNAs were originally determined by comparative sequence analysis in the 1980s (Noller et al., 1981; Noller & Woese, 1981). When crystal structures were determined 20 years later, they agreed almost perfectly; over 97% of predicted pairs were confirmed (Ban et al., 2000; Gutell, Lee, & Cannone, 2002; Wimberly et al., 2000). Comparative sequence analysis has a significant drawback, however; in order to achieve the best possible accuracy, it requires a large investment of labor by a skilled human. Although automated methods to predict a common structure for multiple sequences exist (Bernhart, Hofacker, Will, Gruber, & Stadler, 2008; Do, Foo, & Batzoglou, 2008; Harmanci, Sharma, & Mathews, 2011; Holmes, 2005; Kiryu, Tabei, Kin, & Asai, 2007; Lindgreen, Gardner, & Krogh, 2007; Steffen, Voss, Rehmsmeier, Reeder, & Giegerich, 2006; Torarinsson, Havgaard, & Gorodkin, 2007; Wei, Alpert, & Lawrence, 2011; Will, Reiche, Hofacker, Stadler, & Backofen, 2007; X. Xu, Ji, & Stormo, 2007; Z. Xu & Mathews, 2011), no automated method currently exists that performs as well as a human expert.

Computational secondary structure prediction methods attempt to predict structure in an automated fashion, without human intervention. The most popular method, implemented in mFold/UNAfold (Zuker, 2003), the ViennaRNA Package (Lorenz et al., 2011), and RNAstructure (Bellaousov, Reuter, Seetin, & Mathews, 2013; Reuter & Mathews, 2010), relies on the hypothesis that RNA secondary structures in solution are at equilibrium, and therefore the RNA secondary structure with the lowest Gibbs free energy of folding will be the most prevalent. A model that estimates the folding free energy change of an RNA based on nearest neighbor interactions between adjacent loops and base pair stacks has been derived from optical melting experiments on short RNA molecules (Andronescu, Condon, Turner, & Mathews, 2014; Mathews et al., 2004; Xia et al., 1998). The Nearest Neighbor Database Web site (Turner & Mathews, 2010) provides the current set of energy rules. A search procedure using dynamic programming can then find the structure with the lowest free energy of folding (Nussinov & Jacobson, 1980; Zuker & Stiegler, 1981). Alternative approaches to secondary structure prediction based on machine learning methods provide comparable accuracy (Do, Woods, & Batzoglou, 2006; Lu, Gloor, & Mathews, 2009; Rivas, Lang, & Eddy, 2012). Overall, benchmarks demonstrate that about two-thirds of predicted base pairs are in the accepted structure for RNA sequences with known structure, as determined by comparative sequence analysis.

A notable limitation of computational methods is that the most popular programs ignore the possibility of pseudoknots, i.e., non-nested sets of base pairs (Liu, Mathews, & Turner, 2010). Methods that do attempt to predict pseudoknots typically do so with poor accuracy (Bellaousov & Mathews, 2010).

RNA secondary structure can be probed using small molecules, enzymes, or hydroxyl radical cleavage. These methods modify or cleave an RNA molecule in a way that is dependent on the local secondary structure, such as modifying the Watson–Crick faces of bases that are not involved in a base pair. These modifications or cleavages were traditionally detected by reverse transcription and electrophoresis on a sequencing gel, but are now increasingly detected by capillary electrophoresis or massively parallel sequencing methods. Although these methods by themselves provide no information about which nucleotides actually pair with each other, they increase the accuracy of structure prediction because the information they provide dramatically reduces the number of possible wrong answers that the algorithm could consider.

2. OVERVIEW OF PROBING METHODS

2.1. Overview

The RNA secondary structure probing methods have commonalities. In general, the probing reagent must be used at a concentration such that each RNA molecule is modified no more than once. Once the RNA is modified, the labeled product of the assay (either the labeled RNA itself or a reverse-transcribed DNA complement) is run on a sequencing gel or capillary electrophoresis instrument, generating a readout for each nucleotide where the strength of the signal is correlated with the frequency of modification or cleavage at that nucleotide. These methods have been collectively called Single Nucleotide Resolution Nucleic Acid Structure Mapping experiments, or SNRNASMs, because they all localize probing information to a single nucleotide position (Rocca-Serra et al., 2011). Sites of modification or cleavage by the various popular structure probing molecules are shown in Fig. 1.

2.2. Chemical probing

The oldest methods to chemically probe RNA structure use small molecules that modify the Watson–Crick face of the nucleotide base. Nucleotides that are not involved in a canonical base pair are more accessible to the chemical probe and are modified more frequently. The modified bases block subsequent reverse transcription of the RNA, which can be read out by electrophoresis or by high-throughput sequencing. Therefore, bases that show higher reactivity with the probe, as measured by reverse transcriptase stops,

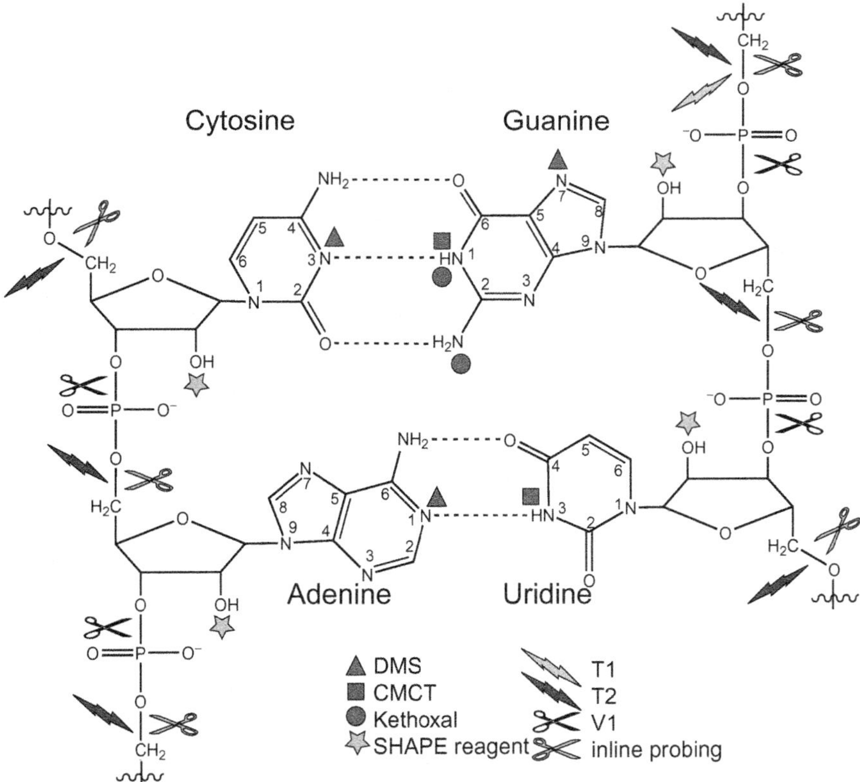

Figure 1 Sites of modification or cleavage by chemical probes, SHAPE reagent, enzymes, and inline probing (Ehresmann et al., 1987). A schematic of a helix containing canonical Watson–Crick base pairs is depicted, with sites of modification marked by shapes and bond cleavage marked by lightning bolts.

are more likely to be unpaired. Small molecule probes include dimethyl sulfate (DMS), which methylates N1 of adenine and N3 of cytosine; kethoxal, which reacts with the exocyclic amine and N1 of guanines; and CMCT, which modifies N3 of uridine and N1 of guanine. DMS also modifies the N7 of guanines. This modification, which is not indicative of secondary structure but is indicative of tertiary contacts, can be detected by aniline dye cleavage of the backbone (Ehresmann et al., 1987).

2.3. SHAPE

SHAPE (Selective 2′ Hydroxyl Acylation followed by Primer Extension) chemistry provides readout of backbone flexibility (Merino, Wilkinson, Coughlan, & Weeks, 2005). SHAPE reagents modify 2′ hydroxyl moieties with a bulky adduct that blocks reverse transcriptase extension of a labeled primer. The reactivity of a given 2′ hydroxyl toward the SHAPE reagent strongly correlates with the local backbone flexibility. Because the backbone at an unpaired nucleotide is typically highly flexible, while the backbone of a nucleotide involved in base pair is less flexible, SHAPE reactivity provides a nucleotide-by-nucleotide readout of base-pairing. The reaction is largely sequence-independent because it occurs on the backbone.

Protocols for probing RNA secondary structure using SHAPE are available (Low & Weeks, 2010; McGinnis, Duncan, & Weeks, 2009; Wilkinson, Merino, & Weeks, 2006). Several SHAPE reagents are available. Historically, the most commonly used agent is *N*-methylisatoic anhydride (NMIA). This has now been largely superceded with 1-methyl-7-nitroisatoic anhydride (1M7) (Mortimer & Weeks, 2007). Recently, it was reported that performing reactions with multiple reagents can provide additional information (Rice, Leonard, & Weeks, 2014).

2.4. Inline probing

Instead of using a probing molecule, the inline probing method relies on RNA self-cleavage in the presence of divalent cations. Nucleophilic attack on the backbone phosphate by the 2′ hydroxyl causes cleavage of the backbone, leaving a phosphate bound to both the 2′ and 3′ hydroxyl groups. This reaction proceeds fast at positions where the backbone is more flexible, so more cleavage occurs at unpaired nucleotides (Regulski & Breaker, 2008; Soukup & Breaker, 1999).

2.5. Enzymatic cleavage

Enzymatic cleavage of the RNA backbone by nucleases gives information on secondary structure. Various nucleases have specificity for different secondary structure elements. Commonly used molecules include RNase V1 (Lockard & Kumar, 1981), which cleaves helices at any base, leaving a 3′ hydroxyl; RNase T1, which cleaves after an unpaired guanine nucleotide (Wrede, Wurst, Vournakis, & Rich, 1979), leaving a 3′ phosphate; and RNase T2 (Vary & Vournakis, 1984), which cleaves after any unpaired nucleotide, leaving a 3′ phosphate. A review article by Knapp (1989) outlines the procedures for enzymatic mapping of RNA using these enzymes.

2.6. Comparison of methods

Methods that modify the pairing faces, methods that read the backbone flexibility, and enzymatic cleavage methods each have advantages and disadvantages. The base-pairing faces of nucleotides in or adjacent to G-U wobble pairs or on the ends of helices are accessible to modification by DMS, CMCT, or kethoxal, and often appear reactive even if they are paired (Mathews et al., 2004). Methods that read backbone flexibility do not report directly on base-pairing, so a nucleotide with low SHAPE or inline probing reactivity might be involved in a tertiary contact or noncanonical interaction rather than a base pair.

One distinct advantage of DMS is that it penetrates cells, allowing structure probing *in vivo* (Y. Ding et al., 2014; Mathews et al., 2004; Zaug & Cech, 1995), although new SHAPE reagents have been developed that also penetrate cells. (Spitale et al., 2013). Enzymes have an advantage in that they directly cleave the RNA, obviating the need for reverse transcription.

2.7. Adaptation to high-throughput technologies

The original implementation of all of these methods probed only a single molecule at a time. High-throughput methods, however, have now been developed for chemical probing (Y. Ding et al., 2014), SHAPE (Aviran et al., 2011; Lucks et al., 2011; Seetin, Kladwang, Bida, & Das, 2014; Siegfried, Busan, Rice, Nelson, & Weeks, 2014; Spitale et al., 2013), and enzymatic cleavage (Kertesz et al., 2010; Underwood et al., 2010), allowing the simultaneous probing of many thousands of different sequences simultaneously, both *in vivo* and *in vitro*. These high-throughput probing methods, in concert with computational structure prediction, have recently resulted in several revelations about secondary structure that might not have been available otherwise, including ATP-dependent unfolding of many RNAs *in vivo* (Rouskin, Zubradt, Washietl, Kellis, & Weissman, 2014), a global profile of plant mRNA secondary structure revealing interesting regulatory features including a triplet pattern of structure in coding sequences (Y. Ding et al., 2014), the discovery of new functionally important pseudoknots in the HIV-1 genome (Siegfried et al., 2014), and a database of SNPs that alter secondary structure of ncRNAs (Wan et al., 2014).

3. IMPROVING THE ACCURACY OF SECONDARY STRUCTURE PREDICTION USING PROBING DATA

3.1. Constraint methods for enzymatic cleavage and chemical modification

Data from enzymatic probing have been used to guide minimum free energy structure prediction algorithm as a hard constraint (Mathews, Sabina, Zuker, & Turner, 1999). Here, nucleotides are forced to be in base pairs (RNase V1) or forced to be unpaired (RNase T1 or T2) if they are accessible to enzymatic cleavage. Structures that are contradictory to the enzymatic data are forbidden to occur and hence the predicted lowest free energy structure will be entirely consistent with the entered data (Mathews et al., 1999). A nucleotide can reliably be considered paired or unpaired only if enzymatic cleavage occurred both 3′ and 5′ of that nucleotide. This rule is needed because, in spite of the selectivity of the enzymes, spurious cleavages, i.e., cleavages apparently inconsistent with the accepted structure, are observed.

To use chemical modification data to improve secondary structure prediction, the free energy minimization and partition function dynamic programming algorithms were modified to treat those data as constraints on the possible structures, where structures contradictory to the data are not allowed (Mathews, 2004; Mathews et al., 2004). Here, nucleotides specified by the user to be accessible to modification are required to be either unpaired, in a G-U wobble pair, or located at the end of a helix. These constraints match empirical observations of the reactivity of nucleotides toward chemical mapping reagents.

Accuracy of RNA secondary structure prediction is customarily scored by sensitivity and positive predictive value (PPV), where sensitivity is the fraction of pairs in the accepted structure that are correctly predicted and PPV is the fraction of predicted pairs in the accepted structures. Lower sensitivity reflects pairs that are missing in the model, and lower PPV reflects pairs that are incorrectly incorporated into the model. For the minimum free energy structure without any experimental data, the average sensitivity score is 69.3% and the average PPV is 61.3% (Bellaousov & Mathews, 2010). While this is sufficiently accurate to develop hypotheses about the structure, the models are far from perfect.

Constraints from enzymatic cleavage or chemical modification provide only a small increase in accuracy, if any, for structures that are already well predicted by free energy minimization alone. These constraints, however, provide a large improvement in the accuracy of structure prediction for sequences that are poorly predicted by energy minimization alone (*sensitivity* $< 50\%$). The presence of hard constraints appears to prevent the algorithm from finding egregiously wrong structures, so the constraints provide a baseline of average performance (Mathews et al., 2004, 1999).

3.2. Restraints for structure prediction using SHAPE and DMS

Quantitative data from probing with SHAPE chemistry has been used to provide restraints on folding, rather than hard constraints. Here, data are collected by capillary electrophoresis or by massively parallel sequencing methods, deconvoluted, normalized, and used to provide a free energy bonus or penalty for each nucleotide in a base stack of:

$$\Delta G_{\mathrm{SHAPE}} = m \times \ln(\text{Normalized reactivity} + 1) + b \tag{1}$$

where m and b are constants that provide the best structure prediction accuracy on a set of training data. b is a negative number, so $\Delta G_{\mathrm{SHAPE}}$ will be negative (favorable) for each pair in the structure where the reactivity is low, and positive (unfavorable) for each highly reactive pair. This formulation treats base pairs at the ends of helices differently from base pairs with stacking pairs both $5'$ and $3'$. Nucleotides paired at the end of helices receive the $\Delta G_{\mathrm{SHAPE}}$ term once, while the other paired nucleotides are in two base pair stacks and receive the $\Delta G_{\mathrm{SHAPE}}$ term twice. This matches the observation that paired nucleotides at the ends of helices tend to be more reactive than paired nucleotides in the middle of helices. Parameters for the pseudoenergy calculation were determined empirically by performing a grid search for the parameters that resulted in the best secondary structure prediction for a large subunit ribosomal RNA (Deigan, Li, Mathews, & Weeks, 2009), and these parameters were subsequently revised using a jackknife optimization procedure on a larger dataset (Hajdin et al., 2013). Predictions using SHAPE data resulted in excellent accuracy, with a sensitivity and PPV of around 90% (Deigan et al., 2009).

The ΔG_{SHAPE} restraint formulation in Eq. (1) empirically weights structures according to the degree in which they match SHAPE data. Subsequent works provide principled methods for using SHAPE data (Eddy, 2014; Sükösd, Swenson, Kjems, & Heitsch, 2013; Washietl, Hofacker, Stadler, & Kellis, 2012; Zarringhalam, Meyer, Dotu, Chuang, & Clote, 2012). These approaches were reviewed in detail by Eddy (2014). Eddy observed that, when the pseudo free energy is treated as the likelihood ratio for nucleotide pairing, Eq. (1) and the fitted parameters specify a likelihood ratio for pairing that is not borne out by observation. Alternative approaches, however, do not at this time exceed the performance of the pseudo free energy formulation.

Quantitative probing data from DMS have also been implemented as a restraint, rather than a constraint, with the Gibbs free energy penalty for an inconsistent structure, i.e., a structure where a reactive A or C nucleotide is not in an unpaired region or helix end (Cordero, Kladwang, VanLang, & Das, 2012). In this work, the log likelihood of a nucleotide being paired with a specific DMS reactivity was used to determine the free energy penalty, rather than the empirical functional form of Eq. (1). This was incorporated into the RNAstructure Fold program, and made available as a web server, http://rmdb.stanford.edu/structureserver/. Interestingly, DMS mapping, despite providing data on fewer nucleotides than SHAPE, still produces secondary structures with similarly high accuracy (Cordero et al., 2012). It may be that data are not needed for each nucleotide, or that DMS has some advantage in probing secondary structure because it probes the Watson–Crick pairing face of nucleotides.

The RNAstructure software package includes an option to use arbitrary free energy restraints provided by the user. This was used to incorporate enzymatic cleavage data as a restraint in RNA structure prediction as part of the FragSeq method, which probes structure at a transcriptome-wide scale (Underwood et al., 2010).

4. USING SHAPE DATA ON A SINGLE SEQUENCE TO IMPROVE SECONDARY STRUCTURE PREDICTION ACCURACY

4.1. Experimental design

The most useful information for improving the accuracy of secondary structure prediction is quantified using SHAPE or DMS mapping data (Cordero et al., 2012; Deigan et al., 2009; Hajdin et al., 2013). This supersedes previous approaches that graded nucleotide reactivity to chemical agents as weak, moderate, or strong using gel electrophoresis bands. This is because the quantified data can be used to restrain structure prediction, and the restraints are more robust to experimental uncertainty and variations in reactivity. Prior work had constrained structure prediction, forcing nucleotides into user-specified configurations (Mathews, 2004; Mathews et al., 1999).

Experimental protocols are available for collecting SHAPE data (Wilkinson et al., 2006). Capillary electrophoresis data can be quantified using the software QuSHAPE (Karabiber, McGinnis, Favorov, & Weeks, 2013). For gel electrophoresis, the SAFA package can be used (Laederach et al., 2008). Normalization should be performed to determine reactivity on a scale from 0 to 2 using the boxplot approach (Hajdin et al., 2013); QuSHAPE does this automatically.

The data are normalized by the following procedure:

1. Find the candidate outliers
   - Let R_i be the raw reactivity of the nucleotide at index i.
   - Sort the R_i values from lowest to highest.
   - Calculate the upper and lower quartiles Q_U and Q_L, which are the median values of the upper and lower halves of the data, respectively.
   - Calculate the interquartile distance $d = Q_U - Q_L$
   - Candidate outliers are nucleotides where the reactivity $R_i > Q_U + 1.5 * d$
2. Remove the candidate outliers from the dataset, highest values first
   - Remove no more than 10% of the data if the sequence is longer than 100 nucleotides. Otherwise, remove no more than 5% of the data.
3. Normalize the data
   - Calculate R_{max}, the mean reactivity of the 10% most reactive nucleotides after the outliers are removed.
   - Divide the reactivity of each nucleotide (including the outliers) by R_{max}.

On this scale, nucleotides with reactivity greater than 0.85 are considered highly reactive and nucleotides with reactivity less than 0.4 are relatively unreactive.

4.2. Using RNAstructure

Secondary structure prediction without pseudoknots can be performed using the web servers at http://rna.urmc.rochester.edu/RNAstructureWeb/ (Bellaousov et al., 2013), the Fold command line program, or the graphical interface (Reuter & Mathews, 2010). RNAstructure can be downloaded from http://rna.urmc.rochester.edu/RNAstructure.html. To predict structures that can contain pseudoknots, the ShapeKnots command line program can be used (Hajdin et al., 2013). Web servers and graphical interfaces for ShapeKnots are forthcoming. The computational cost for ShapeKnots is considerably higher than for Fold; therefore it is recommended that ShapeKnots be used only for sequences of fewer than 750 nucleotides at this time.

Step-by-step protocols for using RNAstructure are available (Mathews, 2014). An extensive online help is also available (http://rna.urmc.rochester.edu/RNAstructureHelp.html). Normalized SHAPE reactivities are read as plain text by RNAstructure, and the file format is specified in the online help (http://rna.urmc.rochester.edu/Text/File_Formats.html). It is important to recognize that a reactivity of 0 (zero) indicates a nucleotide that is not accessible to SHAPE and that this is different than not having reactivity information. For nucleotides for which there is no information available, the file can indicate a reactivity of less than -500 or those nucleotides can be left out of the file. For most applications, the default SHAPE pseudo free energy parameters of $m = 1.8$ kcal/mol and $b = -0.6$ kcal/mol (Eq. (1)) should be used.

The command line tool (draw) and the graphical user interface can both color-annotate predicted secondary structures with SHAPE reactivity data. This information is helpful for interpreting the results of structure prediction. If the web server is used, SHAPE reactivities are automatically used to color-annotate the predicted structures. Additionally, the partition function algorithm (partition) can be used to predict base-pairing probabilities restrained by SHAPE data. Highly probable base pairs (e.g., > 0.9) are more likely to be correctly predicted than less probable base pairs (e.g., < 0.5) (Mathews, 2004). This is also true when SHAPE data are used to restrain the calculation, and the use of SHAPE data generally increases the pairing probabilities of nucleotides predicted to be base paired.

4.3. Example

To illustrate the prediction of secondary structure, an example using the RNAstructure web server (Bellaousov et al., 2013) is provided with the *Escherichia coli* 5S rRNA (Andronescu, Bereg, Hoos, & Condon, 2008). Fig. 2A shows the predicted secondary structure in the absence of experimental restraints. This is a poorly predicted secondary structure, as illustrated by the CircleCompare plots in Fig. 2B, which compares the prediction to

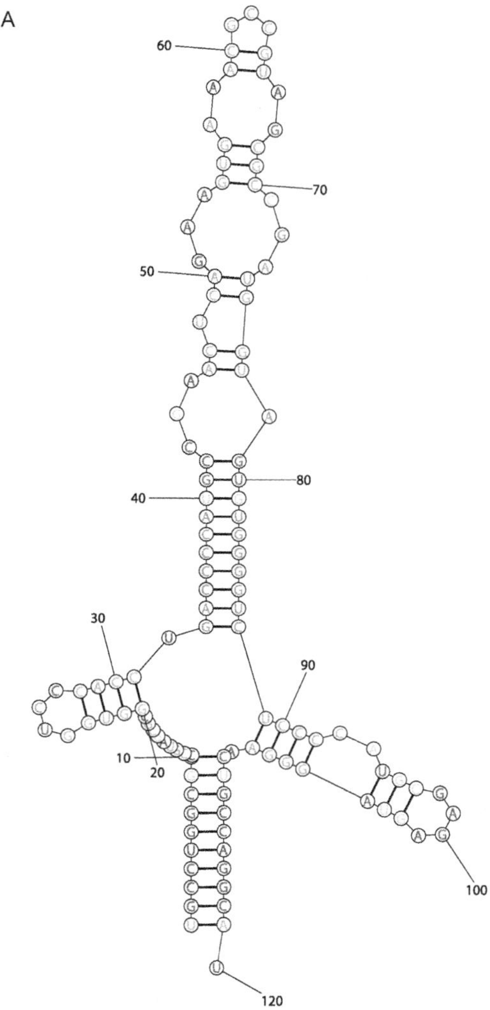

Probability >= 99%

99% > Probability >= 95%

95% > Probability >= 90%

90% > Probability >= 80%

80% > Probability >= 70%

70% > Probability >= 60%

60% > Probability >= 50%

50% > Probability

ENERGY = −51.5 07/31/14 01:02:24

Figure 2 Predicted *E. coli* 5S rRNA secondary structure without SHAPE data. Panel (A) shows the predicted structure, as displayed by the RNAstructure predict a secondary structure web server. The structure is color-annotated by estimated base-pairing probabilities.

(Continued)

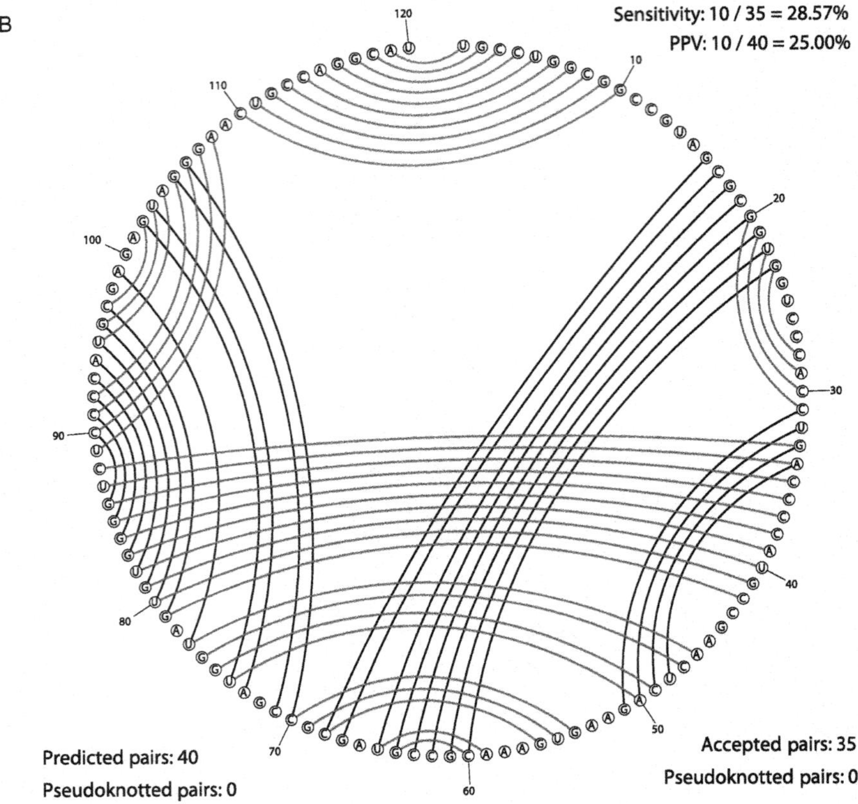

Pair present in both predicted and accepted structure (Green).
Pair present in predicted structure only (Red).
Pair present in accepted structure only (Black).

Figure 2—Cont'd Panel (B) shows a CircleCompare plot from the RNAstructure CircleCompare web server, which compares the predicted structure to the accepted structure. The sequence is drawn around a circle and lines indicate base pairs. Pairs present in both the predicted and accepted structure are green; pairs in the predicted structure only are red; and pairs in the accepted structure only are black. The upper-left shows that the sensitivity (fraction of accepted pairs correctly predicted) is 29% and the PPV (fraction of predicted pairs in the accepted structure) is 25%. (See the color plate.)

the accepted structure. One reason for the poor prediction is that the loop E internal loop motif, involving nucleotides 72–78 and 98–104, is more stable than the nearest neighbor parameters estimate (Serra et al., 2002). Fig. 2A shows the predicted structure with base-pairing probability color-annotation (Mathews, 2004). This provides some clues that this structure is poorly predicted because there is a large fraction of pairs predicted to have base-pairing probabilities below 50%. Fig. 3A shows the secondary structure predicted with SHAPE restraints (Hajdin et al., 2013), and annotated with base-pairing probabilities. The same predicted structure with SHAPE color-annotation is in Fig. 3B, and this shows that the predicted structure is guided by the SHAPE reactivities. This prediction is now accurate, as shown by the CirclePlot in Fig. 3C, with 86% sensitivity and 77% PPV. The probability color-annotation shows a predominance of pairs with greater than 95% pairing probability. The incorrectly predicted helices are composed of pairs with less than 50% base-pairing probability, which would suggest that those predicted pairs are unreliable. The SHAPE-guided structure prediction in this case is not as good as the average performance of SHAPE-guided structure prediction, but this structure prediction would be good enough to develop testable hypotheses and it demonstrates that estimated pairing probabilities can provide guidance about the accuracy of a predicted base pair.

5. OPEN QUESTIONS

Structure probing combined with computational structure prediction has proved useful in modeling RNA secondary structures. A number of questions remain about the best ways to use these data and also about new ways these data can inform structure modeling.

Reactivity data from SHAPE and DMS probing have been quantified and used as restraints, rather than hard constraints, for secondary structure prediction, and produce highly accurate secondary structure models (Cordero et al., 2012; Deigan et al., 2009). It would be instructive to see if the other common secondary structure probing agents, such as kethoxal, could likewise be used in this fashion. CMCT does not work as well as DMS or SHAPE to restrain secondary structure prediction (Cordero et al., 2012).

Another open question is whether there is synergy between different chemical probing methods. DMS and SHAPE together produce a slightly better secondary structure model than DMS or SHAPE alone (Cordero et al., 2012). Combining other types of mapping data may likewise produce further improved structure predictions.

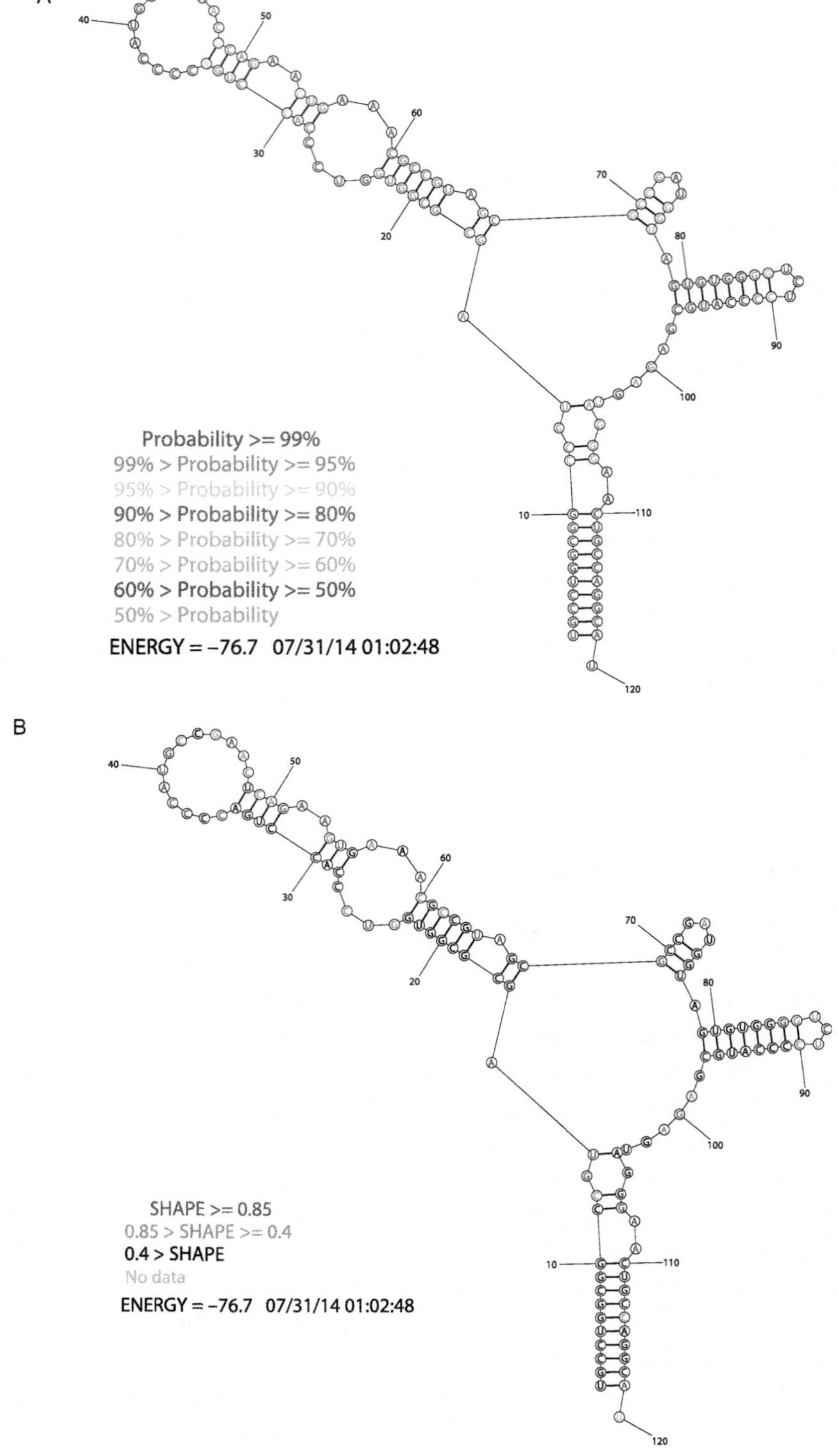

Figure 3 See legend on opposite page.

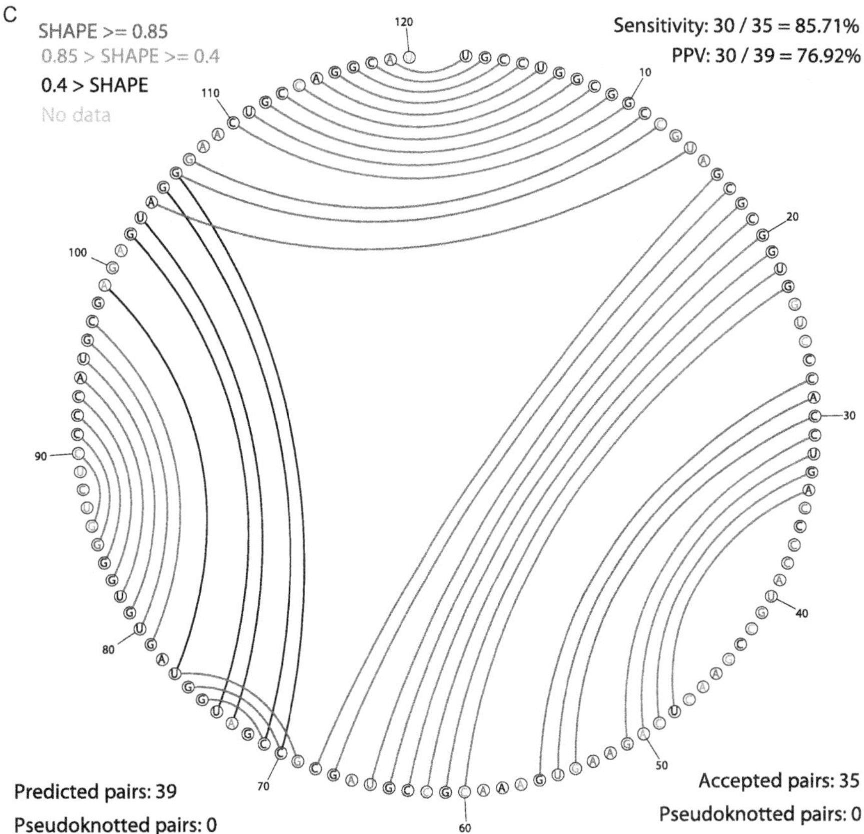

Pair present in both predicted and accepted structure (Green).
Pair present in predicted structure only (Red).
Pair present in accepted structure only (Black).

Figure 3 Predicted *E. coli* 5S rRNA secondary structure with restraints from SHAPE data. Panel (A) shows the predicted structure, as displayed by the RNAstructure predict a secondary structure web server. The structure is color-annotated by estimated base-pairing probabilities. Panel (B) shows the predicted structure with color-annotation by normalized SHAPE reactivities. These plots are both generated by default. Panel (C) shows a CircleCompare plot from the RNAstructure CircleCompare web server, which compares the predicted structure to the accepted structure. The sequence is drawn around a circle and lines indicate base pairs. Pairs present in both predicted and accepted structure are green; pairs in the predicted structure only are red; and pairs in the accepted structure only are black. The upper-left shows that the sensitivity (fraction of accepted pairs correctly predicted) is 86% and the PPV (fraction of predicted pairs in the accepted structure) is 77%. The sequence is color-annotated by SHAPE normalized SHAPE reactivities. (See the color plate.)

Inline probing is another method that has not, to our knowledge, been used to inform computational secondary structure prediction. A comparison between SHAPE data and inline probing data suggested that the data from inline probing are noisier and may produce lower quality structures than SHAPE (Zarringhalam et al., 2012). Because the method requires only a small amount of RNA and needs no other reagents, it might be an attractive method to incorporate into existing secondary structure prediction software.

Studies focusing on using probing data to improve RNA secondary structure prediction have to date focused on RNA sequences with a single, well-defined structure. As transcriptome data are becoming available, new methods should be developed for modeling structures when more than one configuration is expected. This is important for modeling RNA sequences such as riboswitches, which populate multiple states *in vivo*, or for protein coding regions of mRNA that are likely to have heterogeneous secondary structure that may still be important for function.

Another open question is whether 3D structural motifs have reactivity "signatures" which could be identified and incorporated into structure prediction. For example, if the six nucleotides (four "unpaired" nucleotides and the closing base pair) of a stable tetraloop such as GNRA had a characteristic set of reactivities that could be identified, then the structure prediction algorithm could search for this signature and either constrain or restrain the structure to contain this hairpin. This type of information would also be informative for tertiary structure modeling and could be incorporated into a number of the existing approaches (Cao & Chen, 2011; Das, Karanicolas, & Baker, 2010; F. Ding et al., 2008; Parisien & Major, 2008; Rother, Rother, Puton, & Bujnicki, 2011; Seetin & Mathews, 2011).

6. CONCLUSIONS

We are entering a new age of capabilities for probing RNA structure. High-throughput technologies make it feasible to probe transcriptomes either *ex vivo* or *in vivo* and to quantify the extent of reactivity. Quantified SHAPE and DMS data can be used as restraints for secondary structure prediction. These approaches currently yield the most accurate predicted structures, on average. Other methods exist for constraining structure prediction with chemical and enzymatic data, read from gels, and these methods are known to be useful when secondary structure prediction would have otherwise been inaccurate. Continued advances are expected as unsolved problems are studied. This is especially important as new high-throughput experimental methods are developed.

REFERENCES

Andronescu, M., Bereg, V., Hoos, H. H., & Condon, A. (2008). RNA STRAND: The RNA secondary structure and statistical analysis database. *BMC Bioinformatics, 9*, 340.

Andronescu, M., Condon, A., Turner, D. H., & Mathews, D. H. (2014). The determination of RNA folding nearest neighbor parameters. *Methods in Molecular Biology, 1097*, 45–70.

Aviran, S., Trapnell, C., Lucks, J. B., Mortimer, S. A., Luo, S., Schroth, G. P., et al. (2011). Modeling and automation of sequencing-based characterization of RNA structure. *Proceedings of the National Academy of Sciences of the United States of America, 108*(27), 11069–11074.

Ban, N., Nissen, P., Hansen, J., Moore, P. B., & Steitz, T. A. (2000). The complete atomic structure of the large ribosomal subunit at 2.4Å resolution. *Science, 289*, 905–920.

Bellaousov, S., & Mathews, D. H. (2010). ProbKnot: Fast prediction of RNA secondary structure including pseudoknots. *RNA, 16*, 1870–1880.

Bellaousov, S., Reuter, J. S., Seetin, M. G., & Mathews, D. H. (2013). RNAstructure: Web servers for RNA secondary structure prediction and analysis. *Nucleic Acids Research, 41*(Web Server issue), W471–W474.

Bernhart, S. H., Hofacker, I. L., Will, S., Gruber, A. R., & Stadler, P. F. (2008). RNAalifold: Improved consensus structure prediction for RNA alignments. *BMC Bioinformatics, 9*(1), 474.

Cao, S., & Chen, S. J. (2011). Physics-based de novo prediction of RNA 3D structures. *The Journal of Physical Chemistry B, 115*(14), 4216–4226.

Cordero, P., Kladwang, W., VanLang, C. C., & Das, R. (2012). Quantitative dimethyl sulfate mapping for automated RNA secondary structure inference. *Biochemistry, 51*(36), 7037–7039.

Das, R., Karanicolas, J., & Baker, D. (2010). Atomic accuracy in predicting and designing noncanonical RNA structure. *Nature Methods*, 7(4), 291–294.

Deigan, K. E., Li, T. W., Mathews, D. H., & Weeks, K. M. (2009). Accurate SHAPE-directed RNA structure determination. *Proceedings of the National Academy of Sciences of the United States of America, 106*(1), 97–102.

Ding, F., Sharma, S., Chalasani, P., Demidov, V. V., Broude, N. E., & Dokholyan, N. V. (2008). Ab initio RNA folding by discrete molecular dynamics: From structure prediction to folding mechanisms. *RNA, 14*(6), 1164–1173.

Ding, Y., Tang, Y., Kwok, C. K., Zhang, Y., Bevilacqua, P. C., & Assmann, S. M. (2014). In vivo genome-wide profiling of RNA secondary structure reveals novel regulatory features. *Nature, 505*(7485), 696–700.

Do, C. B., Foo, C. S., & Batzoglou, S. (2008). A max-margin model for efficient simultaneous alignment and folding of RNA sequences. *Bioinformatics, 24*(13), i68–i76.

Do, C. B., Woods, D. A., & Batzoglou, S. (2006). CONTRAfold: RNA secondary structure prediction without physics-based models. *Bioinformatics, 22*(14), e90–e98.

Eddy, S. R. (2014). Computational analysis of conserved RNA secondary structure in transcriptomes and genomes. *Annual Review of Biophysics, 43*, 433–456.

Ehresmann, C., Baudin, F., Mougel, M., Romby, P., Ebel, J., & Ehresmann, B. (1987). Probing the structure of RNAs in solution. *Nucleic Acids Research, 15*, 9109–9128.

Fica, S. M., Tuttle, N., Novak, T., Li, N.-S., Lu, J., Koodathingal, P., et al. (2013). RNA catalyses nuclear pre-mRNA splicing. *Nature, 503*(7475), 229–234.

Fire, A., Xu, S., Montgomery, M. K., Kostas, S. A., Driver, S. E., & Mello, C. C. (1998). Potent and specific genetic interference by double-stranded RNA in Caenorhabditis elegans. *Nature, 391*(6669), 806–811.

Garcia-Martin, J. A., Clote, P., & Dotu, I. (2013). RNAiFOLD: A constraint programming algorithm for RNA inverse folding and molecular design. *Journal of Bioinformatics and Computational Biology, 11*(2), 1350001.

Gorodkin, J., Hofacker, I. L., Torarinsson, E., Yao, Z., Havgaard, J. H., & Ruzzo, W. L. (2010). De novo prediction of structured RNAs from genomic sequences. *Trends in Biotechnology*, *28*(1), 9–19.

Gruber, A. R., Findeiss, S., Washietl, S., Hofacker, I. L., & Stadler, P. F. (2010). RNAz 2.0: Improved noncoding RNA detection. *Pacific Symposium on Biocomputing*, *15*, 69–79.

Gutell, R. R., Lee, J. C., & Cannone, J. J. (2002). The accuracy of ribosomal RNA comparative structure models. *Current Opinion in Structural Biology*, *12*, 301–310.

Hajdin, C. E., Bellaousov, S., Huggins, W., Leonard, C. W., Mathews, D. H., & Weeks, K. M. (2013). Accurate SHAPE-directed RNA secondary structure modeling, including pseudoknots. *Proceedings of the National Academy of Sciences of the United States of America*, *110*(14), 5498–5503.

Harmanci, A. O., Sharma, G., & Mathews, D. H. (2011). TurboFold: Iterative probabilistic estimation of secondary structures for multiple RNA sequences. *BMC Bioinformatics*, *12*(1), 108.

Heale, B. S., Soifer, H. S., Bowers, C., & Rossi, J. J. (2005). siRNA target site secondary structure predictions using local stable substructures. *Nucleic Acids Research*, *33*(3), e30.

Hofacker, I. L., Fontana, W., Stadler, P. F., Bonhoeffer, L. S., Tacker, M., & Schuster, P. (1994). Fast folding and comparison of RNA secondary structures. *Monatshefte fuer Chemie*, *125*, 167–168.

Holmes, I. (2005). Accelerated probabilistic inference of RNA structure evolution. *BMC Bioinformatics*, *6*(1), 73.

Karabiber, F., McGinnis, J. L., Favorov, O. V., & Weeks, K. M. (2013). QuShape: Rapid, accurate, and best-practices quantification of nucleic acid probing information, resolved by capillary electrophoresis. *RNA*, *19*(1), 63–73.

Kertesz, M., Wan, Y., Mazor, E., Rinn, J. L., Nutter, R. C., Chang, H. Y., et al. (2010). Genome-wide measurement of RNA secondary structure in yeast. *Nature*, *467*(7311), 103–107.

Kiryu, H., Tabei, Y., Kin, T., & Asai, K. (2007). Murlet: A practical multiple alignment tool for structural RNA sequences. *Bioinformatics*, *23*(13), 1588–1598.

Klein, R. J., & Eddy, S. R. (2003). RSEARCH: Finding homologs of single structures RNA sequences. *BMC Bioinformatics*, *4*, 44.

Knapp, G. (1989). Enzymatic approaches to probing RNA secondary and tertiary structure. *Methods in Enzymology*, *180*, 192–212.

Kruger, K., Grabowski, P. J., Zaug, A. J., Sands, J., Gottschling, D. E., & Cech, T. R. (1982). Self-splicing RNA: Autoexcision and autocyclization of the ribosomal RNA intervening sequence of Tetrahymena. *Cell*, *31*(1), 147–157.

Laederach, A., Das, R., Vicens, Q., Pearlman, S. M., Brenowitz, M., Herschlag, D., et al. (2008). Semiautomated and rapid quantification of nucleic acid footprinting and structure mapping experiments. *Nature Protocols*, *3*(9), 1395–1401.

Lee, J., Kladwang, W., Lee, M., Cantu, D., Azizyan, M., Kim, H., et al. (2014). RNA design rules from a massive open laboratory. *Proceedings of the National Academy of Sciences of the United States of America*, *111*(6), 2122–2127.

Lindgreen, S., Gardner, P. P., & Krogh, A. (2007). MASTR: Multiple alignment and structure prediction of non-coding RNAs using simulated annealing. *Bioinformatics*, *23*(24), 3304–3311.

Liu, B., Mathews, D. H., & Turner, D. H. (2010). RNA pseudoknots: Folding and finding. *F1000 Biology Reports*, *2*, 8.

Lockard, R. E., & Kumar, A. (1981). Mapping tRNA structure in solution using double-strand-specific ribonuclease V1 from cobra venom. *Nucleic Acids Research*, *9*(19), 5125–5140.

Lorenz, R., Bernhart, S. H., Honer Zu Siederdissen, C., Tafer, H., Flamm, C., & Stadler, P. F. (2011). ViennaRNA package 2.0. *Algorithms for Molecular Biology*, *6*, 26.

Low, J. T., & Weeks, K. M. (2010). SHAPE-directed RNA secondary structure prediction. *Methods, 52*(2), 150–158.

Lu, Z. J., Gloor, J. W., & Mathews, D. H. (2009). Improved RNA secondary structure prediction by maximizing expected pair accuracy. *RNA, 15*, 1805–1813.

Lu, Z. J., & Mathews, D. H. (2007). Efficient siRNA selection using hybridization thermodynamics. *Nucleic Acids Research, 36*, 640–647.

Lucks, J. B., Mortimer, S. A., Trapnell, C., Luo, S., Aviran, S., Schroth, G. P., et al. (2011). Multiplexed RNA structure characterization with selective 2'-hydroxyl acylation analyzed by primer extension sequencing (SHAPE-Seq). *Proceedings of the National Academy of Sciences of the United States of America, 108*(27), 11063–11068.

Macke, T., Ecker, D., Gutell, R., Gautheret, D., Case, D. A., & Sampath, R. (2001). RNAMotif: A new RNA secondary structure definition and discovery algorithm. *Nucleic Acids Research, 29*, 4724–4735.

Mathews, D. H. (2004). Using an RNA secondary structure partition function to determine confidence in base pairs predicted by free energy minimization. *RNA, 10*(8), 1178–1190.

Mathews, D. H. (2014). RNA secondary structure analysis using RNAstructure. *Current Protocols in Bioinformatics, 46*, 12.6.1–12.6.25.

Mathews, D. H., Disney, M. D., Childs, J. L., Schroeder, S. J., Zuker, M., & Turner, D. H. (2004). Incorporating chemical modification constraints into a dynamic programming algorithm for prediction of RNA secondary structure. *Proceedings of the National Academy of Sciences of the United States of America, 101*, 7287–7292.

Mathews, D. H., Sabina, J., Zuker, M., & Turner, D. H. (1999). Expanded sequence dependence of thermodynamic parameters provides improved prediction of RNA secondary structure. *Journal of Molecular Biology, 288*, 911–940.

McGinnis, J. L., Duncan, C. D. S., & Weeks, K. M. (2009). High-throughput SHAPE and hydroxyl radical analysis of RNA structure and ribonucleoprotein assembly. *Methods in Enzymology, 468*, 67–89.

Merino, E. J., Wilkinson, K. A., Coughlan, J. L., & Weeks, K. M. (2005). RNA structure analysis at single nucleotide resolution by selective 2'-hydroxyl acylation and primer extension (SHAPE). *Journal of the American Chemical Society, 127*(12), 4223–4231.

Mortimer, S. A., & Weeks, K. M. (2007). A fast-acting reagent for accurate analysis of RNA secondary and tertiary structure by SHAPE chemistry. *Journal of the American Chemical Society, 129*(14), 4144–4145.

Nahvi, A., Sudarsan, N., Ebert, M. S., Zou, X., Brown, K. L., & Breaker, R. R. (2002). Incorporating chemical modification constraints into a dynamic programming algorithm for prediction of RNA secondary structure. *Chemistry & Biology, 9*(9), 1043.

Nawrocki, E. P., Kolbe, D. L., & Eddy, S. R. (2009). Infernal 1.0: Inference of RNA alignments. *Bioinformatics, 25*(10), 1335–1337.

Noller, H. F., Hoffarth, V., & Zimniak, L. (1992). Unusual resistance of peptidyl transferase to protein extraction procedures. *Science, 256*(5062), 1416–1419.

Noller, H. F., Kop, J., Wheaton, V., Brosius, J., Gutell, R. R., Kopylov, A. M., et al. (1981). Secondary structure model for 23S ribosomal RNA. *Nucleic Acids Research, 9*(22), 6167–6189.

Noller, H. F., & Woese, C. R. (1981). Secondary structure of 16S ribosomal RNA. *Science, 212*(4493), 403–411.

Nussinov, R., & Jacobson, A. B. (1980). Fast algorithm for predicting the secondary structure of single-stranded RNA. *Proceedings of the National Academy of Sciences of the United States of America, 77*, 6309–6313.

Pace, N. R., Thomas, B. C., & Woese, C. R. (1999). Probing RNA structure, function, and history by comparative analysis. In R. F. Gesteland, T. R. Cech, & J. F. Atkins (Eds.), *The RNA world* (2nd ed., pp. 113–141). Cold Spring Harbor, New York: Cold Spring Harbor Laboratory Press.

Parisien, M., & Major, F. (2008). The MC-Fold and MC-Sym pipeline infers RNA structure from sequence data. *Nature*, *452*(7183), 51–55.

Regulski, E. E., & Breaker, R. R. (2008). In-line probing analysis of riboswitches. *Methods in Molecular Biology*, *419*, 53–67.

Reuter, J. S., & Mathews, D. H. (2010). RNAstructure: Software for RNA secondary structure prediction and analysis. *BMC Bioinformatics*, *11*, 129.

Rice, G. M., Leonard, C. W., & Weeks, K. M. (2014). RNA secondary structure modeling at consistent high accuracy using differential SHAPE. *RNA*, *20*(6), 846–854.

Rivas, E., Lang, R., & Eddy, S. R. (2012). A range of complex probabilistic models for RNA secondary structure prediction that includes the nearest-neighbor model and more. *RNA*, *18*(2), 193–212.

Rocca-Serra, P., Bellaousov, S., Birmingham, A., Chen, C., Cordero, P., Das, R., et al. (2011). Sharing and archiving nucleic acid structure mapping data. *RNA*, *17*(7), 1204–1212.

Rother, M., Rother, K., Puton, T., & Bujnicki, J. M. (2011). ModeRNA: A tool for comparative modeling of RNA 3D structure. *Nucleic Acids Research*, *39*(10), 4007–4022.

Rouskin, S., Zubradt, M., Washietl, S., Kellis, M., & Weissman, J. S. (2014). Genome-wide probing of RNA structure reveals active unfolding of mRNA structures in vivo. *Nature*, *505*(7485), 701–705.

Seetin, M. G., Kladwang, W., Bida, J. P., & Das, R. (2014). Massively parallel RNA chemical mapping with a reduced bias MAP-seq protocol. *Methods in Molecular Biology*, *1086*, 95–117.

Seetin, M. G., & Mathews, D. H. (2011). Automated RNA tertiary structure prediction from secondary structure and low-resolution restraints. *Journal of Computational Chemistry*, *32*(10), 2232–2244.

Serganov, A., & Nudler, E. (2013). A decade of riboswitches. *Cell*, *152*(1–2), 17–24.

Serra, M. J., Baird, J. D., Dale, T., Fey, B. L., Retatagos, K., & Westhof, E. (2002). Effects of magnesium ions on the stabilization of RNA oligomers of defined structures. *RNA*, *8*(3), 307–323.

Siegfried, N. A., Busan, S., Rice, G. M., Nelson, J. A. E., & Weeks, K. M. (2014). RNA motif discovery by SHAPE and mutational profiling (SHAPE-MaP). *Nature Methods*, *advance on*.

Soukup, G. A., & Breaker, R. R. (1999). Relationship between internucleotide linkage geometry and the stability of RNA. *RNA*, *5*(10), 1308–1325.

Spitale, R. C., Crisalli, P., Flynn, R. A., Torre, E. A., Kool, E. T., & Chang, H. Y. (2013). RNA SHAPE analysis in living cells. *Nature Chemical Biology*, *9*(1), 18–20.

Steffen, P., Voss, B., Rehmsmeier, M., Reeder, J., & Giegerich, R. (2006). RNAshapes: An integrated RNA analysis package based on abstract shapes. *Bioinformatics*, *22*(4), 500–503.

Sükösd, Z., Swenson, M. S., Kjems, J., & Heitsch, C. E. (2013). Evaluating the accuracy of SHAPE-directed RNA secondary structure predictions. *Nucleic Acids Research*, *41*(5), 2807–2816.

Tafer, H., Ameres, S. L., Obernosterer, G., Gebeshuber, C. A., Schroeder, R., Martinez, J., et al. (2008). The impact of target site accessibility on the design of effective siRNAs. *Nature Biotechnology*, *26*(5), 578–583.

Tinoco, I., Jr., & Bustamante, C. (1999). How RNA folds. *Journal of Molecular Biology*, *293*(2), 271–281.

Torarinsson, E., Havgaard, J. H., & Gorodkin, J. (2007). Multiple structural alignment and clustering of RNA sequences. *Bioinformatics*, *23*(8), 926–932.

Torarinsson, E., Sawera, M., Havgaard, J. H., Fredholm, M., & Gorodkin, J. (2006). Thousands of corresponding human and mouse genomic regions unalignable in primary sequence contain common RNA structure. *Genome Research*, *16*(7), 885–889.

Turner, D. H., & Mathews, D. H. (2010). NNDB: The nearest neighbor parameter database for predicting stability of nucleic acid secondary structure. *Nucleic Acids Research*, *38*(Database issue), D280–D282.

Underwood, J. G., Uzilov, A. V., Katzman, S., Onodera, C. S., Mainzer, J. E., Mathews, D. H., et al. (2010). FragSeq: Transcriptome-wide RNA structure probing using high-throughput sequencing. *Nature Methods*, 7(12), 995–1001.

Uzilov, A. V., Keegan, J. M., & Mathews, D. H. (2006). Detection of non-coding RNAs on the basis of predicted secondary structure formation free energy change. *BMC Bioinformatics*, 7(1), 173.

Vary, C. P., & Vournakis, J. N. (1984). RNA structure analysis using T2 ribonuclease: Detection of pH and metal ion induced conformational changes in yeast tRNAPhe. *Nucleic Acids Research*, *12*(17), 6763–6778.

Wan, Y., Qu, K., Zhang, Q. C., Flynn, R. A., Manor, O., Ouyang, Z., et al. (2014). Landscape and variation of RNA secondary structure across the human transcriptome. *Nature*, *505*(7485), 706–709.

Washietl, S., Hofacker, I. L., Stadler, P. F., & Kellis, M. (2012). RNA folding with soft constraints: Reconciliation of probing data and thermodynamic secondary structure prediction. *Nucleic Acids Research*, *40*(10), 4261–4272.

Wei, D., Alpert, L. V., & Lawrence, C. E. (2011). RNAG: A new Gibbs sampler for predicting RNA secondary structure for unaligned sequences. *Bioinformatics*, *27*(18), 2486–2493.

Wilkinson, K. A., Merino, E. J., & Weeks, K. M. (2006). Selective 2'-hydroxyl acylation analyzed by primer extension (SHAPE): Quantitative RNA structure analysis at single nucleotide resolution. *Nature Protocols*, *1*(3), 1610–1616.

Will, S., Reiche, K., Hofacker, I. L., Stadler, P. F., & Backofen, R. (2007). Inferring non-coding RNA families and classes by means of genome-scale structure-based clustering. *PLoS Computational Biology*, *3*(4), e65.

Wimberly, B. T., Brodersen, D. E., Clemons, W. M., Jr., Morgan-Warren, R. J., Carter, A. P., Vonrhein, C., et al. (2000). Structure of the 30S ribosomal subunit. *Nature*, *407*, 327–329.

Winkler, W., Nahvi, A., & Breaker, R. R. (2002). Thiamine derivatives bind messenger RNAs directly to regulate bacterial gene expression. *Nature*, *419*(6910), 952–956.

Wrede, P., Wurst, R., Vournakis, J., & Rich, A. (1979). Conformational changes of yeast tRNAPhe and E. coli tRNA2Glu as indicated by different nuclease digestion patterns. *Journal of Biological Chemistry*, *254*(19), 9608–9616.

Xia, T., SantaLucia, J., Jr., Burkard, M. E., Kierzek, R., Schroeder, S. J., Jiao, X., et al. (1998). Thermodynamic parameters for an expanded nearest-neighbor model for formation of RNA duplexes with Watson-Crick pairs. *Biochemistry*, *37*, 14719–14735.

Xu, X., Ji, Y., & Stormo, G. D. (2007). RNA Sampler: A new sampling based algorithm for common RNA secondary structure prediction and structural alignment. *Bioinformatics*, *23*(15), 1883–1891.

Xu, Z., & Mathews, D. H. (2011). Multilign: An algorithm to predict secondary structures conserved in multiple RNA sequences. *Bioinformatics*, *27*(5), 626–632.

Yao, Z., Weinberg, Z., & Ruzzo, W. L. (2006). CMfinder—A covariance model based RNA motif finding algorithm. *Bioinformatics*, *22*(4), 445–452.

Zadeh, J. N., Steenberg, C. D., Bois, J. S., Wolfe, B. R., Pierce, M. B., Khan, A. R., et al. (2010). NUPACK. Analysis and design of nucleic acid systems. *Journal of Computational Chemistry*, *32*(3), 439–452.

Zarringhalam, K., Meyer, M. M., Dotu, I., Chuang, J. H., & Clote, P. (2012). Integrating chemical footprinting data into RNA secondary structure prediction. *PLoS One*, 7(10), e45160.

Zaug, A. J., & Cech, T. R. (1995). Analysis of the structure of Tetrahymena nuclear RNAs in vivo: Telomerase RNA, the self-splicing rRNA Intron, and U2 snRNA. *RNA*, *1*, 363–374.

Zuker, M. (2003). Mfold web server for nucleic acid folding and hybridization prediction. *Nucleic Acids Research*, *31*(13), 3406–3415.

Zuker, M., & Stiegler, P. (1981). Optimal computer folding of large RNA sequences using thermodynamics and auxiliary information. *Nucleic Acids Research*, *9*, 133–148.

CHAPTER FIVE

Computational Prediction of Riboswitch Tertiary Structures Including Pseudoknots by RAGTOP: A Hierarchical Graph Sampling Approach

Namhee Kim, Mai Zahran, Tamar Schlick[1]

Department of Chemistry and Courant Institute of Mathematical Sciences, New York University, New York, USA

[1]Corresponding author: e-mail address: schlick@nyu.edu

Contents

Abstract

The modular organization of RNA structure has been exploited in various computational and theoretical approaches to identify RNA tertiary (3D) motifs and assemble RNA structures. Riboswitches exemplify this modularity in terms of both structural and functional adaptability of RNA components. Here, we extend our computational approach based on tree graph sampling to the prediction of riboswitch topologies by defining additional edges to mimick pseudoknots. Starting from a secondary (2D) structure, we construct an initial graph deduced from predicted junction topologies by our data-mining algorithm RNAJAG trained on known RNAs; we sample these graphs in 3D space guided by knowledge-based statistical potentials derived from bending and torsion measures

Methods in Enzymology, Volume 553
ISSN 0076-6879
http://dx.doi.org/10.1016/bs.mie.2014.10.054

of internal loops as well as radii of gyration for known RNAs. We present graph sampling results for 10 representative riboswitches, 6 of them with pseudoknots, and compare our predictions to solved structures based on global and local RMSD measures. Our results indicate that the helical arrangements in riboswitches can be approximated using our combination of modified 3D tree graph representations for pseudoknots, junction prediction, graph moves, and scoring functions. Future challenges in the field of riboswitch prediction and design are also discussed.

1. INTRODUCTION

1.1. Riboswitch structure and function

Many noncoding RNAs have important regulatory and catalytic roles in various cells and viruses. Riboswitches represent a common type of noncoding RNA that is present in the 5′ UTRs of certain mRNAs (Serganov & Nudler, 2013). They offer important specialized components involved in the regulation of cellular function and operate through a conformational switch upon binding to a ligand (Barrick & Breaker, 2007; Breaker, 2012; Serganov & Patel, 2007). The regulatory mechanisms involved include, for example, formation or deletion of transcription terminator (Peselis & Serganov, 2012; Proshkin, Mironov, & Nudler, 2014), sequestration of ribosome-binding sites (Winkler & Breaker, 2005), and emergence of alternative cleavage sites (Cheah, Wachter, Sudarsan, & Breaker, 2007). The ligand can be a small molecule or ion, and the binding interaction triggers a conformational change in the RNA and subsequent altered expression of the open reading frame located within its mRNA. Typically, each riboswitch consists of an aptamer domain that forms the binding pocket for the target metabolite or ion and an expression platform, which overlaps with the aptamer region of the riboswitch and exerts genetic control by one of several possible mechanisms (Breaker, 2012).

Riboswitches exhibit a diverse range of secondary (2D) and tertiary (3D) structures (Montange & Batey, 2008; Peselis & Serganov, 2012), but they generally contain a natural ligand binding or "aptamer domain" (Gold, Polisky, Uhlenbeck, & Yarus, 1995) and an "expression platform" (Winkler & Breaker, 2003; see Fig. 1). The thiamine pyrophosphate (TPP) riboswitch is a classic example. The TPP riboswitch can adopt two structural conformations upon binding to TPP. TPP can only bind upon formation of a thi-box domain between the aptamer domain and the expression platform. The binding triggers a switch in the entire RNA structure

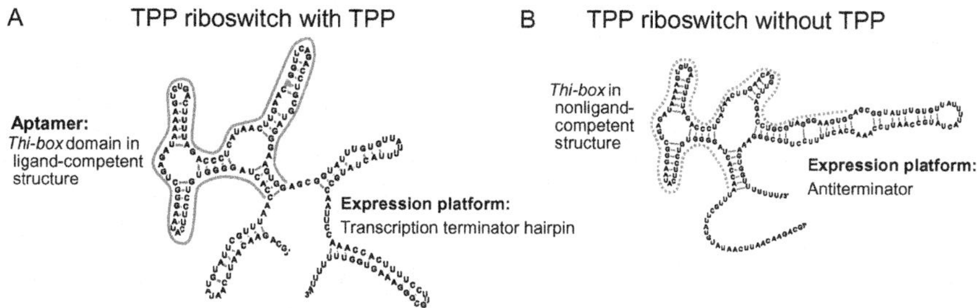

Figure 1 Riboswitch structure and function illustrated for the TPP riboswitch. (A) TPP riboswitch bound to TPP. In this state, the thi-box and terminator are formed to terminate transcription elongation. (B) An alternative riboswitch structure without TPP-box domain. In this alternative state, the antiterminator is formed which occurs when TPP is not bound.

from an active "on" state to an inactive "off" state that causes the formation of a terminator hairpin (Edwards, Klein, & Ferre-D'Amare, 2007; Serganov, Polonskaia, Phan, Breaker, & Patel, 2006).

Based on solved 3D structures, riboswitches have been classified into two types—Type I and Type II (Montange & Batey, 2008). Type I riboswitches are characterized by a single binding pocket supported by a largely pre-established global fold. This arrangement limits ligand-induced conformational changes in the RNA to a small region. The purine riboswitches, the glmS riboswitch, and the *S*-adenosylmethionine (SAM)-II riboswitch are of this type. Type II riboswitches contain binding pockets split into at least two spatially distinct sites. As a result, binding induces both local changes to the binding pocket as well as global rearrangements to the architecture of the RNA. This latter class includes the TPP riboswitch, SAM-I riboswitch, and M-box magnesium riboswitch (Montange & Batey, 2008). Similar features are found in other noncoding RNAs, making it possible to begin to build a hierarchical classification of RNA structure based on the spatial organization of their active sites and associated 2D structural elements.

1.2. Riboswitch motifs

Riboswitch selectivity is encoded within their conserved sensing domains. These domains can vary in the size and complexity of their 2D and 3D structures. All major riboswitch classes are determined at high resolution in complex with their ligands (see Table 1 for examples). Even though they adopt different conformations, most riboswitch structures can be classified depending on the motif they contain: junctions or pseudoknots. Other motifs

Table 1 List of 10 representative riboswitches from the PDB database

PDB	Class	Organism	*L*	*J*	IL	PK	Reference
4ENB	Fluoride	*T. petrophila*	52	–	1	1	Ren, Rajashankar, and Patel (2012)
2KZL	T-box	*B. subtilis*	55	–	2	0	Wang and Nikonowicz (2011)
2G9C	Purine	Artificial	67	3WJ	0	1	Gilbert, Mediatore, and Batey (2006)
3RKF	Guanine	Artificial	67	3WJ	0	1	Buck et al. (2011)
3Q3Z	c-di-GMP-II	*C. acetobutylicum*	77	–	1	1	Smith, Shanahan, Moore, Simon, and Strobel (2011)
3D2G	TPP	*A. thaliana*	77	3WJ	2	0	Thore, Frick, and Ban (2008)
2GDI	TPP	*E. coli*	80	3WJ	3	0	Thore et al. (2008)
2HOJ	TPP	*E. coli*	83	3WJ	2	0	Edwards and Ferre-D'Amare (2006)
2GIS	SAM-I	*T. tengcongensis*	94	4WJ	2	1	Montange and Batey (2006)
4B5R	SAM-I	*H. marismortui*	95	4WJ	2	1	Daldrop and Lilley (2013)

Each riboswitch is classified by ligand, organism, sequence length (*L*), junction type (*J*), number of internal loops (IL), number of pseudoknots (PK), and reference of its experimental structure.

such as kink-turn (k-turn) motifs (Lilley, 2014; Wang, Daldrop, Huang, & Lilley, 2014) recur in many riboswitches as well.

Junctions include single-stranded regions with three or more helical arms. These junctions form long-distance 3D interactions stabilizing the overall conformation. For example, junctions in the Mg^{2+} class II riboswitches (Serganov & Nudler, 2013) are positioned far from the regulatory helix, but one of the helices of the junction folds back toward the regulatory helix and stabilizes it through long-range 3D interactions. Ligands can bind to the RNA in the junction region or close to the regulatory helix, thus stabilizing the global conformation and 3D interactions.

Some riboswitch architectures, including the SAM-I and fluoride riboswitches, are governed by small pseudoknots, which are formed when two single-stranded regions flanked by a stem are base paired. Some junction riboswitches like *glmS* riboswitch and SAM-I riboswitch also contain pseudoknots that are particularly crucial in the formation of ligand-binding pockets and long-distance 3D contacts.

Many riboswitches contain recurrent structural motifs, which are present in other natural and artificial RNAs, such as k-turn motifs. A k-turn motif is a bulge, which generates a kink between two helices with an angle of ~50° (Lilley, 2014). For example, the cyclic-diGMP, cobalamine, and T-box riboswitches have k-turn motifs. As this motif introduces a tight bend into the axis of the duplex RNA, k-turn motifs serve as key architectural elements that help generate specific ligand-binding pockets (Lilley, 2014). Like other functional RNAs, riboswitches employ these motifs as basic building blocks in their complex spatial conformations.

1.3. Advances in computational approaches for RNA structure prediction

Understanding the mechanisms behind RNA functions requires RNA 3D structural knowledge. Multidisciplinary approaches in biology have been commonly used in the past decades and are particularly valuable in the study of RNA molecules. Indeed, RNA's regulatory roles combined with its modular architecture makes it a suitable subject for systematic computational approaches (see our recent reviews, Kim, Fuhr, & Schlick, 2013; Kim, Petingi, & Schlick, 2013; Laing & Schlick, 2011). Theoretical contributions to the prediction of RNA structure have been made from the prediction of 2D structure to the prediction of 3D folds. For example, programs for predicting RNA 2D structures such as Mfold (Zuker, 2003), RNAfold (Hofacker, 2003), ContextFold (Dowell & Eddy, 2006), and PKNOT (Rivas & Eddy, 1999) for pseudoknot folding, are widely used. Programs to fold 3D structures of small RNAs, such as NAST (Jonikas et al., 2009), FARNA (Das & Baker, 2007; Das, Karanicolas, & Baker, 2010), Vfold (Cao & Chen, 2011; Xu, Zhao, & Chen, 2014), and MC-Sym (Parisien & Major, 2008), have also been developed using coarse-grained models, free energy minimization, and fragment assembly approaches, respectively. These programs can predict 3D structures of small RNAs up to ~40 nucleotides within ~6 Å root-mean-square-deviation (RMSD) of atomic positions from native structures (Laing & Schlick, 2011). Other programs to annotate motifs in 2D and 3D structures such as FR3D (Petrov, Zirbel, & Leontis, 2011, 2013), MC-Annotate (Gendron, Lemieux, & Major, 2001), and RNAVIEW (Yang et al., 2003) provide useful tools to extract 2D information from 3D structures (Antczak et al., 2014; Kim, Laing, et al., 2014; Laing et al., 2013) and expand our knowledge of recurrent 3D interactions in RNA structures (Kim et al., 2013).

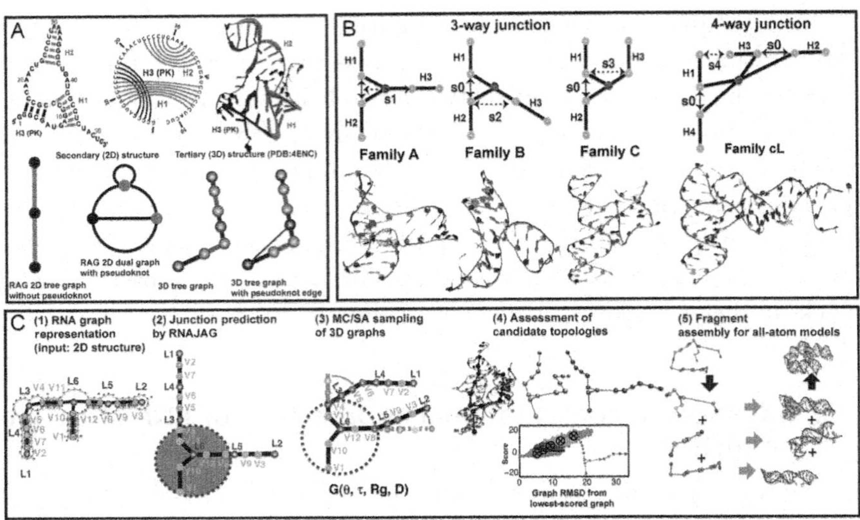

Figure 2 Our hierarchical graph folding approach. (A) RNA-As-Graph (RAG) representation (e.g., fluoride riboswitch, PDB entry 4ENC): RNA 2D and 3D structure including a pseudoknot structure, with corresponding 2D tree graph, dual graph, 3D tree, and refined 3D tree graph; the pseudoknot structure is represented by an additional edge in the 3D tree graph (red). (B) Junction families of 3- and 4-way junctions in riboswitches (family types A, B, C for 3-way junction and cL for 4-way junction) where the helical arm has different helical arrangements (perpendicular, diagonal, or parallel) with respect to the coaxially stacked helices. (C) Our hierarchical graph folding approach RAGTOP, from 2D graph to all-atom models (e.g., TPP riboswitch, PDB entry 2GDI): (1) given a 2D structure, a 2D tree topology is annotated. (2) An initial planar graph is constructed by junction prediction by RNAJAG and edge size estimation. (3) Graphs are sampled by Monte Carlo sampling with two moves (random or restricted) and guided by knowledge-based statistical potentials for bending and torsion angles of internal loops, radii of gyration, and size of pseudoknot edge. (4) Candidate graphs are assessed by the lowest RMSD (P1), the lowest score (P2, for restricted moves), or the lowest cluster representatives (P3, for random moves), and compared with reference graphs translated from solved riboswitches based on local and global RMSD measures. (5) All-atom models are constructed by graph partitioning, fragment search, and assembly of corresponding all-atom modules in RAG-3D. See details in Kim, Laing, et al. (2014) and Laing et al. (2013). (See the color plate.)

We have contributed to the field of RNA structure modeling by developing RNA-As-Graphs (RAG), a resource for RNAs modeled as planar tree and dual graphs to assist the cataloging, analyzing, and designing of RNA structures (see Fig. 2A; Fera et al., 2004; Gan et al., 2004; Izzo, Kim, Elmetwaly, & Schlick, 2011). The simplification and abstraction of RNA structures as graphs drastically reduces the conformational space and allows enumeration and classification of RNA structures according to essential but simplified topological aspects, such as helical arrangements and loop connectivity. RAG has been applied to the prediction of RNA-like topologies (Izzo et al., 2011; Koessler, Knisley, Knisley, & Haynes, 2010), *in silico* modeling of *in vitro* selection (Kim, Izzo, Elmetwaly, Gan, & Schlick, 2010), analysis of large viral RNA (Gopal, Zhou, Knobler, & Gelbart, 2012), analysis and design of riboswitches (Quarta, Kim, Izzo, & Schlick, 2009), prediction of RNA junction topology (Laing et al., 2013), and graph partitioning for the discovery of RNA modularity (Kim, Zheng, Elmetwaly, & Schlick, 2014).

Recently, we have extended RNA graph representations from 2D to 3D graphs and developed a hierarchical sampling approach (which we term "RAGTOP" here for RNA-As-Graph-Topologies) to predict global 3D topologies compatible with a given RNA 2D structure (see Fig. 2; Kim, Laing, et al., 2014). RNA 3D graphs represent both topological connectivity of 2D structures and geometrical aspects of helical arrangements in 3D. Our overall approach exploits graph representations to accelerate conformational sampling and generate a graph approximation to an RNA 3D structure. We utilize the modular and hierarchical features of RNA structures in two steps. First, we predict junction topologies based on our data-mining program called RNAJAG (RNA-Junction-As-Graphs; Laing et al., 2013). Second, we sample 3D graphs guided by knowledge-based statistical potentials derived from bending and torsion measures of internal loops as well as radii of gyration for known RNAs (Kim, Laing, et al., 2014). This graph sampling approach RAGTOP has demonstrated significant improvements over current approaches for characterizing 3D global helical arrangements in large RNAs from a given 2D structure (Kim, Laing, et al., 2014).

Here, we modify our 3D tree graph representation to represent riboswitches with pseudoknots and apply our hierarchical graph sampling tool for the prediction of representative riboswitch structures. In Section 2, we characterize how to represent RNA 2D and 3D structures including both pseudoknot-free and pseudoknot structures by extended tree graphs. We also describe our hierarchical graph folding approach for the prediction of riboswitch structures. Section 3 presents our graph sampling and prediction results for 10 different riboswitches, 6 with pseudoknots, and compares them to solved structures based on global and local RMSD measures. We conclude in Section 4 with summary and future challenges.

2. HIERARCHICAL GRAPH FOLDING APPROACH

Here, we modify RAGTOP described as in Kim, Laing, et al. (2014) to handle pseudoknots by altering the 3D graph representations by additional edges for pseudoknot interactions, adding a term for pseudoknots in our scoring function, and treating MC pivot moves at pseudoknots differently (see below).

2.1. RNA 2D and 3D graph representation

RNA graph representations provide the basis of our hierarchical folding approach RAGTOP (see Fig. 2A). Recently, we developed 3D tree graph representations which preserve the rules for 2D tree graph representation and can further represent parallel and antiparallel helical arrangements in 3D space: (1) unpaired RNA 2D building blocks (hairpin loops, internal loops, junctions) and the helix ends are translated to vertices; (2) helices in pseudoknot-free structures are translated to edges; (3) edges are also set to represent the connection between vertices of unpaired regions and helix ends; (4) 3D coordinates are assigned for each vertex at the centers of helices and loops (see Kim, Laing, et al., 2014 for full equations). Here, to represent pseudoknot structures, we modify our 3D graph representation by additional edges for pseudoknot interactions: (5) the pseudoknot interactions are translated into edges. The pseudoknot edge is formed by the connection of two loop vertices, which interact via pseudoknot base pairing. Figure 2A shows an example of a modified 3D graph for a pseudoknot in the fluoride riboswitch (PDB entry 4ENC), formed by base pairs connecting one internal loop and the other dangling end, represented by a pseudoknot edge.

2.2. Junction prediction by RNAJAG

RNAJAG predicts helical arrangements of RNA junction structures as tree graphs from a given 2D structure. RNAJAG can indicate the family type and stacking orientation of a given junction by a data-mining approach based on the random forests (decision tree) procedure trained using loop length, adenine base content, and free energy estimates of two base pairs in junction helix ends (Laing et al., 2013). The 3- and 4-way junctions are classified into three families (called A, B, and C; Lescoute & Westhof, 2006) and nine families (H, cH, cL, cK, π, cW, Ψ, cX, and X), respectively, according to coaxial stacking and helical configuration (Laing & Schlick, 2009). RNAJAG is used in RAGTOP to generate initial graph structures with specific helical arrangements (Laing et al., 2013). Four recurrent junction family types (A, B, C, and cL) in riboswitch structures are shown in Fig. 2B.

2.3. Monte Carlo simulated annealing (MC/SA) graph sampling

Starting from initial graph setup by size measures and junction prediction, we perform MC/SA sampling of RNA 3D graphs at the flexible vertices in internal loops (Kim, Laing, et al., 2014). We use two types of moves—restricted pivot moves, by reciprocally decreasing angle ranges from 360° to 10° along MC steps, and random pivot moves (Kim, Laing, et al., 2014). Here, in addition to internal loops, we allow pivot moves for hairpin loop vertices if they are involved in a pseudoknot interaction. To score our graphs, we developed knowledge-based statistical potentials from statistical analyses of geometrical features of solved RNA structures, including bending and torsion angles between two helices of internal loops, and radii of gyration of the entire RNAs (Kim, Laing, et al., 2014). Here, to model pseudoknots, we modify the scoring function by adding a term for the pseudoknot edge ($G_{pk} = D - \overline{D}$ where D is the length of an pseudoknot edge of each sampled graph and $\overline{D}$ is the "equilibrium" length of a pseudoknot edge observed from known pseudoknots (between 10 and 15 Å)). Thus, the scoring function for a pseudoknot structure (G) is the sum of scores for pseudoknot-free tree graph ($G_{internal} + G_{Rg}$, see Kim, Laing, et al., 2014 for details) plus G_{pk}. The scores guide the conformational sampling: if the score for a new conformation is lower than that of the old conformation, the new conformation is accepted. If the new score is higher, the SA sampling proceeds: some moves with higher score at each step are accepted with decreasing probability along the MC steps (for details, see Kim, Laing, et al., 2014).

2.4. Assessment of sampled graphs

After MC/SA, the candidate graphs are compared to the 3D graphs translated from solved RNAs by three procedures (P1–P3; Kim, Laing, et al., 2014). P1 directly compares the candidate graph with the lowest RMSD score among the final pool of accepted graphs to the reference graph translated from the solved structure. P2 compares our lowest-scored graph among accepted graphs to the reference graph. For random moves, conformational space is more globally sampled compared with restricted moves, and additional clustering is required to select a representative graph from among five clusters (P3). Thus, for P3 we compare the cluster representatives to the reference graph. We compare resulting graphs to reference graphs translated from solved structures by the average RMSDs. Graph-based RMSD provides a valid measure to compare global topological similarities with positive correlation to atomic RMSD (Kim, Laing, et al., 2014). However, as elaborated in Parisien, Cruz, Westhof, and Major (2009), RMSD spreads the structural dissimilarities and does not specify local errors such as base interactions and local helical arrangements. As we start with given 2D structures, base interactions are same for all sampled graphs. Thus, in addition to graph RMSDs, we compare resulting graphs in terms of vertex-to-vertex distances, which account for the specificity of local helical arrangements.

2.5. All-atom building by RAG-3D

For the all-atom model building, we use a threading-like procedure based on a search for graph similarities with the 3D graphs classified in our database RAG-3D (Zahran, Elmetwaly, & Schlick, 2014), an extension of the original RAG database containing 2D planar graphs (Fera et al., 2004; Izzo et al., 2011). RAG-3D contains 3D atomic models extracted from RNA structures present in the PDB database, and linked to corresponding 3D graphs. In RAG-3D, the 3D graphs are classified based on the original RAG motif IDs, which reflect topological properties of 2D structural elements. All-atom models are constructed in three steps: (i) identifying the motif ID of the target graph; (ii) comparing the target graph to all 3D graphs catalogued with the same motif ID in RAG-3D based on a standard RMSD calculation; and (iii) selecting the lowest-score RMSD graph, extracting its all-atom 3D coordinates, and adjusting base content to match that of the target sequence, while keeping the backbone intact.

3. APPLICATION TO RIBOSWITCH STRUCTURE PREDICTION

We apply the RAGTOP procedure as modified here for pseudoknots to a set of 10 representative riboswitches in Table 1. These 10 riboswitches have diverse structural features in terms of sequence length and diverse combinations of structural elements, namely internal loops, junctions with 3- and 4-way junctions, and pseudoknots. In particular, among these 10 riboswitches, 7 structures have junctions and 6 have pseudoknots.

Table 2 and Fig. 3 provide our graph results for junction prediction and graphs after Monte Carlo sampling for these 10 representative riboswitches.

Table 2 Graph results for 10 riboswitches

PDB	Class	L	J	Native		RNAJAG		RMSD after MC/SA		
				Coaxial	Family	Coaxial	Family	P1	P2	P3
4ENB	Fluoride	52	No	N/A	N/A	N/A	N/A	2.45	2.76	3.08
2KZL	T-box	55	No	N/A	N/A	N/A	N/A	6.06	11.05	12.50
2G9C	Purine	67	3WJ	H1H3	C	H1H3	C	3.88	4.64	4.07
3RKF	Guanine	67	3WJ	H1H3	C	H1H3	C	3.67	4.59	6.23
3Q3Z	c-di-GMP-II	77	No	N/A	N/A	N/A	N/A	2.22	8.63	3.61
3D2G	TPP	77	3WJ	H1H2	A	H1H2	A	3.62	15.74	18.34
2GDI	TPP	80	3WJ	H1H2	A	H2H1	A	4.25	17.98	17.90
2HOJ	TPP	83	3WJ	H1H2	A	H1H2	A	4.70	14.93	13.01
2GIS	SAM	94	4WJ	H1H4 and H2H3	cL	H1H4 and H2H3	cL	11.41	14.94	15.64
4B5R	SAM	95	4WJ	H1H4 and H2H3	cL	H1H4 and H2H3	cL	10.77	16.40	13.98

For riboswitches with junctions, each junction is listed with its junction family and coaxial-stacking arrangement from native structures and predicted structures by RNAJAG. N/A indicates that the structure does not include junction. After MC/SA sampling, graph RMSDs between reference graphs from solved structures and our sampled graphs by MC/SA—the lowest (P1, random moves), the lowest score (P2, restricted moves), and the lowest cluster representative (P3, random moves) after MC/SA—are shown. See Figs. 3 and 4 for vertex-to-vertex distance measures between corresponding vertices of reference and sampled graphs (P1–P3).

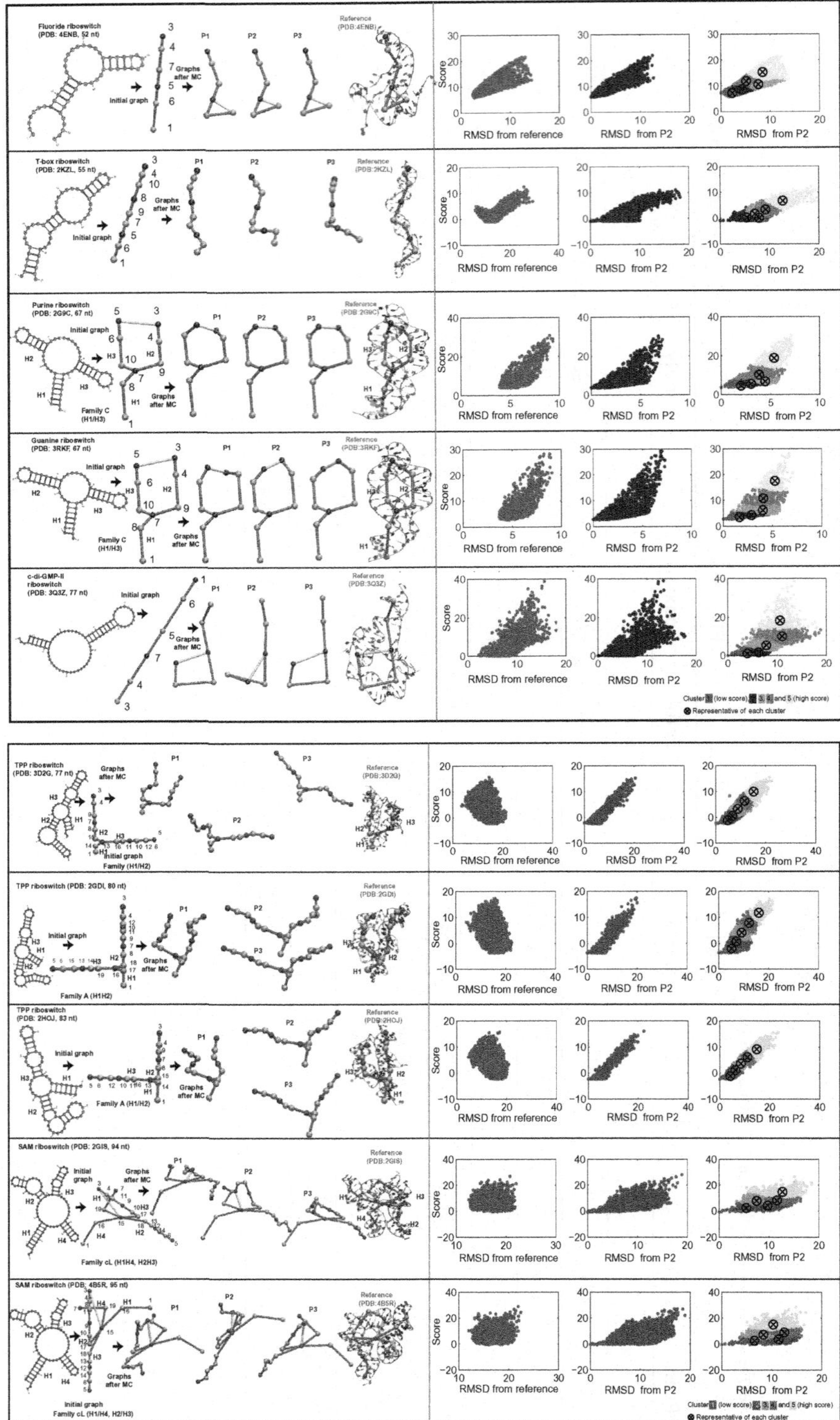

Figure 3 See legend on opposite page.

We assess candidate graphs with respect to predicted graphs (translated from solved structures) rather than predicted atomic models versus solved atomic models, by junction types, coaxial stacking, and graph RMSD. Our analysis has shown that graph and all-atom RMSDs are positively correlated with positive correlation coefficient (~0.89); thus, overall similarity between structures can be captured by graph RMSD (Kim, Laing, et al., 2014). Figure 4 shows graph results by vertex-to-vertex distance specifying local errors by helical arrangements.

For RNAs containing junctions (7 of 10 riboswitches in Table 1), junction families and coaxial stacking are perfectly predicted by RNAJAG based on a collective training set of 244 junctions. The purine and guanine riboswitches (PDB entries 2G9C and 3RKF) have 3-way junctions with Family C having coaxial stacking of H1 and H3 and a parallel helix H2, which is predicted correctly by RNAJAG. Similarly, for TPP riboswitches (PDB entries 3D2G, 2HOJ, and 2GDI), the 3-way junctions are classified and predicted as Family A with coaxial stacking of H1 and H2 and a perpendicular helix H3. For SAM riboswitches (PDB entries 2GIS and 4B5R), the 4-way junctions are classified and predicted as cL, which is an expanded version of Family A 3-way junction, with coaxial stacking of H1 and H4 and H2 and H3. Thus, RNAJAG can predict junction classes and coaxial stacking for riboswitches very well, with prediction accuracy much higher than overall prediction accuracy for all RNAs: 95%/92% in 3-/4-way junctions for coaxial tacking and 94%/87% for family type in 3-/4-way junctions. The abundance of junction structures in riboswitches increases the prediction accuracy for junction classifications.

Starting from an initial graph constructed from the RNAJAG junction prediction, we sample riboswitch graphs using 10^4 steps for restricted pivot moves, which converge to one region of conformational space, as well as random pivot moves, which explore multiple regions of space and thus require clustering analysis. Graph sampling improves orientation of loops, pseudoknots, and overall 3D topology geometries. For RMSDs relative to reference graphs (P1 in Table 2), the lowest values range from 2.22 Å

Figure 3—cont'd Graph results for riboswitch structures—best graph with the lowest RMSD from reference graph based on random pivot moves (P1), the lowest-scored graph based on restricted pivot moves (P2), and the lowest cluster representative of landscapes with respect to the lowest-scored graph based on random pivot moves (P3). Pseudoknot edges in the graphs are indicated in yellow. Landscapes with respect to reference structure and the lowest-scored graph based on restricted and random pivot moves are also shown. (See the color plate.)

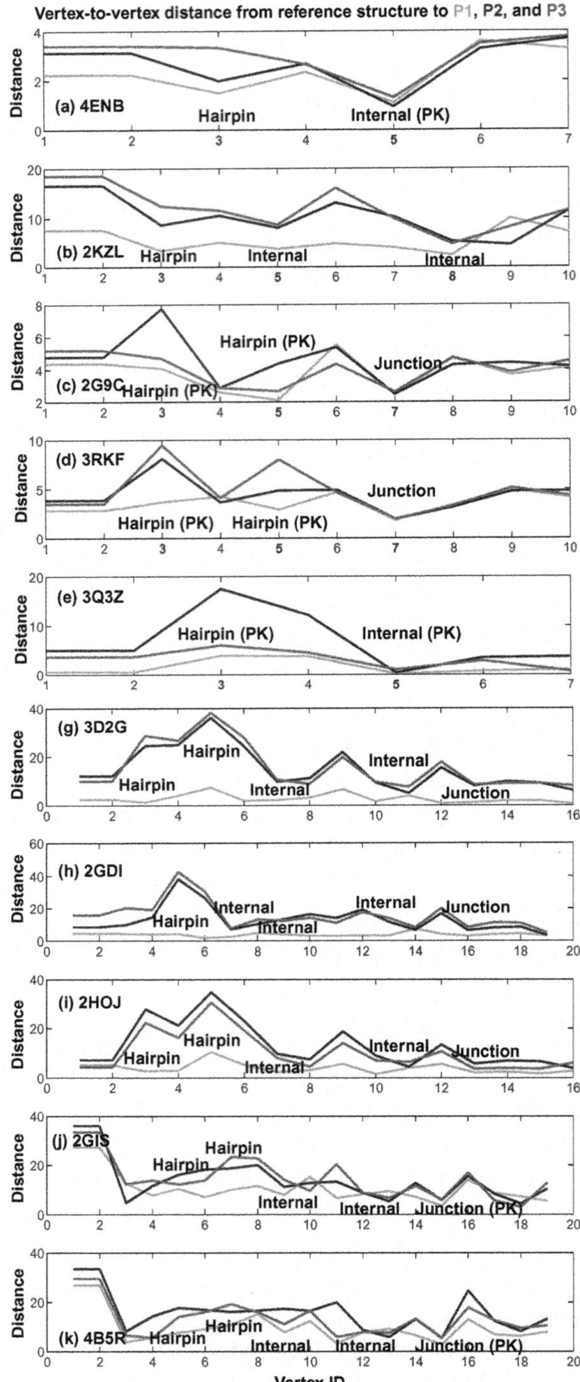

Figure 4 Local geometry analysis by vertex-to-vertex distance measures from reference graphs to the lowest RMSD (P1, green line (light gray in the print version)), the lowest-scored (P2, blue line (dark gray in the print version)), and the lowest cluster representative (P3, red line (gray in the print version)) graphs for 10 riboswitches in Tables 1 and 2. Vertex ID matches with graphs in Fig. 3.

(c-di-GMP-II riboswitch, PDB entry 3Q3Z) to 11.41 Å (SAM riboswitch, PDB entry 2GIS) using random moves. For most riboswitches except the SAM riboswitch (PDB entries 2GIS and 4B5R), the RMSD with respect to known structure (P1) is less than or close to 6 Å. For SAM riboswitches with 4-way junctions, the RMSD is high (11.41 Å for PDB entry 2GIS and 10.77 Å for PDB entry 4B5R). This is because the initial junction geometries are held rigid during MC sampling. Although the 4-way junction class for the initial junction geometry is predicted correctly (family cL and coaxial stacking of H1H4 and H2H3), related distances are imperfect. For example, the distance between helices H1 and H3 should be much closer than that between H2 and H4, but the generated graph has the longer distance between H2 and H4.

When the reference graph is not known, we consider both the lowest-scored graph based on restricted moves (P2) and the lowest-scored graph representatives among five clusters based on random moves (P3). For P2, graph RMSDs range from 2.76 Å (fluoride riboswitch, PDB entry 4ENB) to 17.98 Å (TPP riboswitch, PDB entry 2GDI). For random graphs, representative graphs from five clusters sorted by score from low- to high-offer candidate 3D topologies in the absence of solutions (Fig. 3). Representative graphs from cluster 1 have RMSDs ranging from 3.08 Å (fluoride riboswitch, PDB entry 4ENB) to 18.34 Å (TPP riboswitch, PDB entry 3D2G; P3 in Table 1), similar to the lowest-scored graphs (P2).

The modified 3D tree graphs with additional edges handle pseudoknots effectively. For fluoride riboswitch (PDB entry 4ENB), an edge connecting the internal loop vertex ID 5 and the dangling end vertex 1 represent its pseudoknot (see Fig. 3 for vertex ID). For the purine and guanine riboswitches (PDB entries 2G9C and 3RKF), two distant hairpin loops are interconnected to form a pseudoknot as represented by an edge connecting vertices 5 and 3. For the c-di-GMP-II riboswitch (PDB entry 3Q3Z), a pseudoknot is formed by the connection between one hairpin (vertex ID 3) and the other internal loop (vertex ID 5). For the SAM riboswitches (PDB entries 2GIS and 4B5R), a pseudoknot edge is formed by the connection between a junction (vertex ID 15) and a hairpin loop (vertex ID 3). The sampled graphs approximate global interactions due to pseudoknots by closer distances between the two interconnected strands estimated by the size of pseudoknot edges. After MC/SA, the size of pseudoknot edge in the resulting graphs (P1–P3) is similar with that in reference riboswitches, which ranges from 10 Å (pseudoknot edge connecting vertices 3 and 5 in PDB entries 2G9C and 3RKF) to 20 Å (pseudoknot edge connecting vertices

5 and 15). All RMSD measures between resulting graphs (P1–P3) and reference graphs are relatively small for these pseudoknot structures. Even for the SAM riboswitches (PDB entries 2GIS and 4B5R) whose initial graph geometries are poor, the overall graph RMSDs are reduced by around 2 Å compared to graph results without pseudoknot edges (from 17.43 Å (Kim, Laing, et al., 2014) to 15.64 Å as shown in P3 in Table 2). For P3, RMSD ranges from 3.08 Å (PDB entry 4ENB) to 15.64 Å (PDB entry 2GIS).

As an alternative to the global RMSD measures, we also analyze local geometrical features. Figure 4 shows the results of local distances between corresponding vertices in aligned resulting graphs and reference graphs, which is developed in a similar spirit in local base interactions whose dissimilarities are spread in RMSD measures (Parisien et al., 2009). These vertex-to-vertex distance measures indicate the local similarities/dissimilarities. We see that the distance for hairpins is larger than that for junctions and internal loops (for example, see hairpin vertices ID 3 and 5 in TPP riboswitch, PDB entry 3D2G). This is because we locate the hairpin edge based on only size estimated by the sequence length of a hairpin sequence, without consideration of the angles about a hairpin edge.

Even though our graph prediction provides the information about the overall helical arrangements of riboswitches, atomic models are ultimately required. We applied our build-up approach to the 3-way junction structure of guanine riboswitch (PDB entry 3RKF; Laing et al., 2013). This 3-way junction guanine riboswitch RNA contains 53 nucleotides. RNAJAG was able to correctly predict both the junction family type and the coaxial stacking and predicted a graph with RMSD value of 4.32 Å with respect to the graph of its native structure translated from the PDB structure. We superimposed the predicted graph against all the graphs of the same motif ID family (namely, 4_2) available in the RAG-3D database, and ranked them based on their RMSDs with respect to the target graph. We extracted the all-atom coordinates of the lowest-RMSD graph found (4.41 Å), and created a model by mutating the bases to match the query sequence. We obtained an RMSD value of 5.09 Å for the all-atom model junction region compared to its native structure, as shown in figure 10 in Laing et al. (2013).

4. FUTURE CHALLENGES AND PERSPECTIVES

We have extended and applied our hierarchical computational approach (named RAGTOP here) to predict riboswitch tertiary structure with pseudoknots by combining our coarse-grained graph sampling approach (Kim, Laing, et al., 2014), which utilizes RNAJAG for initial junction prediction and knowledge-based scoring functions for MC/SA sampling, with modified features for pseudoknots. Three added features for pseudoknots include modified 3D graphs with pseudoknot edges, graph moves for hairpin loop vertices involving a pseudoknot formation, and updated scoring function with an additional term for pseudoknot edge lengths. Our sampling based on geometric statistical potentials produces graphs whose 3D shapes resemble native structures, and the lowest-scored graphs are also reasonably selected without knowledge of reference graphs both in our restricted (P2) and random (P3) move protocols. The application of our approach to 10 riboswitches with internal loops, 3- and 4-way junctions, and pseudoknots shows that graph-based sampling can reasonably predict the junction structures of riboswitches, and provide a good approximation for global helical arrangements. Graph-based structure prediction is expected to continue to a useful tool to predict and design riboswitch structures.

Our current graph sampling approach is primarily applicable to the prediction of riboswitch structures with 3- and 4-way junctions. However, riboswitches can have higher-order junctions (e.g., lysine riboswitch, 5-way junctions). RNAJAG can potentially be extended to predict higher-order junctions. For example, 5-way junctions can be partitioned into 3- and 4-way subjunctions. Our prediction of RNA junctions could also be extended from the current three discrete models (parallel, perpendicular, and diagonal helical arrangements) to model continuous helical orientations.

Our 3D graph representations might also be modified and expanded to represent long-range interactions. Here, we added an edge to represent a pseudoknot formed by intertwined and long-range base pair interactions between two hairpin loops and between a hairpin loop and a junction/internal loop. This representation provides a reasonable framework for pseudoknot topology sampling. The corresponding pivot moves for hairpin loops involving pseudoknots and the modified scoring function term to target a pseudoknot edge length guide our graph sampling towards native-like pseudoknot topologies. In the future, dual graphs could also be explored to model pseudoknot structures rigorously.

Our scoring function does not account for k-turn motifs, hairpin angles, and ligand-binding cases. However, since k-turn motifs have highly conserved sequence contents and 2D structures, it is possible to identify them based on primary and 2D structures. Our scoring function could discern internal loops where k-turn can potentially occur and score them differently. In addition, the shape of hairpin loops could also be considered, so as to locate nonplanar hairpin edges. A separation of self-folding RNA parameters from those for substrate-binding RNAs could also be envisioned.

The application of RAGTOP to riboswitch design appears promising. Combined with other bioinformatics searches of sequences and accurate 2D folding algorithms, our approach could be applied to design new riboswitch structures. Furthermore, since riboswitches are potential targets for the construction of artificial genetic circuits that could be controlled by nonnatural compounds, understanding the regulatory and structural principles exploited by natural riboswitches could help in the development of synthetic riboswitches that respond to a particular ligand. Computational structure prediction of riboswitch candidates with unknown 3D structures could ultimately help identify and design new riboswitch classes.

ACKNOWLEDGMENTS

This work is supported by the National Institutes of Health (GM100469, GM081410). We also gratefully acknowledge the Telluride Science Research Center (TSRC) and the NIH conference grant (GM112216-01) which brought together RNA scientists and stimulated this work.

REFERENCES

Antczak, M., Zok, T., Popenda, M., Lukasiak, P., Adamiak, R. W., Blazewicz, J., et al. (2014). RNApdbee—A webserver to derive secondary structures from pdb files of knotted and unknotted RNAs. *Nucleic Acids Research*, *42*(Web Server issue), W368–W372.

Barrick, J. E., & Breaker, R. R. (2007). The distributions, mechanisms, and structures of metabolite-binding riboswitches. *Genome Biology*, *8*(11), R239.

Breaker, R. R. (2012). Riboswitches and the RNA world. *Cold Spring Harbor Perspectives in Biology*, *4*(2), a003566.

Buck, J., Wacker, A., Warkentin, E., Wohnert, J., Wirmer-Bartoschek, J., & Schwalbe, H. (2011). Influence of ground-state structure and Mg^{2+} binding on folding kinetics of the guanine-sensing riboswitch aptamer domain. *Nucleic Acids Research*, *39*(22), 9768–9778.

Cao, S., & Chen, S. J. (2011). Physics-based de novo prediction of RNA 3D structures. *The Journal of Physical Chemistry B*, *115*, 4216–4226.

Cheah, M. T., Wachter, A., Sudarsan, N., & Breaker, R. R. (2007). Control of alternative RNA splicing and gene expression by eukaryotic riboswitches. *Nature*, *447*(7143), 497–500.

Daldrop, P., & Lilley, D. M. (2013). The plasticity of a structural motif in RNA: Structural polymorphism of a kink turn as a function of its environment. *RNA*, *19*(3), 357–364.

Das, R., & Baker, D. (2007). Automated de novo prediction of native-like RNA tertiary structures. *Proceedings of the National Academy of Sciences of the United States of America*, *104*(37), 14664–14669.

Das, R., Karanicolas, J., & Baker, D. (2010). Atomic accuracy in predicting and designing noncanonical RNA structure. *Nature Methods*, 7(4), 291–294.

Dowell, R. D., & Eddy, S. R. (2006). Efficient pairwise RNA structure prediction and alignment using sequence alignment constraints. *BMC Bioinformatics*, 7, 400.

Edwards, T. E., & Ferre-D'Amare, A. R. (2006). Crystal structures of the thi-box riboswitch bound to thiamine pyrophosphate analogs reveal adaptive RNA-small molecule recognition. *Structure*, *14*(9), 1459–1468.

Edwards, T. E., Klein, D. J., & Ferre-D'Amare, A. R. (2007). Riboswitches: Small-molecule recognition by gene regulatory RNAs. *Current Opinion in Structural Biology*, *17*(3), 273–279.

Fera, D., Kim, N., Shiffeldrim, N., Zorn, J., Laserson, U., Gan, H. H., et al. (2004). RAG: RNA-As-Graphs web resource. *BMC Bioinformatics*, *5*, 88.

Gan, H. H., Fera, D., Zorn, J., Shiffeldrim, N., Tang, M., Laserson, U., et al. (2004). RAG: RNA-As-Graphs database—Concepts, analysis, and features. *Bioinformatics*, *20*(8), 1285–1291.

Gendron, P., Lemieux, S., & Major, F. (2001). Quantitative analysis of nucleic acid three-dimensional structures. *Journal of Molecular Biology*, *308*(5), 919–936.

Gilbert, S. D., Mediatore, S. J., & Batey, R. T. (2006). Modified pyrimidines specifically bind the purine riboswitch. *Journal of the American Chemical Society*, *128*(44), 14214–14215.

Gold, L., Polisky, B., Uhlenbeck, O., & Yarus, M. (1995). Diversity of oligonucleotide functions. *Annual Review of Biochemistry*, *64*, 763–797.

Gopal, A., Zhou, Z. H., Knobler, C. M., & Gelbart, W. M. (2012). Visualizing large RNA molecules in solution. *RNA*, *18*(2), 284–299.

Hofacker, I. L. (2003). Vienna RNA secondary structure server. *Nucleic Acids Research*, *31*(13), 3429–3431.

Izzo, J. A., Kim, N., Elmetwaly, S., & Schlick, T. (2011). RAG: An update to the RNA-As-Graphs resource. *BMC Bioinformatics*, *12*, 219.

Jonikas, M. A., Radmer, R. J., Laederach, A., Das, R., Pearlman, S., Herschlag, D., et al. (2009). Coarse-grained modeling of large RNA molecules with knowledge-based potentials and structural filters. *RNA*, *15*(2), 189–199.

Kim, N., Fuhr, N., & Schlick, T. (2013). Graph applications to RNA structure and function. In R. Russell (Ed.), *Biophysics of RNA folding, Biophysics for the life sciences: Vol. 3.* (pp. 23–51). Berlin and Heidelberg: Springer Verlag.

Kim, N., Izzo, J. A., Elmetwaly, S., Gan, H. H., & Schlick, T. (2010). Computational generation and screening of RNA motifs in large nucleotide sequence pools. *Nucleic Acids Research*, *38*(13), e139.

Kim, N., Laing, C., Elmetwaly, S., Jung, S., Curuksu, J., & Schlick, T. (2014). Graph-based sampling for approximating global helical topologies of RNA. *Proceedings of the National Academy of Sciences of the United States of America*, *111*(11), 4079–4084.

Kim, N., Petingi, L., & Schlick, T. (2013). Network theory tools for RNA modeling. *WSEAS Transaction on Mathematics*, *12*(9), 941.

Kim, N., Zheng, Z., Elmetwaly, S., & Schlick, T. (2014). RNA graph partitioning for the discovery of RNA modularity: A novel application of graph partition algorithm to biology. *PLoS One*, *9*(9), e106074.

Koessler, D. R., Knisley, D. J., Knisley, J., & Haynes, T. (2010). A predictive model for secondary RNA structure using graph theory and a neural network. *BMC Bioinformatics*, *11*(Suppl. 6), S21.

Laing, C., Jung, S., Kim, N., Elmetwaly, S., Zahran, M., & Schlick, T. (2013). Predicting helical topologies in RNA junctions as tree graphs. *PLoS One*, *8*(8), e71947.

Laing, C., & Schlick, T. (2009). Analysis of four-way junctions in RNA structures. *Journal of Molecular Biology*, *390*(3), 547–559.

Laing, C., & Schlick, T. (2011). Computational approaches to RNA structure prediction, analysis, and design. *Current Opinion in Structural Biology*, *21*(3), 306–318.

Lescoute, A., & Westhof, E. (2006). Topology of three-way junctions in folded RNAs. *RNA*, *12*, 83–93.

Lilley, D. M. (2014). The K-turn motif in riboswitches and other RNA species. *Biochimica et Biophysica Acta, 1839*, 995–1004.
Montange, R. K., & Batey, R. T. (2006). Structure of the S-adenosylmethionine riboswitch regulatory mRNA element. *Nature, 441*(7097), 1172–1175.
Montange, R. K., & Batey, R. T. (2008). Riboswitches: Emerging themes in RNA structure and function. *Annual Review of Biophysics, 37*, 117–133.
Parisien, M., Cruz, J. A., Westhof, E., & Major, F. (2009). New metrics for comparing and assessing discrepancies between RNA 3D structures and models. *RNA, 15*(10), 1875–1885.
Parisien, M., & Major, F. (2008). The MC-Fold and MC-Sym pipeline infers RNA structure from sequence data. *Nature, 452*(7183), 51–55.
Peselis, A., & Serganov, A. (2012). Structural insights into ligand binding and gene expression control by an adenosylcobalamin riboswitch. *Nature Structural & Molecular Biology, 19*(11), 1182–1184.
Petrov, A. I., Zirbel, C. L., & Leontis, N. B. (2011). WebFR3D—A server for finding, aligning and analyzing recurrent RNA 3D motifs. *Nucleic Acids Research, 39*(Web Server issue), W50–W55.
Petrov, A. I., Zirbel, C. L., & Leontis, N. B. (2013). Automated classification of RNA 3D motifs and the RNA 3D Motif Atlas. *RNA, 19*(10), 1327–1340.
Proshkin, S., Mironov, A., & Nudler, E. (2014). Riboswitches in regulation of Rho-dependent transcription termination. *Biochimica et Biophysica Acta, 1839*, 974–977.
Quarta, G., Kim, N., Izzo, J. A., & Schlick, T. (2009). Analysis of riboswitch structure and function by an energy landscape framework. *Journal of Molecular Biology, 393*(4), 993–1003.
Ren, A., Rajashankar, K. R., & Patel, D. J. (2012). Fluoride ion encapsulation by Mg^{2+} ions and phosphates in a fluoride riboswitch. *Nature, 486*(7401), 85–89.
Rivas, E., & Eddy, S. R. (1999). A dynamic programming algorithm for RNA structure prediction including pseudoknots. *Journal of Molecular Biology, 285*(5), 2053–2068.
Serganov, A., & Nudler, E. (2013). A decade of riboswitches. *Cell, 152*(1–2), 17–24.
Serganov, A., & Patel, D. J. (2007). Ribozymes, riboswitches and beyond: Regulation of gene expression without proteins. *Nature Reviews. Genetics, 8*(10), 776–790.
Serganov, A., Polonskaia, A., Phan, A. T., Breaker, R. R., & Patel, D. J. (2006). Structural basis for gene regulation by a thiamine pyrophosphate-sensing riboswitch. *Nature, 441*(7097), 1167–1171.
Smith, K. D., Shanahan, C. A., Moore, E. L., Simon, A. C., & Strobel, S. A. (2011). Structural basis of differential ligand recognition by two classes of bis-(3′-5′)-cyclic dimeric guanosine monophosphate-binding riboswitches. *Proceedings of the National Academy of Sciences of the United States of America, 108*(19), 7757–7762.
Thore, S., Frick, C., & Ban, N. (2008). Structural basis of thiamine pyrophosphate analogues binding to the eukaryotic riboswitch. *Journal of the American Chemical Society, 130*(26), 8116–8117.
Wang, J., Daldrop, P., Huang, L., & Lilley, D. M. (2014). The k-junction motif in RNA structure. *Nucleic Acids Research, 42*(8), 5322–5331.
Wang, J., & Nikonowicz, E. P. (2011). Solution structure of the K-turn and specifier loop domains from the Bacillus subtilis tyrS T-box leader RNA. *Journal of Molecular Biology, 408*(1), 99–117.
Winkler, W. C., & Breaker, R. R. (2003). Genetic control by metabolite-binding riboswitches. *Chembiochem, 4*(10), 1024–1032 (Review).
Winkler, W. C., & Breaker, R. R. (2005). Regulation of bacterial gene expression by riboswitches. *Annual Review of Microbiology, 59*, 487–517.

Xu, X. J., Zhao, P. N., & Chen, S. J. (2014). Vfold: A web server for RNA structure and folding thermodynamics prediction. *PLoS One*, *9*(9), e107504.

Yang, H., Jossinet, F., Leontis, N., Chen, L., Westbrook, J., Berman, H., et al. (2003). Tools for the automatic identification and classification of RNA base pairs. *Nucleic Acids Research*, *31*(13), 3450–3460.

Zahran, M., Elmetwaly, S., & Schlick, T. (2014). 3D-RAG: A Tool for RNA 3D Substructuring. In *Preparation*.

Zuker, M. (2003). Mfold web server for nucleic acid folding and hybridization prediction. *Nucleic Acids Research*, *31*(13), 3406–3415.

SECTION II

RNA Dynamics and Thermodynamics

CHAPTER SIX

Using Reweighted Pulling Simulations to Characterize Conformational Changes in Riboswitches

Francesco Di Palma, Francesco Colizzi, Giovanni Bussi[1]
Scuola Internazionale Superiore di Studi Avanzati (SISSA), Trieste, Italy
[1]Corresponding author: e-mail address: bussi@sissa.it

Contents

Abstract

Riboswitches are RNA sequences located in noncoding portions of mRNA that can sense specific ligands and subsequently control gene expression. The ligand-binding event induces conformational changes in the riboswitch that are then transmitted to the gene expression apparatus. Probing the mechanisms of such a fine regulation at atomic resolution is very difficult experimentally and molecular dynamics (MD) could be used to quantify the ligand-dependent behavior of a riboswitch. However, since the accessible time scale of fully atomistic simulations is limited, this can only be done using enhanced sampling techniques. Here, we discuss the application of steered MD to

Methods in Enzymology, Volume 553
ISSN 0076-6879
http://dx.doi.org/10.1016/bs.mie.2014.10.055

the characterization of the ligand-dependent stability of the aptamer terminal helix in the *add* adenine-sensing riboswitch. The employed techniques are discussed in detail and sample input files are provided. We show that with a limited computational effort it is possible to quantify, in terms of free energy, the stacking interaction between the ligand and the terminal helix, obtaining results in agreement with thermodynamic experiments.

1. INTRODUCTION

Since their first description in 2002 (Mironov et al., 2002; Nahvi et al., 2002), riboswitches have come into the limelight in RNA biology as an important and widespread cis-acting gene regulatory mechanism. Riboswitches are structured RNA domains located in the untranslated region (UTR) of the transcript mRNA they regulate. Riboswitches control gene expression mainly at the transcriptional and translational levels by means of allosteric structural changes induced by the binding of the sensed ligand (Mandal & Breaker, 2004b). They selectively respond to fundamental metabolites such as coenzymes, nucleobases and their derivatives, amino acids, ions, and tRNAs (Serganov & Nudler, 2013) to control a broad range of genes, including those involved in the metabolism or in the uptake of the ligand itself. A classification of riboswitches based either on the sensed ligand or the organization of the binding pocket can be found in the literature (Breaker, 2012; Montagne & Batey, 2008). In bacteria, they reside at the mRNA 5′-UTR and have a general common architecture (Winkler & Breaker, 2005). Namely, they are composed of an aptamer region, recognizing and binding the ligand, and an expression platform that directly interacts with the gene expression apparatus. Upon ligand binding a structural shift occurs between the ON and the OFF states, stabilizing one conformation to the detriment of the other. Many pathogenic bacteria use riboswitches to control essential metabolic pathways and structure-based ligand design approaches are currently used to explore the possibility of developing novel antibacterial agents (Colizzi, Lamontagne, Lafontaine, & Bussi, 2014; Daldrop et al., 2011; Sund, Lind, & Aqvist, 2014; Warner et al., 2014).

The *add* adenine riboswitch is a translational regulator of the adenosine deaminase gene and contributes to regulate the adenine concentration in the cellular environment (Mandal & Breaker, 2004a). It acts using a negative feedback mechanism (Kim & Breaker, 2008). In the absence of the adenine the Shine-Dalgarno (SD) sequence of the *add* gene is paired with a complementary sequence (anti-SD) and is not available for ribosome binding. Conversely, when the ligand is present, the anti-SD becomes an integral part of the aptamer terminal helix, so that the SD sequence is unpaired and translation can be initiated. The X-ray crystal structure of the *add* aptamer domain was solved in 2004 and was found to have a "tuning-fork"-like three-dimensional structure, wherein the adenine binding occurs in a pocket formed at the junction of three stems (P1, P2, and P3), two of which are hairpins interacting via kissing loops (L2–L3) (Serganov et al., 2004). Neupane, Yu, Foster, Wang, and Woodside (2011) have used mechanical manipulation with optical tweezers to show that the aptamer folds rapidly before the expression platform is transcribed. Adenine can bind the aptamer and the ON state is stabilized with respect to the OFF state both during and after transcription (Lemay et al., 2011). The ligand-dependent folding mechanism of the aptamer domain has been also investigated with real-time multidimensional NMR, showing a joint interplay between the different secondary structure elements and the ligand (Lee, Gal, Frydman, & Varani, 2010). The dynamics of tertiary interactions, the terminal helix pairing, and the organization of the binding pocket are structural changes that one needs to characterize for an in-depth understanding of the riboswitch behavior. However, the motion picture that can be inferred from this kind of experiments lacks the atomistic details about the multiple conformations sampled during the process.

To this aim, molecular dynamics (MD) at atomistic resolution (Tuckerman, 2010), where all the atoms of a system are evolved with physics based equations, could provide important insights on RNA conformational changes. In principle, one could imagine to perform a long MD simulation where the riboswitch is initialized in its holo form (ligand bound) and the ligand is subsequently removed. On the appropriate time scale, one could expect the riboswitch to spontaneously convert into the correct apo form (ligand unbound). Unless using a special-purpose molecular simulation machine (Dror, Young, & Shaw, 2011), this is not possible since the typical timescales required for these conformational changes are on the order of the millisecond, whereas MD is still limited to the microsecond time scale. Yet, the prediction of thermodynamic information such as the ligand-induced stabilization of the holo form is still very difficult with straightforward MD. Several techniques have been developed in the last decades to tackle this issue and to allow changes that should happen on a long time scale to be observed in practical simulation times (see, e.g., Abrams & Bussi, 2014; Dellago & Hummer, 2014).

In this chapter, we show the results of a study on the *add* riboswitch and illustrate how it is possible to use steered MD (Grubmüller, Heymann, & Tavan, 1996) to enhance the sampling of conformational changes in a riboswitch. In this context, steered MD can be used to address specific questions: which are and how strong are the interactions responsible for the ligand-induced stabilization of the terminal helix? Similarly to other methods, steered MD acts on an *a priori* chosen collective variable (CV). We present and compare two applications where the steering is applied on CVs of growing complexity. First, the full terminal helix is disrupted upon pulling its termini and we investigate the effect of the ligand. This study can provide a qualitative insight on the role of the ligand. Second, we analyze the ligand-induced stabilization of the first base pair (bp) of the terminal stem, which is a good proxy for the stem stability. We also describe how data from steered MD can be postprocessed to provide quantitative thermodynamic information. Some of the results presented here have been obtained by analyzing in more details the simulations described in a previous paper of ours (Di Palma, Colizzi, & Bussi, 2013). We also provide practical instructions and include several example input files that could be used to apply the same protocol to similar problems.

2. METHODS AND THEORY

Several techniques can be used to accelerate MD. In some of them *a priori* information about the process under investigation is required. These methods require the definition of descriptors named CVs, which are arbitrary functions of the microscopic coordinates. A bias potential is then added to modify the observed probability distribution of these CVs. If these CVs include the degrees of freedom that are responsible for the conformational change under investigation, a bias can be built to favor the transition state and thus to greatly enhance the transition rate. CVs should thus not only be able to distinguish reactants from products but also to properly drive the system through the transition. As an example, a natural CV to describe a binding process could be the distance between two molecules. However, if the binding involves some important rearrangement, it will be difficult to accelerate it by just biasing this CV. The comparison of several CVs and the choice of a relevant one is itself an active component of the investigation process that helps understanding the dynamics of the system of interest.

Methods where a bias potential acts on a preselected CV have their root in the umbrella sampling method (Torrie & Valleau, 1977). Some examples are steered MD (Grubmüller et al., 1996), adaptive biasing force (Darve & Pohorille, 2001), and metadynamics (Laio & Parrinello, 2002). We here focus on steered MD used in combination with a reweighting scheme that we have recently developed (Colizzi & Bussi, 2012). In the following sections, we briefly explain the theory behind steered MD and the reweighting of the probability distribution and we describe the actual protocol used to prepare and analyze our simulations.

2.1 Steered MD

Steered MD can be considered as an *in silico* representation of mechanical manipulation experiments such as pulling with AFM or optical tweezers. However, whereas mechanical manipulation techniques can only be used to pull the distance between two atoms (typically, the end-to-end distance of a polymer), steered MD can be used to pull any conceivable function of the microscopic coordinates. Two main protocols are employed, namely constant force or constant velocity (Isralewitz, Gao, & Schulten, 2001). In the first one, a force is directly applied to one or more atoms, and the extension or the displacement is monitored throughout the simulation. We focus here on the second one, where a moving harmonic potential is used to pull the system along a predefined CV $s(x)$. Here x represents the microscopic coordinates of the system. The bias potential thus reads

$$V(x) = \frac{k}{2}(s(x) - s_0 - vt)^2$$

Here, k is the spring stiffness, s_0 is the initial position of the restraint, and v is its velocity. This procedure allows the system to be forced to move along a CV with a known schedule. For example, one can pull a torsional angle to enforce an isomerization process to happen in a prefixed amount of time. This makes the method particularly convenient when a single CV has been identified to be relevant for the process under investigation and limited computational resources are available.

2.2 Jarzynski equality

Once a CV and a pulling protocol, able to induce the desired conformational change, have been determined, it is necessary to interpret the resulting trajectories. We notice here that typical simulations are in the nanosecond to microsecond time scale. Thus, conformational changes are artificially induced in a time that is orders of magnitude faster than the natural time required for a spontaneous process. Special analysis techniques are required to recover the correct thermodynamic properties from a simulation that is very far from thermodynamic equilibrium (Minh & McCammon, 2008).

If pulling is done in a quasi-static limit, the work performed corresponds to the free-energy change. In the more realistic case of finite velocity, the work provides an upper estimate of the free-energy change. Since steered MD is a stochastic process, pulling should be performed several times and multiple trajectories should be averaged. According to the Jarzynski equality (Jarzynski, 1997a), the free-energy change can be obtained from an exponential average of the work

$$e^{-\beta\Delta F} = \lim_{N\to\infty} \frac{1}{N} \sum_{i=1}^{N} e^{-\beta W_i}$$

Here, N is the total number of trajectories, W_i is the work performed on the ith trajectory, and $\beta = 1/k_B T$ is the inverse of the thermal energy. This identity is exact only in the limit of infinite number of realizations. Averages computed with a finite number of trajectories are known to overestimate the free-energy change as well (Gore, Ritort, & Bustamante, 2003). However, this expression is very useful since it can correct for the presence of unphysical trajectories in the ensemble. Those are indeed expected to exhibit a large performed work and thus contribute to the exponential average with a negligible weight.

We remark that the value of the work can also be used as a guideline for the choice of the CV: if two different CVs are capable to induce the same conformational change, the one for which the work is lower is the one that allows a transition to happen closer to equilibrium. With such a CV, the number of trajectories required to converge the results is expected to be lower.

2.3 Reweighting scheme

The Jarzynski relationship allows one to pull a CV and compute the free-energy landscape projected on that CV. We have recently shown (Colizzi & Bussi, 2012) that the scheme can be modified so as to allow the calculation of the free-energy landscape projected on another, *a posteriori* chosen CV. In practice, using this method one can, for instance, enforce an easy-to-pull distance and then estimate the free-energy difference of the process by looking at the rupture or formation of interactions among neighboring residues or at changes in the number of surrounding water molecules so as to quantify solvation.

The method is based on the combination of an identity first suggested by Jarzynski (1997b) and of weighted-histogram analysis method (Kumar, Rosenberg, Bouzida, Swendsen, & Kollman, 1992). It is also related to a method introduced by Hummer and Szabo (2001). The algorithm allows weights to be computed for the configuration at time t from the ith trajectory as

$$w_i(t) = \frac{e^{-\beta(W_i(t)-F(t))}}{\int e^{-\beta(V(x_i(t),t')-F(t'))}dt'}$$

Here, $W_i(t)$ is the work performed up to time t on the ith replica and $F(t)$ is a term that corresponds to the free energy of the restrained system at time t and can be obtained solving

$$e^{-\beta F(t)} = \sum_i \int w_i(t')e^{-\beta V(x_i(t'),t)}dt'$$

These two equations must be solved self-consistently to obtain the weights $w_i(t)$ that can then be used to compute the free energy with respect to any *a posteriori* chosen CV $s'(x)$ as

$$F(s') = -k_B T \log \sum_i \int w_i(t)\delta(s' - s'(x))dt$$

Here, the δ function selects all the configurations for which the value of the new CV is exactly s'. In combination with the maximum likelihood approach introduced by Minh and Adib (2008), this method has also been generalized to bidirectional pulling (Do, Carloni, Varani, & Bussi, 2013).

2.4 Error estimate

To compute the error in the estimated free energies, one can use a bootstrapping procedure that takes advantage of the fact that several independent estimates of the same value are computed. Given N the number of trajectories to be analyzed, the simplest approach consists of re-sampling randomly N trajectories with replacement from this ensemble. Weighting factors can then be re-computed using only these trajectories in the self-consistent procedure, potentially including duplicate ones. The standard deviation of the results provides an estimate of the error. With this procedure, the error can be computed for weights w or directly for any desired observable quantity O.

Namely, one can compute the free-energy difference between two relevant metastable conformations for every bootstrap sample and compute the error from their standard deviation.

We notice that the combination of the trajectories here is a nonlinear procedure, so that the average of the bootstrap samples is not necessary identical to the result obtained including all the trajectories. We thus here define the error as

$$\varepsilon = \sqrt{\frac{1}{M-1}\sum_{m=1}^{M}\left(O_m - O_{\text{tot}}\right)^2}$$

Here, M is the number of bootstrap samples, O_m is the observable computed from the mth bootstrap sample, and O_{tot} is the observable computed using all the available trajectories.

2.5 Protocols

We simulated the holo and the apo form of the A-riboswitch aptamer domain, both composed of 71 nucleotides (see Fig. 1A and C). The apo form was generated by adenine removal from the ligand-bound (holo) crystal structure (PDB ID 1Y26; Serganov et al., 2004). This deletion is justified by the fact that the apo and holo forms of the aptamer alone have been shown experimentally to share an overall similar secondary structure (Lemay et al., 2011). We note that this is not the case when the expression platform is present, as the riboswitch is expected to undergo a conformational change towards the ON state in the presence of the ligand.

MD simulations were performed using the Amber99 force field (Wang, Cieplak, & Kollman, 2000) refined with the parmbsc0 corrections (Pérez et al., 2007) and the more recent refinement on the χ torsional parameters (Zgarbová et al., 2011).

A first set of steered MD simulations aimed at inducing the opening of the whole P1 stem was performed in a rhombic dodecahedral simulation box including $\approx$39,500 TIP3P water molecules (Jorgensen, Chandrasekhar, Madura, Impey, & Klein, 1983) and NaCl at [0.15 M]. Temperature and pressure were controlled using stochastic velocity rescaling (Bussi, Donadio, & Parrinello, 2007) and Berendsen barostat (Berendsen, Postma, van Gunsteren, DiNola, & Haak, 1984). After equilibration, a steered MD was performed using as a CV the distance between the centers of mass of the terminal nucleotides (C1 and G71). By incrementing this distance, the rupture of the P1 stem is enforced. In our protocol, we pull it from

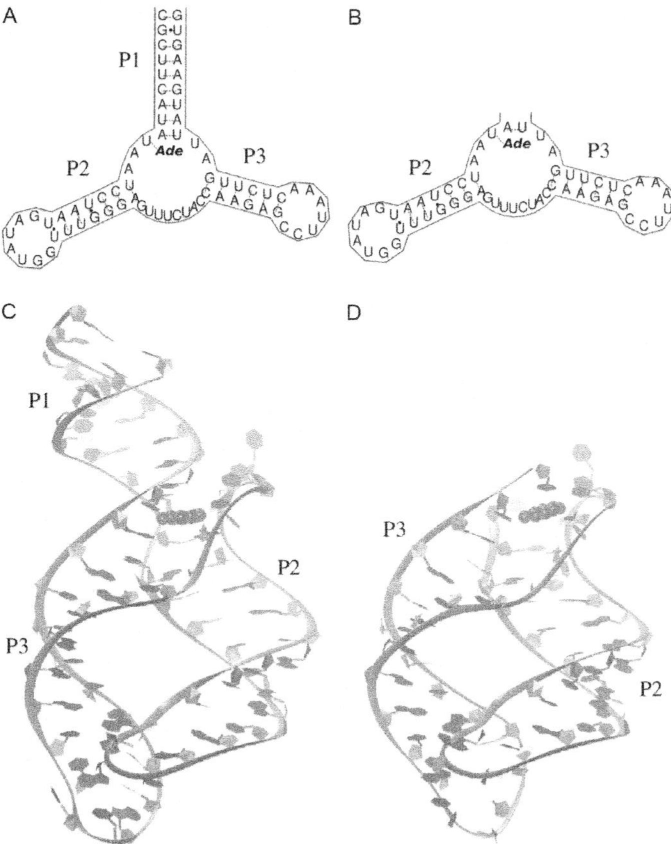

Figure 1 Secondary and three-dimensional structures of the A-riboswitch aptamer domain. Both the representations are given for the whole system (A and C) and for the *ad hoc* modified system lacking the first 8 bp of the P1 stem (ΔP1, B and D). The three stems (P1, P2, and P3) are labeled. The ligand is labeled in the secondary structure (Ade) and shown as ball-and-stick in the 3D structures.

1.05 to 12.3 nm. The latter value is sufficient to completely unfold the 9 bps of the helix. Two different velocities v were used to evaluate the influence of the pulling velocity, namely 0.5625 nm/ns (fast, corresponding to a 20 ns simulation) and 0.225 nm/ns (slow, corresponding to a 50 ns simulation). In both cases, the spring constant was chosen based on the CV fluctuations in a free simulation and set to 7.6×10^3 kJ/mol/nm^2.

A second set of steered MD simulations aimed at inducing the rupture of the A9–U63 pair was performed in a smaller hexagonal prism simulation box including ≈ 13,100 TIP3P water molecules (Jorgensen et al., 1983) and NaCl at [0.15 M]. The first 8 bps of the P1 stem (i.e., from C1 to U8 and from A64 to G71) were cut in both systems to generate the ΔP1 holo and ΔP1 apo systems (see Fig. 1B and D). This procedure reduces the noise during the pulling and allows the calculation to be focused on the influence of the ligand on the A9–U63 pairing. A moving harmonic restraint on the root-mean-square deviation (RMSD) from the reference crystal structure was used to induce the A9–U63 pairing rupture. This bp is the one directly interacting with the ligand, when it is present in the binding site. The RMSD was chosen as a CV, because it can unambiguously discriminate between the native conformation (low RMSD) and any other possible conformation. The steered CV was pulled at constant velocity of 0.175 nm/ns from 0 to 0.35 nm of RMSD in 2 ns setting the spring constant to 3.9×10^4 kJ/mol/nm^2. Such a pulling scheme induced the complete unbinding of the A9–U63 bp in the presence and the absence of the ligand (see Fig. 4). Initial conformations were taken from a restrained ensemble at equilibrium. We here chose equally spaced starting frames from a preliminary 8.192 ns simulation where the RMSD was restrained at 0.

Simulations were carried out with the Gromacs 4.6 program package (Pronk et al., 2013) combined with the PLUMED plugin (Bonomi et al., 2009; Tribello, Bonomi, Branduardi, Camilloni, & Bussi, 2014). Other simulation details can be found in Di Palma et al. (2013).

2.6 Structural analysis

The P1 unfolding in the presence and in the absence of the ligand was analyzed from the structural point of view. To monitor the behavior of the 9 bp forming the stem, we computed the RMSD of each bp from its native conformation during the simulation. The value of this RMSD is expected to start from zero and to increase when the specific bp is broken. We also analyzed the trajectories with a recently developed approach (Bottaro, Di Palma, & Bussi, 2014) where a local coordinate system is constructed in the center of the six-membered ring of each nucleobase and pairings are identified based on the relative position of the two bases. This procedure allows pairs to be annotated using the Leontis–Westhof nomenclature (Leontis & Westhof, 2001).

2.7 Sample input files

For the pulling simulations, we used the PLUMED plugin. We provide here sample input files for the latest PLUMED version (2.1). This is a sample input file for the pulling simulations where the entire P1 stem is disrupted by pulling the distance between the terminal nucleotides:

```
# make the molecule whole across periodic boundary condition
WHOLEMOLECULES ENTITY0=1-2257
# compute center of mass of nt C1
c1: COM ATOMS= 1-29
# compute center of mass of nt G71
c2: COM ATOMS= 2223-2257
# compute the distance between the two nt
d: DISTANCE ATOMS=c1,c2 NOPBC
# apply a moving restraint (fast pulling)
r: ...
  MOVINGRESTRAINT
  ARG=d KAPPA0=7630 AT0=1.05 AT1=12.30
  STEP0=0 STEP1=10000000
# the line above should be replaced with this for slow pulling
# STEP0=0 STEP1=25000000
...
# print the results for subsequent analysis
# namely: distance, center of the restraint, performed work
PRINT ARG=d,r.d_cntr,r.d_work FILE=COLVAR
```

This is a sample input file for the pulling simulations where the A9–U63 bp is disrupted by pulling its RMSD from the native structure:

```
# make the molecule whole across periodic boundary condition
WHOLEMOLECULES ENTITY0=1-1746
# compute the RMSD from the native structure
# ref.pdb only contains atoms from the relevant nucleotides
rmsd: RMSD REFERENCE=ref.pdb TYPE=OPTIMAL
# apply a moving restraint (fast pulling)
r: ...
  MOVINGRESTRAINT
  ARG=d KAPPA0=39150 AT0=0 AT1=0.35
  STEP0=0 STEP1=1000000
...
# additionally, monitor hydrogen bonds count
hb: ...
  COORDINATION GROUPA=17,17,20,23 GROUPB=1734,1738,1735,1735
  D_0 0.35 R_0 0.0000001 NN 6 MM 12 PAIR
...
```

```
# print the results for subsequent analysis
PRINT ARG=d,rmsd.d_cntr,rmsd.d_kappa,rmsd.d_work,hb FILE=COLVAR
```

To choose the starting conformation for multiple steered MD simulations, one should perform a restrained simulation replacing the `MOVINGRESTRAINT` command above with `RESTRAINT`:

```
RESTRAINT ARG=d KAPPA=39150 AT=0
```

Starting points should be then selected as equally spaced from the resulting trajectory.

For the annotation of steered MD trajectories where the P1 stem was disrupted, we used the baRNAba tool (http://github.com/srnas/barnaba) with the following command:

```
baRNAba ANNOTATE -f trajectory.pdb
```

For the reweighting of steered MD trajectories where the bp A9–U63 was disrupted we used an *in house* program that is available online (http://github.com/srnas/smd-reweight). The multiple COLVAR files obtained from the steered MD simulations should be concatenated and redirected to the smd-reweight command as in this example:

```
cat COLVAR* | smd-reweight --nbins int
```

The COLVAR files here are expected to contain one row per frame, and the following ordered columns: time, value of the pulled CV, position of the restraint, stiffness of the restraint, work, and the analyzed CV.

3. RESULTS AND DISCUSSION

3.1 Pulling the P1 stem

We first analyzed the rupture of the P1 stem pulling on the distance between the terminal bases. This was done at two different pulling rates, namely $\nu = 0.5625$ nm/ns (fast, corresponding to a 20 ns simulation) and $\nu = 0.225$ nm/ns (slow, corresponding to a 50 ns simulation) for both the apo and the holo forms. A movie showing the slow trajectory for the holo form can be found as Supplementary data (http://dx.doi.org/10.1016/bs.mie.2014.10.055). Here, one can qualitatively observe that the unzipping mechanism for the P1 stem is consistent with the one discussed in our previous work (Colizzi & Bussi, 2012). At the end of the simulation, the entire P1 stem and part of the binding pocket are disrupted.

The rupture of the individual bps is shown in Figs. 2 and 3 by monitoring the RMSD from native and the occupation of the bp, respectively. The bp close to the ligand (A9–U63) was broken at a later stage in the holo form when compared with the apo form. Table 1 reports the time corresponding to the

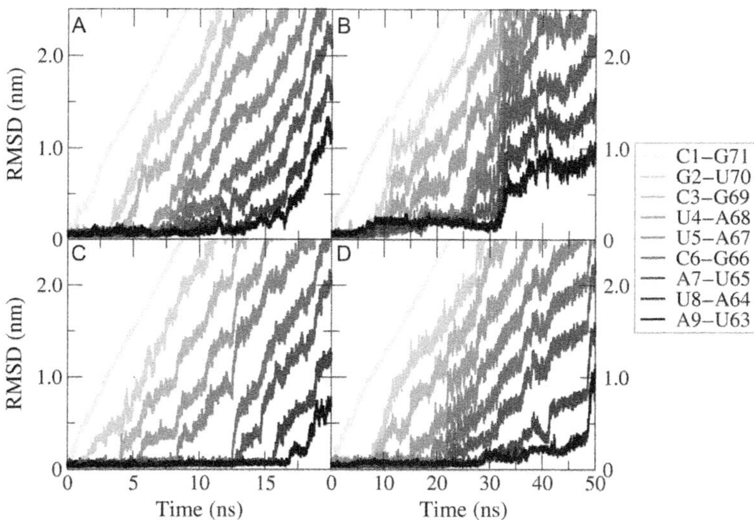

Figure 2 RMSD from the native conformation of each base pair of the P1 stem (A, apo fast; B, apo slow; C, holo fast; and D, holo slow). Each base pair (gray scale) can be considered disrupted at RMSD > 0.5. In both fast and slow runs, the last base pair (A9–U63, black) was disrupted at a later stage in the holo form.

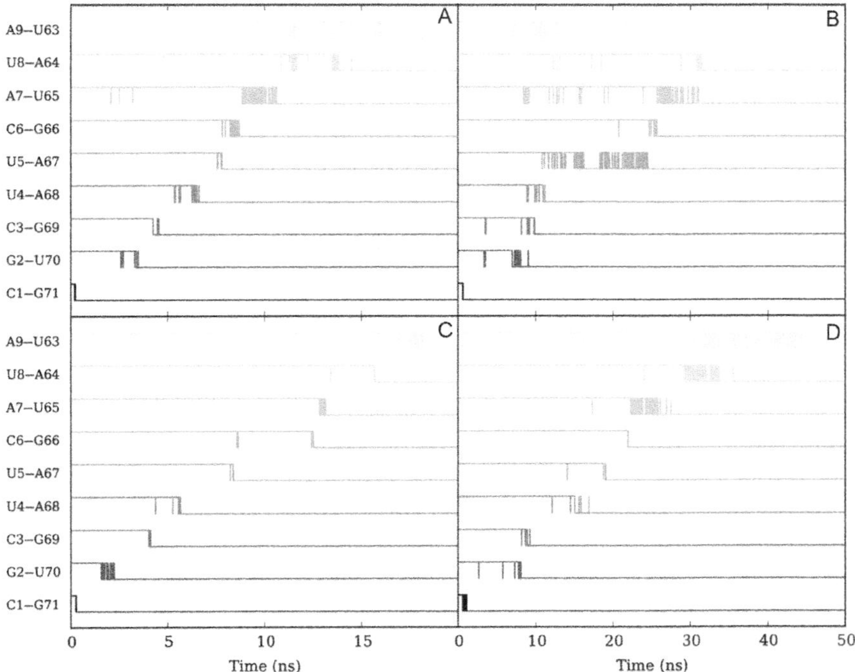

Figure 3 Monitoring of the P1 stem bp ruptures. Individual base pairing of the 9 bp of the P1 stem (gray scale) along the four simulations has been identified using baRNAba (A, apo fast; B, apo slow; C, holo fast; and D, holo slow). Refer to Table 1 for more details about the times at which the base pairs are definitively unpaired.

Table 1 Time corresponding to the definitive disruption of P1 stem base pairs during pulling, as in Fig. 3

Base pair	Fast apo	Fast holo	Slow apo	Slow holo
C1–G71	0.3	0.3	0.7	1.1
G2–U70	3.5	2.2	9.0	9.0
C3–G69	4.5	4.2	9.9	10.0
U4–A68	6.6	6.0	11.2	16.9
U5–A67	7.8	8.4	24.5	19.0
C6–G66	8.7	12.5	25.7	23.0
A7–U65	10.6	13.2	31	27.5
U8–A64	14.5	15.7	31.5	35.5
A9–U63	16.6	18.2	32	44.5

All values are expressed in ns.

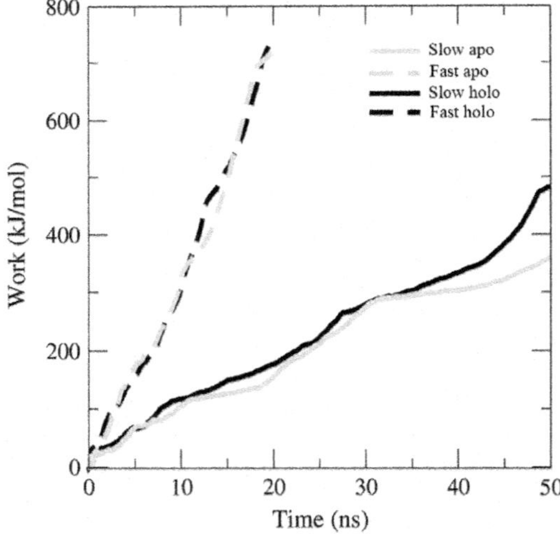

Figure 4 Mechanical work performed during the unfolding process of the whole P1 stem as a function of the simulation time. Profiles for both fast (dashed) and slow (solid) runs are shown. Results for the apo and the holo forms are in gray and black, respectively. Dissipation is larger in the faster simulations.

definitive disruption of each bp for all the cases. The largest discrepancies between holo and apo can be observed in the slow simulation for the A9–U63 pair confirming that this pair is stabilized by the ligand presence.

In Fig. 4, the work performed to open the P1 stem, with and without the ligand, is shown for fast and slow runs. It can be appreciated that the total work is larger in the fast runs. We remind here that the work provides in general an overestimation of the free-energy change, the difference being the dissipated work. The closer the process is to equilibrium the lower the dissipated work. The slow pulling simulations are thus here closer to equilibrium than the fast ones. However, the pulling velocities employed in our calculations are several orders of magnitude faster than the ones typically employed in the mechanical manipulation experiments and the process is highly out of equilibrium also for our slow pulling. In this respect, the values of the work exerted, on the order of hundreds of kJ/mol, have no quantitative meaning and cannot be used to determine reliably the free-energy change.

Apo and holo forms are also compared in Fig. 4. For fast pulling simulations, the apo and holo profiles are hardly distinguishable. At this pulling rate, artifacts caused by the highly out-of-equilibrium pulling procedure likely hide the relatively small differences between the apo and the holo systems. However, at a slow pulling rate, an interestingly informative behavior is observed. The plots of the work exerted on the system in the apo and holo forms are not significantly distinguishable until around the 30th nanosecond of pulling. After this point, the two plots clearly diverge and the holo system appears to be the most stable one. Interestingly, the 30th nanosecond of pulling corresponds to the beginning of the displacement of the bp (A9–U63) that directly interacts with the ligand as shown in Figs. 2 and 3. From this qualitative data, one may speculate that the effect of the ligand stabilization is poorly transmitted along the P1-helix beyond the first interacting bp (A9–U63).

The overall result of these simulations is that the presence of the ligand is stabilizing the P1 stem by means of interactions with its first bp (A9–U63). However, at this level, this is only a qualitative indication. The rupture of an RNA helix is a rather complex process. Pulling on the distance between the termini might be effective in experimental settings or in coarse-grain models (Lin & Thirumalai, 2008) when the slow pulling speed allows the process to remain close to equilibrium. On the contrary, in atomistic MD, it is definitely unlikely to induce such a complex process in a meaningful manner and at a reasonable time scale. As a matter of fact, due to the complexity of the process and the inadequate representation of its determinants by one CV, even using steered MD simulations on the time scale of the unbiased transition the process would likely happen out of equilibrium (Pan, Weinreich, Shan, Scarpazza, & Shaw, 2014).

3.2 Pulling the A9–U63 bp

The pulling of the entire P1 stem suggested a possible role of the ligand in stabilizing the adjacent A9–U63 bp. We here analyze and discuss the results obtained by enforcing the rupture of such a bp directly. We note that these simulations were performed in an *ad hoc* modified system where the rest of the stem P1 was removed (ΔP1). This modification finds its basis in the fact that ligand binding can happen also in aptamers with truncated P1 stem (Lemay & Lafontaine, 2007) and by the weak interactions between the P1 stem and the rest of the aptamer, but for the A9–U63 bp.

Around 512 pulling simulations were performed for both apo and holo forms. The work profiles for the holo form are shown in Fig. 5, where the large variance of the work can be appreciated. We remind that using the Jarzynski equality is very sensitive to low-work realizations. Six outliers for which we observed the transient formation of a cis Hoogsteen/Sugar noncanonical bp at large RMSD values (Leontis, Stombaugh, & Westhof, 2002) are also highlighted in Fig. 5. Similar transient conformations were not observed when pulling the apo form.

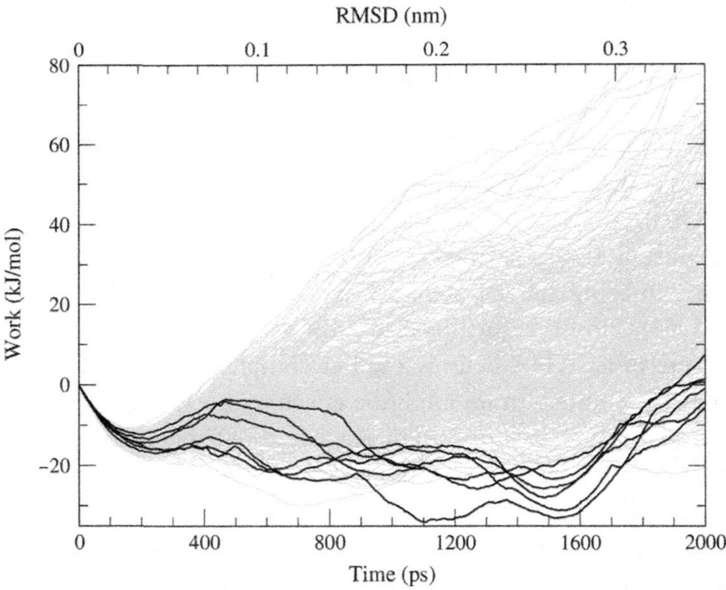

Figure 5 Mechanical work performed during the rupture of the A9–U63 base pair for the ΔP1 holo form (gray). Work is shown as a function of time (lower scale) and of the position of the restraint (upper scale). Around 512 simulations are overlaid. Outlier simulations for which a noncanonical pairing is observed at large RMSD are highlighted in black.

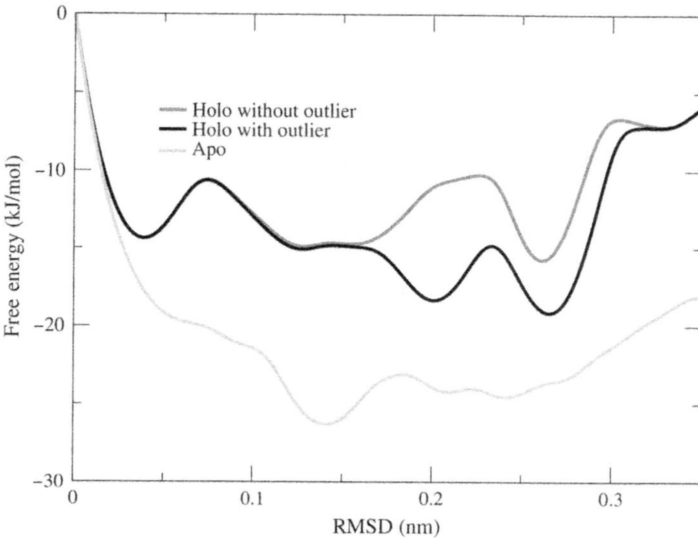

Figure 6 Free-energy profiles of the A9–U63 bp unpairing process resulting from Jarzynski equality as a function of the value of the RMSD from native in the ΔP1 systems. The profiles computed using all the 512 simulations for the holo (black) and the apo (light gray) forms are shown. In dark gray, here we show the profile for the holo form obtained by excluding the six outliers (see also Fig. 5).

The free-energy change computed using the Jarzynski equality is shown in Fig. 6, for both apo and holo forms. After an initial decrease, which is related to the entropy gain when larger values of the RMSD are allowed, the profiles stabilize to values that are visually different in the apo and holo forms. These data qualitatively confirms that the ligand is stabilizing the A9–U63 bp. For the holo form, we show the results both including and excluding the six outliers highlighted in Fig. 5. The inclusion of the outliers does not change the qualitative pictures. However, it is not trivial to quantify, in an objective manner, the difference because the two profiles present several ripples. The behavior of the outliers can be better interpreted by looking at the hydrogen bonds between A9 and U63 along the pulling (Fig. 7). Here one can appreciate that in the selected outlier the hydrogen bonds survive longer, so that low energy states are still possible at relatively large values of the RMSD. This is an example of how ambiguous the RMSD can be as a CV along which the free-energy profile is constructed.

To solve such an ambiguity, we directly looked at the rupture of the interactions describing the opening of the bp and then we reconstructed the free-energy profile by reweighting the probability distribution

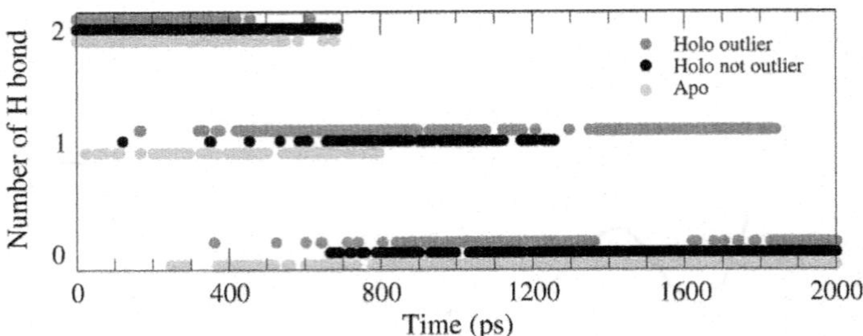

Figure 7 Hydrogen bonds occurrence between A9 and U63 for typical apo (light gray) and holo simulations (black). The different behavior of one among the outliers of the holo form is shown in dark gray.

Table 2 ΔF required for the disruption of the A9–U63 base pairs, in both apo and holo forms

Apo	Holo	Holo w/out outliers
-2.1 ± 1.2	1.3 ± 1.5	3.1 ± 0.9

For the holo form, results obtained without including the six outliers are shown.

accordingly. We thus used the reweighting scheme discussed above to compute the free energy as a function of the number of hydrogen bonds between A9 and U63. In Table 2, we report the free-energy difference between the state where at least one hydrogen bond is formed and the state where there are no hydrogen bonds. Results are slightly different from those reported in a previous work of ours (Di Palma et al., 2013) due to small numerical errors in the solution of the self-consistent reweighting equations. The numerical values are, however, compatible within the reported errors. The values of ΔF reported in Table 2 quantify the stability of the A9–U63 bp in presence and absence of the ligand. These absolute numbers should be interpreted with care because of the intrinsic overestimation of free-energy changes when the Jarzynski equation is applied to a finite sample of trajectories. Additionally, the number of possible conformations with no hydrogen bonds is very large and its entropic contribution could be underestimated here. However, the apo and holo results can be expected to be affected by similar systematic errors so that their difference should be reliably estimated ($\Delta\Delta F = 3.4 \pm 2$kJ/mol). The obtained result is compatible with the stabilization provided by an additional AU bp as measured from thermodynamic experiments (Mathews et al., 2004). Table 2 also reports for the influence of the outliers in the final results. The difference is small and does not affect our conclusion. However, it is larger than the reported error.

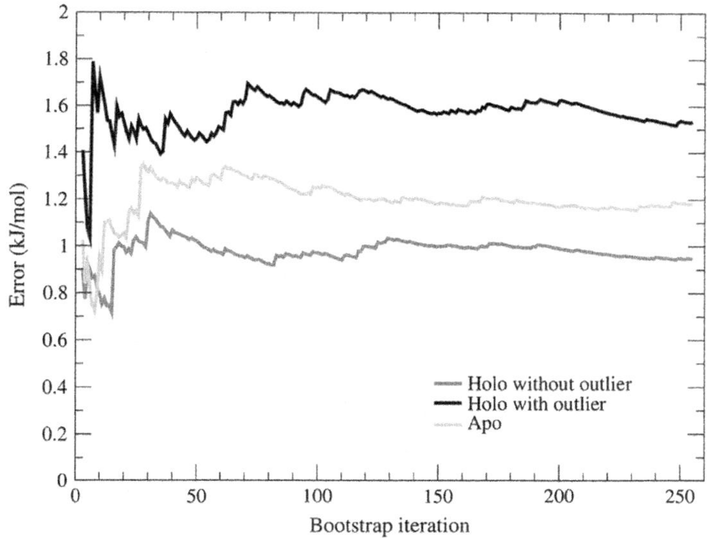

Figure 8 Error in the free-energy difference between states with no hydrogen bonds and states with at least one hydrogen bond, as a function of the bootstrap iteration. Data for holo (black) and apo (light gray) are shown. The error for the holo form obtained excluding the six outliers is shown in dark gray.

We remark that low-work realizations have a large impact in all analysis procedures based on the Jarzynski equality, and that they should be included. We also notice that by removing these trajectories the apparent error evaluated by the bootstrap procedure is artificially decreased.

In Fig. 8, we show how the error estimation on the reported free-energy changes depends on the number of bootstrap iterations. As it can be seen, the error estimate is quite robust and a few bootstrap iterations are enough to converge its calculation. We underline here that this estimate only includes statistical errors that can be detected by looking at the differences between the 512 realizations. As a consequence, if an alternative low-work pathway is never sampled in any of the steered MD simulations then the result can be biased. These errors can be expected to affect in the same manner simulations on similar systems made with the same protocol. Thus, we believe that the method can be reliably used to estimate differential stabilities.

3.3 Comparison with experiments

We comment here that single molecule experiments (Neupane et al., 2011) report a much larger value of $\Delta\Delta F$ (33.5 $\pm$ 9 kJ/mol) for the ligand-induced stabilization of the entire aptamer. In those experiments, the effect of the ligand can only be seen in the very last transition along the folding pathway. The last metastable conformation along the folding pathway is not affected by the ligand. This conformation has been interpreted as one where the stems P2 and P3 are folded and the binding pocket is already organized as in the fully folded conformation. We remark that it is very difficult to access these molecular details in force spectroscopy experiments, and that often one has to infer the conformation from macroscopic information such as the polymer length. Therefore, the difference between the values reported in our work and the values measured from single molecule experiments could be ascribed to the fact that the last transition along the folding pathway also requires a substantial pocket reorganization, possibly coupled with the formation of the L2–L3 kissing complex. This reorganization is ligand dependent and could take into account for the additional coupling between ligand binding and P1 stem folding.

3.4 Possible methodological improvements

Steered MD is one of the most straightforward enhanced sampling techniques. It is grounded on the solid theoretical framework of the Jarzynski equality and allows one to induce a conformational transition in a prefixed time. It, thus, can be used also when limited computational resources are available. Moreover, it is easy to analyze the work distribution to assess the method and possibly optimize the choice of the CV. However, steered MD is limited to the pulling of a single CV at a time and requires one to guess the range of relevant values for the CV. An alternative is umbrella sampling (Torrie & Valleau, 1977), often used in a multiple-restraint approach in combination with weighted-histogram analysis method (Kumar et al., 1992). Umbrella sampling can be used with one or two CVs, but care must be taken since sometime it is not straightforward to recognize when CVs are not sufficient to describe the reaction (see, e.g., Di Palma & Bottaro, submitted for publication, for a critical assessment of umbrella sampling as applied to the kissing loop interaction in the adenine riboswitch). For more complicated conformational changes, where one CV is not enough to describe the reaction, metadynamics (Laio & Parrinello, 2002) could be a better alternative. When used in its well-tempered formulation (Barducci, Bussi, & Parrinello, 2008), metadynamics do not require the boundary of the relevant region in the CV space to be prefixed. If a large number of CV is required, one can use the bias exchange formalism (Piana & Laio, 2007). Alternatively, when finding effective CVs is too difficult, metadynamics can be combined with complementary techniques such as parallel tempering (Bussi, Gervasio, Laio, & Parrinello, 2006).

4. CONCLUSIONS

MD at atomistic resolution in combination with enhanced sampling techniques is a powerful tool and can be used to predict the thermodynamics of conformational changes in riboswitches and to generate experimentally testable hypothesis. In this chapter, we have shown how it is possible to use steered MD to quantify the ligand-induced effect on the stabilization of the terminal helix of an adenine riboswitch aptamer domain. Results are in agreement with available thermodynamic data and can help in the structural interpretation of single molecule experiments. Statistical error can be kept under control with a bootstrap analysis. Employed methods were discussed in details and an effort was dedicated to explain the choice of the fundamental parameters in our simulation and analysis protocols. As a final remark, we observe that by carefully handling the choice of the proper CVs and the simulation protocols the approaches discussed in this work could be effectively applied to other similar systems.

ACKNOWLEDGMENTS

The research leading to these results has received funding from the European Research Council under the European Union's Seventh Framework Programme (FP/2007-2013)/ERC Grant Agreement No. 306662, S-RNA-S. We acknowledge the CINECA award No. HP10B2G6OF, 2012 under the ISCRA initiative for the availability of high performance computing resources. FC thanks the Université de Sherbrooke in Québec, Canada for providing infrastructures.

REFERENCES

Abrams, C., & Bussi, G. (2014). Enhanced sampling in molecular dynamics using metadynamics, replica-exchange, and temperature-acceleration. *Entropy*, *16*, 163–199.

Barducci, A., Bussi, G., & Parrinello, M. (2008). Well-tempered metadynamics: A smoothly converging and tunable free-energy method. *Physical Review Letters*, *100*, 020603.

Berendsen, H. J. C., Postma, J. P. M., van Gunsteren, W. F., DiNola, A., & Haak, J. R. (1984). Molecular dynamics with coupling to an external bath. *The Journal of Chemical Physics*, *81*, 3684.

Bonomi, M., Branduardi, D., Bussi, G., Camilloni, C., Provasi, D., Raiteri, P., et al. (2009). PLUMED: A portable plugin for free-energy calculations with molecular dynamics. *Computer Physics Communications*, *180*, 1961–1972.

Bottaro, S., Di Palma, F., & Bussi, G. (2014). The role of nucleobase interactions in RNA structure and dynamics. *Nucleic Acids Research*, *42*, 13306–13314.

Breaker, R. R. (2012). Riboswitches and the RNA world. *Cold Spring Harbor Perspectives in Biology*, *4*, a003566.

Bussi, G., Donadio, D., & Parrinello, M. (2007). Canonical sampling through velocity rescaling. *The Journal of Chemical Physics*, *126*(1), 014101–014107.

Bussi, G., Gervasio, F. L., Laio, A., & Parrinello, M. (2006). Free-energy landscape for β hairpin folding from combined parallel tempering and metadynamics. *Journal of the American Chemical Society*, *128*, 13435–13441.

Colizzi, F., & Bussi, G. (2012). RNA unwinding from reweighted pulling simulations. *Journal of the American Chemical Society*, *134*, 5173–5179.

Colizzi, F., Lamontagne, A. M., Lafontaine, D. A., & Bussi, G. (2014). Probing riboswitch binding sites with molecular docking, focused libraries, and in-line probing assays. *Methods in Molecular Biology*, *1103*, 141–151.

Daldrop, P., Reyes, F. E., Robinson, D. A., Hammond, C. M., Lilley, D. M., Batey, R. T., et al. (2011). Novel ligands for a purine riboswitch discovered by RNA-ligand docking. *Chemistry & Biology*, *18*, 324–335.

Darve, E., & Pohorille, A. (2001). Calculating free energies using average force. *The Journal of Chemical Physics*, *115*, 9169–9183.

Dellago, C., & Hummer, G. (2014). Computing equilibrium free energies using non-equilibrium molecular dynamics. *Entropy*, *16*, 41–61.

Di Palma, F., Bottaro, S., & Bussi, G. Kissing loop interaction in adenine riboswitch: Insights from umbrella sampling simulations, submitted for publication.

Di Palma, F., Colizzi, F., & Bussi, G. (2013). Ligand-induced stabilization of the aptamer terminal helix in the add adenine riboswitch. *RNA*, *19*, 1517–1524.

Do, T. N., Carloni, P., Varani, G., & Bussi, G. (2013). RNA/peptide binding driven by electrostatics—Insight from bidirectional pulling simulations. *Journal of Chemical Theory and Computation*, *9*, 1720–1730.

Dror, R. O., Young, C., & Shaw, D. E. (2011). *Encyclopedia of parallel computing* (pp. 60–71). New York, NY: Springer.

Gore, J., Ritort, F., & Bustamante, C. (2003). Bias and error in estimates of equilibrium free-energy differences from nonequilibrium measurements. *Proceedings of the National Academy of Sciences of the United States of America*, *100*, 12564–12569.

Grubmüller, H., Heymann, B., & Tavan, P. (1996). Ligand binding: Molecular mechanics calculation of the streptavidin–biotin rupture. *Science*, *271*, 997–999.

Hummer, G., & Szabo, A. (2001). Free energy reconstruction from nonequilibrium single-molecule pulling experiments. *Proceedings of the National Academy of Sciences of the United States of America*, *98*, 3658–3661.

Isralewitz, B., Gao, M., & Schulten, K. (2001). Steered molecular dynamics and mechanical functions of proteins. *Current Opinion in Structural Biology*, *11*, 224–230.

Jarzynski, C. (1997a). Nonequilibrium equality for free energy differences. *Physical Review Letters*, *78*, 2690–2693.

Jarzynski, C. (1997b). Equilibrium free-energy differences from nonequilibrium measurements: A master-equation approach. *Physical Review E*, *56*, 5018–5035.

Jorgensen, W. L., Chandrasekhar, J., Madura, J. D., Impey, R. W., & Klein, M. L. (1983). Comparison of simple potential functions for simulating liquid water. *The Journal of Chemical Physics*, *79*, 926.

Kim, J. N., & Breaker, R. R. (2008). Purine sensing by riboswitches. *Biology of the Cell*, *100*, 1–11.

Kumar, S., Rosenberg, J. M., Bouzida, D., Swendsen, R. H., & Kollman, P. A. (1992). The weighted histogram analysis method for free-energy calculations on biomolecules. I. The method. *Journal of Computational Chemistry*, *13*, 1011–1021.

Laio, A., & Parrinello, M. (2002). Escaping free-energy minima. *Proceedings of the National Academy of Sciences of the United States of America*, *99*, 12562–12566.

Lee, M. K., Gal, M., Frydman, L., & Varani, G. (2010). Real-time multidimensional NMR follows RNA folding with second resolution. *Proceedings of the National Academy of Sciences of the United States of America, 107*, 9192–9197.

Lemay, J. F., Desnoyers, G., Blouin, S., Heppell, B., Bastet, L., St-Pierre, P., et al. (2011). Comparative study between transcriptionally- and translationally-acting adenine riboswitches reveals key differences in riboswitch regulatory mechanisms. *PLoS Genetics, 7*, e1001278.

Lemay, J. C., & Lafontaine, D. A. (2007). Core requirements of the adenine riboswitch aptamer for ligand binding. *RNA, 13*, 339–350.

Leontis, N. B., Stombaugh, J., & Westhof, E. (2002). The non-Watson–Crick base pairs and their associated isostericity matrices. *Nucleic Acids Research, 30*, 3497–3531.

Leontis, N. B., & Westhof, E. (2001). Geometric nomenclature and classification of RNA base pairs. *RNA, 7*, 499–512.

Lin, J.-C., & Thirumalai, D. (2008). Relative stability of helices determines the folding landscape of adenine riboswitch aptamers. *Journal of the American Chemical Society, 130*, 14080–14081.

Mandal, M., & Breaker, R. R. (2004a). Adenine riboswitches and gene activation by disruption of a transcription terminator. *Nature Structural & Molecular Biology, 11*, 29–35.

Mandal, M., & Breaker, R. R. (2004b). Gene regulation by riboswitches. *Nature Reviews. Molecular Cell Biology, 5*, 451–463.

Mathews, D. H., Disney, M. D., Childs, J. L., Schroeder, S. J., Zuker, M., & Turner, D. H. (2004). Incorporating chemical modification constraints into a dynamic programming algorithm for prediction of RNA secondary structure. *Proceedings of the National Academy of Sciences of the United States of America, 101*, 7287–7292.

Minh, D., & Adib, A. (2008). Optimized free energies from bidirectional single-molecule force spectroscopy. *Physical Review Letters, 100*, 180602–180606.

Minh, D., & McCammon, J. A. (2008). Springs and speeds in free energy reconstruction from irreversible single-molecule pulling experiments. *The Journal of Physical Chemistry B, 112*, 5892–5897.

Mironov, A. S., Gusarov, I., Rafikov, R., Lopez, L. E., Shatalin, K., Kreneva, R. A., et al. (2002). Sensing small molecules by nascent RNA: A mechanism to control transcription in bacteria. *Cell, 111*, 747–756.

Montagne, R., & Batey, R. (2008). Riboswitches: Emerging themes in RNA structure and function. *Annual Review of Biophysics, 37*, 117–133.

Nahvi, A., Sudarsan, N., Ebert, M. S., Zou, X., Brown, K. L., & Breaker, R. R. (2002). Genetic control by a metabolite binding mRNA. *Chemistry & Biology, 9*, 1043–1049.

Neupane, K., Yu, H., Foster, D. A., Wang, F., & Woodside, M. T. (2011). Single-molecule force spectroscopy of the add adenine riboswitch relates folding to regulatory mechanism. *Nucleic Acids Research, 39*, 7677–7687.

Pan, A. C., Weinreich, T. M., Shan, Y., Scarpazza, D. P., & Shaw, D. E. (2014). Assessing the accuracy of two enhanced sampling methods using EGFR kinase transition pathways: The influence of collective variable choice. *Journal of Chemical Theory and Computation, 10*, 2860–2865.

Pérez, A., Marchán, I., Svozil, D., Sponer, J., Cheatham, T. E., III, Laughton, C. A., et al. (2007). Refinement of the AMBER force field for nucleic acids: Improving the description of α/γ conformers. *Biophysical Journal, 92*, 3817–3829.

Piana, S., & Laio, A. (2007). A bias-exchange approach to protein folding. *The Journal of Physical Chemistry. B, 111*, 4553–4559.

Pronk, S., Páll, S., Schulz, R., Larsson, P., Bjelkmar, P., Apostolov, R., et al. (2013). GROMACS 4.5: A high-throughput and highly parallel open source molecular simulation toolkit. *Bioinformatics, 29*, 845–854.

Serganov, A., & Nudler, E. (2013). A decade of riboswitches. *Cell*, *152*, 17–24.

Serganov, A., Yuan, Y. R., Pikovskaya, O., Polonskaia, A., Malinina, L., Phan, A. T., et al. (2004). Structural basis for discriminative regulation of gene expression by adenine-and guanine-sensing mRNAs. *Chemistry & Biology*, *11*, 1729–1741.

Sund, J., Lind, C., & Aqvist, J. (2014). Binding site preorganization and ligand discrimination in the purine riboswitch. *The Journal of Physical Chemistry. B*, http://dx.doi.org/10.1021/jp5052358, in press.

Torrie, G. M., & Valleau, J. P. (1977). Nonphysical sampling distributions in Monte Carlo free-energy estimation: Umbrella sampling. *Journal of Computational Physics*, *23*, 187–199.

Tribello, G. A., Bonomi, M., Branduardi, D., Camilloni, C., & Bussi, G. (2014). PLUMED 2: New feathers for an old bird. *Computer Physics Communications*, *185*, 604–613.

Tuckerman, M. E. (2010). *Statistical mechanics: Theory and molecular simulation*. Oxford: Oxford University Press.

Wang, J., Cieplak, P., & Kollman, P. A. (2000). How well does a restrained electrostatic potential (RESP) model perform in calculating conformational energies of organic and biological molecules? *Journal of Computational Chemistry*, *21*(12), 1049–1074.

Warner, K. D., Homan, P., Weeks, K. M., Smith, A. G., Abell, C., & Ferré-D'Amaré, A. R. (2014). Validating fragment-based drug discovery for biological RNAs: Lead fragments bind and remodel the TPP riboswitch specifically. *Chemistry & Biology*, *21*, 591–595.

Winkler, W. C., & Breaker, R. R. (2005). Regulation of bacterial gene expression by riboswitches. *Annual Review of Microbiology*, *59*, 487–517.

Zgarbová, M., Otyepka, M., Šponer, J., Mládek, A., Banáš, P., Cheatham, T. E., III, et al. (2011). Refinement of the Cornell et al. nucleic acids force field based on reference quantum chemical calculations of glycosidic torsion profiles. *Journal of Chemical Theory and Computation*, 7(9), 2886–2902.

CHAPTER SEVEN

Force Field Dependence of Riboswitch Dynamics

Christian A. Hanke, Holger Gohlke[1]
Mathematisch-Naturwissenschaftliche Fakultät, Institut für Pharmazeutische und Medizinische Chemie, Heinrich-Heine-Universität Düsseldorf, Düsseldorf, Germany
[1]Corresponding author: e-mail address: gohlke@uni-duesseldorf.de

Contents

Abstract

Riboswitches are noncoding regulatory elements that control gene expression in response to the presence of metabolites, which bind to the aptamer domain. Metabolite binding appears to occur through a combination of conformational selection and induced fit mechanism. This demands to characterize the structural dynamics of the *apo* state of aptamer domains. In principle, molecular dynamics (MD) simulations can give insights at the atomistic level into the dynamics of the aptamer domain. However, it is unclear to what extent contemporary force fields can bias such insights. Here, we show that the Amber force field ff99 yields the best agreement with detailed experimental observations on differences in the structural dynamics of wild type and mutant aptamer domains of the guanine-sensing riboswitch (Gsw), including a pronounced influence of Mg^{2+}. In contrast, applying ff99 with parmbsc0 and parmχ_{OL} modifications (denoted ff10) results in strongly damped motions and overly stable tertiary loop–loop interactions. These results are based on 58 MD simulations with an aggregate simulation time >11 μs, careful modeling of Mg^{2+} ions, and thorough statistical testing. Our results suggest that the moderate stabilization of the χ-*anti* region in ff10 can have an unwanted damping effect on functionally relevant structural dynamics of marginally stable RNA systems. This suggestion is supported by crystal structure analyses of Gsw aptamer domains that reveal χ torsions with high-*anti* values in the most mobile regions.

Methods in Enzymology, Volume 553
ISSN 0076-6879
http://dx.doi.org/10.1016/bs.mie.2014.10.056

We expect that future RNA force field development will benefit from considering marginally stable RNA systems and optimization toward good representations of dynamics in addition to structural characteristics.

1. INTRODUCTION

Riboswitches are noncoding genetic regulatory elements mostly found in the 5′ untranslated region of bacterial mRNA. Binding of a small molecule metabolite to the aptamer domain results in a large conformational change in the downstream expression platform, which modulates gene expression (Tucker & Breaker, 2005). Structural and related studies have revealed that riboswitches appear to bind their ligands through a combination of conformational selection and induced fit mechanisms (Serganov & Nudler, 2013). This demands to characterize the structural dynamics and energetics of the *apo* state of aptamer domains in order to understand the full process of gene regulation.

The role of the *apo* state is of particular importance for transcriptional gene regulation by riboswitches as there the decision whether the genes should be transcribed or not has to be made already during transcription (Stoddard et al., 2010). The guanine-sensing riboswitch (Gsw) found in the xpt-pbuX operon of *Bacillus subtilis* exerts transcriptional gene regulation and belongs to the class of purine sensing riboswitches (Batey, 2012; Batey, Gilbert, & Montange, 2004). Besides guanine, Gsw binds related ligands such as hypoxanthine (Batey et al., 2004; Gilbert, Stoddard, Wise, & Batey, 2006). Its aptamer domain consists of three paired regions P1, P2, and P3, two loops L2 and L3 capping the P2 and P3 regions, respectively, and three joining regions J1/2, J2/3, and J3/1 connecting the paired regions (Fig. 1A and B). The ligand is tightly bound in the binding pocket formed by the three joining regions (Batey et al., 2004). The aptamer terminal P1 helix connects to the downstream expression platform and is essential for the gene regulation (Batey et al., 2004). Tertiary interactions between the L2 and L3 loops are formed by two base quadruples stacked onto each other (Fig. 1C–F). These interactions were found to be present in the unbound state of the aptamer domain (Noeske, Schwalbe, & Wöhnert, 2007), and have been suggested to restrict the conformational freedom of the domain, thereby promoting ligand binding under physiological conditions (Stoddard, Gilbert, & Batey, 2008). In a mutant without these tertiary interactions, the aptamer's ability to bind

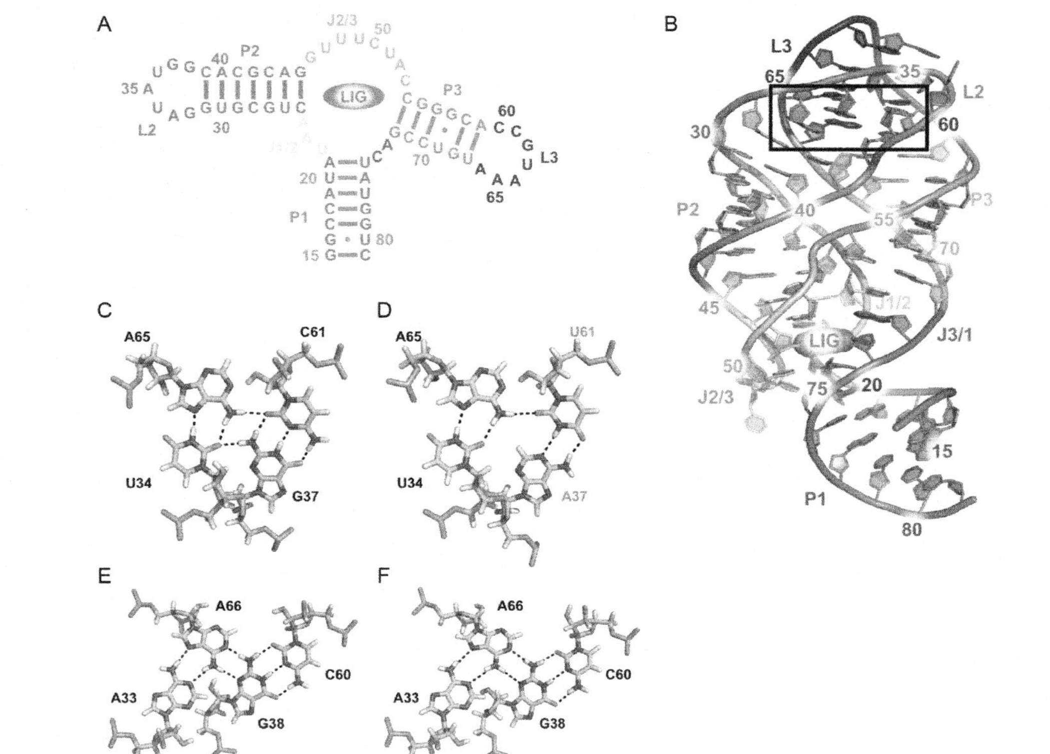

Figure 1 (A) Secondary structure and sequence used in this study of the aptamer domain of Gsw^{apt}; the secondary structure elements are assigned according to Batey et al. (2004). In addition, the ligand-binding site is indicated by the purple oval and denoted by LIG. (B) Tertiary structure of the aptamer domain of the Gsw bound to hypoxanthine (LIG; PDB ID 4FE5, Stoddard et al., 2013). The secondary structure elements are assigned as in (A). The black rectangle indicates the location of the base quadruples. (C and D) Upper base quadruple formed by the L2 and L3 loops for Gsw^{apt} (C) and Gsw^{loop} (D). Here, the G37A/C61U mutation is located (highlighted by the red labels). (E and F) Lower base quadruple formed by the L2 and L3 loops for Gsw^{apt} (E) and Gsw^{loop} (F). (See the color plate.)

hypoxanthine was abolished (Batey et al., 2004). In contrast, the G37A/C61U double mutation, reducing the number of hydrogen bonds within the upper base quadruple from seven in the wild type to five, led to an aptamer domain where formation of loop–loop interactions in the *apo* state and binding of hypoxanthine show a pronounced Mg^{2+} dependence (Buck, Noeske, Wöhnert, & Schwalbe, 2010; Noeske, Buck, et al., 2007). To what extent the loop–loop interactions contribute to the stabilization of the P1 helix, in addition to the ligand (Batey et al., 2004), has not been described so far.

With the long-term goal to gain insights at the atomistic level by molecular dynamics (MD) simulations into the function of the Gsw, we set out here to identify simulation conditions that allow appropriate modeling of the experimentally demonstrated marginal stability of the unbound aptamer domain, in particular when the G37A/C61U double mutation is present, and the pronounced influence of Mg^{2+} on the structural dynamics. With this we address one of the challenges that MD simulations of RNA face, the potential bias due to the applied force field (Sponer et al., 2014).

While contemporary force fields for RNA well describe canonical helices, they may show different successes on more complex RNAs (see references Cheatham & Case, 2013; Sponer et al., 2014; Sponer, Cang & Cheatham, 2012 for recent reviews). Here, we test the Amber (Case et al., 2005) force field ff99 (Wang, Cieplak, & Kollman, 2000) itself and together with the refinements parmbsc0 (Perez et al., 2007) and parmbsc0 + parmχ_{OL3} (Zgarbova et al., 2011). Note that the latter combination, ff99 + parmbsc0 + parmχ_{OL3}, has been termed ff10 in the context of RNA simulations in Amber (as of Amber12, this force field is also referred to as ff12). In parmbsc0, the α/γ dihedrals have been reparameterized in order to prevent the collapse of B-DNA due to the accumulation of irreversible nonnative γ-*trans* dihedral states (Perez et al., 2007; Sponer et al., 2014). As γ-*trans* degradation has never substantially affected canonical A-RNA simulations (Reblova et al., 2006), it was initially not clear if parmbsc0 brings any substantial changes for RNA simulations (Besseova, Otyepka, Reblova, & Sponer, 2009). However, long-timescale MD simulations of RNA revealed sudden irreversible transitions to ladder-like structures (Mlynsky et al., 2010), which required a reparameterization of the χ dihedral to suppress *anti* to high-*anti* χ shifts in RNA (Sponer et al., 2014), leading to the parmχ_{OL3} refinement (Zgarbova et al., 2011). At present, it is recommended to use parmχ_{OL3} together with parmbsc0 (Banas et al., 2010).

Another level of complexity arises in terms of describing the pronounced influence of Mg^{2+} ions on the structural dynamics of the aptamer domain of the Gsw (Buck et al., 2010). Mg^{2+} ions play an important role for the stability of nucleic acid structures (Draper, Grilley, & Soto, 2005; Pyle, 2002; Woodson, 2005). However, Mg^{2+} ions are challenging to treat in MD simulations due to the slow exchange kinetics of their first shell ligands (Ohtaki & Radnai, 1993), the low-diffusion coefficient (Mills & Lobo, 1989), and difficulties in modeling polarization and charge-transfer effects with fixed-charge additive force fields (Auffinger, 2012; Sponer et al., 2014). Here, we test two parameter sets for Mg^{2+} ions available for the use with Amber force fields: the long-used parameters developed by Aqvist (1992) and a set of parameters developed recently with the aim to improve agreement with experimentally determined kinetic properties of Mg^{2+} ions, especially with respect to the exchange of the first solvation shell (Allner, Nilsson, & Villa, 2012). Furthermore, we probe the influence of the method of Mg^{2+} ion placement on the simulation results. In total, we report on MD simulations with an aggregate sampling >11 μs (Table 1).

2. METHODS

2.1. Setup of the simulations

The starting structures for MD simulations of the aptamer domain of the Gsw in the *apo* state were obtained from the respective ligand bound crystal structures after removal of the ligand, ions, and crystal water molecules. The sequence of the wild type structure (PDB ID 4FE5, Stoddard et al., 2013) was adapted to match the sequence of the G37A/C61U mutant (PDB ID 3RKF, Buck et al., 2011). As a result, the wild type and mutant structures only differ in the nucleotides 37 and 61. The modified wild type will be referred to as Gsw^{apt}, and the G37A/C61U mutant will be referred to as Gsw^{loop}, according to Buck et al. (2011).

For both structures, three simulation systems with different Mg^{2+} concentrations (0, 12, and 20 Mg^{2+} ions per RNA molecule) were set up according to investigations on the Mg^{2+} dependence of Gsw^{loop} properties by Buck et al. (2010). For the setup, the *leap* program from the AmberTools suite of programs was used (Case et al., 2012).

For investigating the influence of the initial Mg^{2+} ion placement, one set of simulations was set up using the default placement procedure for ions of *leap*. The other simulations were set up using a modified protocol to place the Mg^{2+} ions with a first hydration shell of six water molecules: initially, a dummy atom was placed by *leap* instead of a Mg^{2+} ion, which had a 2+ charge, but a larger radius of 4 Å. This radius is similar to the one of a

Table 1 List of MD simulations presented in this work

Gsw variant	Number of Mg^{2+} ions[a]	Force field	Simulation time[b]	Mg^{2+} parameters
Gsw^{apt}	0	ff99[c]	3 × 200	Aqvist[d]
	12	ff99[c]	3 × 200	Aqvist[d]
	20	ff99[c]	3 × 200	Aqvist[d]
Gsw^{loop}	0	ff99[c]	3 × 200	Aqvist[d]
	12	ff99[c]	3 × 200	Aqvist[d]
	20	ff99[c]	3 × 200	Aqvist[d]
Gsw^{apt}	0	parmbsc0[e]	3 × 200	Aqvist[d]
	12	parmbsc0[e]	3 × 200	Aqvist[d]
	20	parmbsc0[e]	3 × 200	Aqvist[d]
Gsw^{loop}	0	parmbsc0[e]	3 × 200	Aqvist[d]
	12	parmbsc0[e]	3 × 200	Aqvist[d]
	20	parmbsc0[e]	3 × 200	Aqvist[d]
Gsw^{apt}	0	ff10[f]	3 × 200	Aqvist[d]
	12	ff10[f]	3 × 200	Aqvist[d]
	20	ff10[f]	3 × 200	Aqvist[d]
Gsw^{loop}	0	ff10[f]	3 × 200	Aqvist[d]
	12	ff10[f]	3 × 200	Aqvist[d]
	20	ff10[f]	3 × 200	Aqvist[d]
Gsw^{apt}	20	ff99[c]	3 × 100	Allner et al.[g]
Gsw^{apt}	20	ff99[c]	1 × 100	Aqvist[d,h]

[a]Per RNA molecule.
[b]In ns.
[c]Wang et al. (2000).
[d]Aqvist (1992).
[e]Perez et al. (2007).
[f]Zgarbova et al. (2011).
[g]Allner et al. (2012).
[h]The default placement procedure of *leap* was used for the Mg^{2+} ions in this simulation.

hexahydrated Mg^{2+} ion (Robinson, Gao, Sanishvili, Joachimiak, & Wang, 2000). Afterwards, the dummy atom was replaced by a Mg^{2+} ion with the correct radius and six water molecules surrounding it. Using a larger dummy atom initially in this two-step procedure results in a position of the Mg^{2+} ion that also allows accommodating the first hydration shell. Using a Mg^{2+} ion with the correct radius in the initial step instead would result in the ion be placed too closely to the RNA to accommodate the waters.

Using either placement method, 12 or 20 Mg^{2+} ions with parameters of Aqvist (1992) were added to Gsw^{apt} and Gsw^{loop}. To neutralize the simulation systems, also the ones without Mg^{2+} ions, Na^+ ions (Na^+ parameters for ff99 and ff99 + parmbsc0: Aqvist, 1990 and for ff10: Joung & Cheatham, 2008) were added by *leap*. The systems were then placed in an octahedral box of TIP3P water (Jorgensen, Chandrasekhar, Madura, Impey, & Klein, 1983) such that the distance between the edge of the water box and the closest RNA atom was at least 11 Å.

For each of the six simulation systems (Gsw^{apt} and Gsw^{loop} with 0, 12, or 20 Mg^{2+} ions per RNA molecule), three independent MD simulations were performed. These simulations were performed for each of the three force fields described earlier: ff99 (Wang et al., 2000), ff99 + parmbsc0 (Perez et al., 2007), and ff10 (Zgarbova et al., 2011) (Table 1).

Before the start of the MD simulation, each system was minimized by 200 steps of the steepest descent minimization, followed by 50 steps of conjugate gradient minimization. The particle mesh Ewald method (Darden, York, & Pedersen, 1993) was used to treat long-range electrostatic interactions, and bond lengths involving bonds to hydrogen atoms were constrained using the SHAKE algorithm (Ryckaert, Ciccotti, & Berendsen, 1977). The time step for all MD simulations was 2 fs, with a direct-space nonbonded cutoff of 9 Å. Applying harmonic force restraints with force constants of 5 kcal mol^{-1} $Å^{-2}$ to all solute atoms (and if present, also to the Mg^{2+} ions and their first hydration shell waters), we carried out canonical ensemble (NVT)-MD simulations for 50 ps, during which the system was heated from 100 to 300 K. Subsequent isothermal isobaric ensemble (NPT)-MD simulations were used for 50 ps to adjust the solvent density. Finally, the force constants of the harmonic restraints on positions of RNA, Mg^{2+} ions, and their first hydration shell waters were gradually reduced to 1 kcal mol^{-1} $Å^{-2}$ during 250 ps of NVT-MD simulations. This was followed by 50 ps of NVT-MD simulations without applying positional restraints. From the following 200 ns of NVT-MD simulations at 300 K performed with the GPU version of *pmemd* (Salomon-Ferrer, Gotz, Poole, Le Grand, & Walker, 2013) from the Amber 11 suite of programs (Case et al., 2005, 2010) conformations were extracted every 20 ps.

In order to investigate the influence of the Mg^{2+} parameters on the structural dynamics of the aptamer domain, we also performed MD simulations of Gsw^{apt} in the presence of 20 Mg^{2+} ions now using the recently developed Mg^{2+} parameters (Allner et al., 2012). The simulated system was otherwise identical in terms of the initial coordinates and the force field parameters for the RNA (ff99) and the water model (TIP3P). For the system, three independent MD simulations of 100 ns length each were performed.

2.2. Trajectory analysis

The MD trajectories were analyzed using the *cpptraj* tool (Roe & Cheatham, 2013) from the AmberTools suite of programs (Case et al., 2012). The initial 50 ns of each trajectory was not considered for analysis.

The radius of gyration (R_g), a measure for the compactness of a structure, was calculated for all atoms except the P1 region. Prior to the calculation of root-mean-square fluctuations (RMSF) over all atoms, global translational, and rotational differences between the structures along the trajectory were removed by least-squares fitting onto the average structure. In order to reduce the influence of very mobile regions on the picture of internal motions (Gohlke, Kuhn, & Case, 2004), conformations were root-mean-square fitted on those 80% of the nucleotides with the lowest fluctuations. These 54 nucleotides, which henceforth will be referred to as "core nucleotides", were chosen from an initial calculation of the mean RMSF over the three simulations of Gsw^{apt} in the absence of Mg^{2+} ions. The root-mean-square deviations (RMSD) of atomic positions were calculated as a measure for the structural similarity with respect to the initial structure. The RMSD was calculated after aligning the conformations onto the starting structure using the core nucleotides; the RMSD was then calculated for all atoms of these core nucleotides (Table 2).

The occupancy of hydrogen bonds between bases in the base quadruples of the L2/L3 region was determined using default geometrical parameters (distance, 3.5 Å and angle, 120°). The statistical uncertainty within one trajectory was determined from block averaging over blocks of 10 ns length. Finally, the values reported in Table 3 were averaged over all hydrogen bonds between a pair of bases.

In order to characterize the behavior of Mg^{2+} ions, we calculated diffusion coefficients from the mean-square displacements of all Mg^{2+} ions as moving averages with varying time windows. Mg^{2+} ions with a reduced number of water molecules in the first hydration shell were identified as those that have at least one RNA atom within 3.5 Å.

Table 2 RMSD with respect to the initial structure of the core nucleotides of Gsw^{apt} and Gsw^{loop} using different force fields[a]

Simulated system		ff99	bsc0	ff10
Gsw^{apt}	0 Mg^{2+b}	2.9 ± 0.4	3.5 ± 0.5	2.7 ± 0.3
	12 Mg^{2+b}	2.7 ± 0.4	2.8 ± 0.3	2.3 ± 0.3
	20 Mg^{2+b}	2.1 ± 0.3	3.0 ± 0.4	2.2 ± 0.3
Gsw^{loop}	0 Mg^{2+b}	3.7 ± 0.6	4.1 ± 0.7	2.7 ± 0.3
	12 Mg^{2+b}	2.3 ± 0.3	2.9 ± 0.3	2.5 ± 0.3
	20 Mg^{2+b}	2.5 ± 0.3	3.6 ± 0.3	2.6 ± 0.2

[a]In Å; mean ± SEM over three trajectories of 200 ns length. Nucleotides 5–21, 23–33, 36–47, and 51–64 are taken as core nucleotides. All atoms of the aptamer domain were considered.
[b]Number of Mg^{2+} ions per RNA molecule.

Table 3 Hydrogen bond occupancy in the L2/L3 loop region of Gsw^{apt} and Gsw^{loop} using different force fields[a]

RNA	Pair of nucleotides	ff99	bsc0	ff10
Gsw^{apt}	U34–A65	80.9 ± 10.6	96.9 ± 0.7	95.7 ± 1.0
	U34–G37	86.1 ± 10.5	99.6 ± 0.2	99.2 ± 0.3
	C61–A65	74.9 ± 12.0	98.4 ± 0.6	99.3 ± 0.2
	C61–G37	99.6 ± 0.3	99.9 ± 0.0	99.9 ± 0.0
	A33–A66	69.4 ± 12.6	88.0 ± 7.5	97.1 ± 1.4
	G38–A66	57.5 ± 9.1	63.9 ± 3.4	80.2 ± 2.6
	G38–C60	97.8 ± 0.8	97.2 ± 0.5	97.8 ± 0.5
Gsw^{loop}	U34–A65	66.1 ± 11.9	16.0 ± 10.4	87.6 ± 10.1
	U61–A65	13.1 ± 11.1	28.7 ± 14.0	56.9 ± 11.1
	U61–A37	70.2 ± 7.9	95.9 ± 1.6	67.9 ± 6.1
	A33–A66	50.0 ± 15.0	1.5 ± 1.7	85.2 ± 10.1
	G38–A66	41.6 ± 15.1	54.8 ± 10.3	84.5 ± 7.7
	G38–C60	77.0 ± 10.0	98.4 ± 1.0	99.2 ± 0.4

[a]In %; mean ± SEM over three trajectories of 200 ns length in the absence of Mg^{2+} ions. The nucleotides 34, 37, 61, and 65 belong to the upper base quadruple (Fig. 1C and D) where the G37A/C61U mutation is located; the nucleotides 33, 38, 60, and 66 form the lower base quadruple (Fig. 1E and F).

Circular variances of dihedral angles were first calculated on a per-nucleotide basis and afterwards averaged over all nucleotides. The circular variance for a torsion angle θ was calculated following MacArthur and Thornton (1993) as:

$$\mathrm{Var}(\theta) = 1 - \overline{R} \tag{1}$$

where $\overline{R} = R/n$, n is the number of values, and R is given as

$$R^2 = \left(\sum\nolimits_{i=1}^{n} \cos\theta_i\right)^2 + \left(\sum\nolimits_{i=1}^{n} \sin\theta_i\right)^2 \tag{2}$$

Reported mean values (±standard error in the mean, SEM) were calculated over the independent simulations that were performed for each system. The overall SEM was calculated from the SEMs of the single trajectories according to the laws of error propagation. For testing the equality of two mean values a Welch two-sample *t*-test (Snedecor & Cochran, 1989; Welch, 1947) was applied, which does not require that the two samples have equal variances, as implemented in the program R.

3. RESULTS AND DISCUSSION

3.1. Placement of Mg^{2+} ions

Mg^{2+} ions have a strong influence on the structure and stability of RNA molecules (Draper et al., 2005; Woodson, 2005). However, only few MD simulations of RNA systems have been carried out using Mg^{2+} ions (Auffinger, 2012; Saini, Homeyer, Fulle, & Gohlke, 2013). This is a consequence of the challenges that persist in modeling these cations (see above), which result in poor sampling properties and the difficulty with current MD techniques to simulate losing of a ligand (e.g., a water or a nucleotide atom) from the inner-coordination sphere as a prerequisite to form a new inner-sphere complex (e.g., with a(nother) nucleotide atom) (Auffinger, 2012). This necessitates to pay particular attention to the initial placement of Mg^{2+} ions when setting up a simulation system, particularly if no direct experimental information on the location of Mg^{2+} is available as in the case of the aptamer domain of Gsw.

Applying the default placement procedure of the *leap* program (Schafmeister, Ross, & Romanovski, 1995) of the Amber software (Case et al., 2005) ions are added to favorable positions near the solute according to gas-phase electrostatics. For RNA, this is usually close to the negatively charged phosphate groups of the backbone, as shown for Gsw^{apt} in Fig. 2A for 20 Mg^{2+} ions after the thermalization step involving 500 ps of MD simulations. Already then all ions possess an inner-sphere hydration shell of at most five water molecules, i.e., they have at least one inner-sphere contact with an oxygen atom of the RNA backbone, likely due to the lack of a complete hydration shell around the ions after the placement. As a consequence, none of the ions moves away from its initial position during the first 100 ns of the MD simulations (Fig. 2B; note that a broader ion cloud particularly in the P1 region results from ion movements coupled to RNA motions).

A particularly favorable position exists if two phosphate groups are close together in the starting structure, as given for G27 in the P2 region and G68 in the P3 region (Fig. 2C). The Mg^{2+} ion placed there practically acts as a distance restraint between the two groups even over simulation times of 200 ns due to the slow exchange kinetics of inner-sphere ligands: the ion–phosphate oxygen interactions never break during that time. Such stable ion–phosphate oxygen interactions cause compaction of the surrounding structure (Fig. 2D). In addition, we observed that the connected nucleotides move in a correlated manner, which extends to sequentially neighboring nucleotides (data not shown). Even if placed as a hexahydrated Mg^{2+} bound initially as an outer-sphere complex at such a position, the Mg^{2+} ion can lose a water molecule from the inner sphere and directly chelate to the phosphate groups (Fig. 3A). Such an exchange is fostered by the force field bias toward inner-shell binding of Mg^{2+} ions in RNA simulations (Auffinger, 2012; Reblova et al., 2006, 2003).

As an alternative to the default placement, we developed a procedure in order to place Mg^{2+} ions with a first hydration shell by *leap*. Initially, we added dummy ions to the RNA with a charge equal to Mg^{2+} but a radius of 4 Å, which is close to the radius of a hexahydrated Mg^{2+} ion (Robinson et al., 2000). Subsequently, these dummy ions were replaced by Mg^{2+} ions with a first hydration shell of six water molecules, and the whole system was solvated in a box of TIP3P water. Harmonic restraints were then applied to the RNA, the Mg^{2+} ions, and the first hydration shell waters during the thermalization in order to counteract ion–RNA backbone interactions. As a result, it took at least 5 ns during the production runs after removal of all restraints before the first Mg^{2+} ion lost an inner-sphere water; further water losses of other ions were observed throughout the trajectory such that after 200 ns between 60% and 80% of the Mg^{2+} ions were still hexahydrated (Fig. 3A). If restraints on the ions and first hydration shell waters were omitted in the thermalization phase, occasionally Mg^{2+} ions lost one water

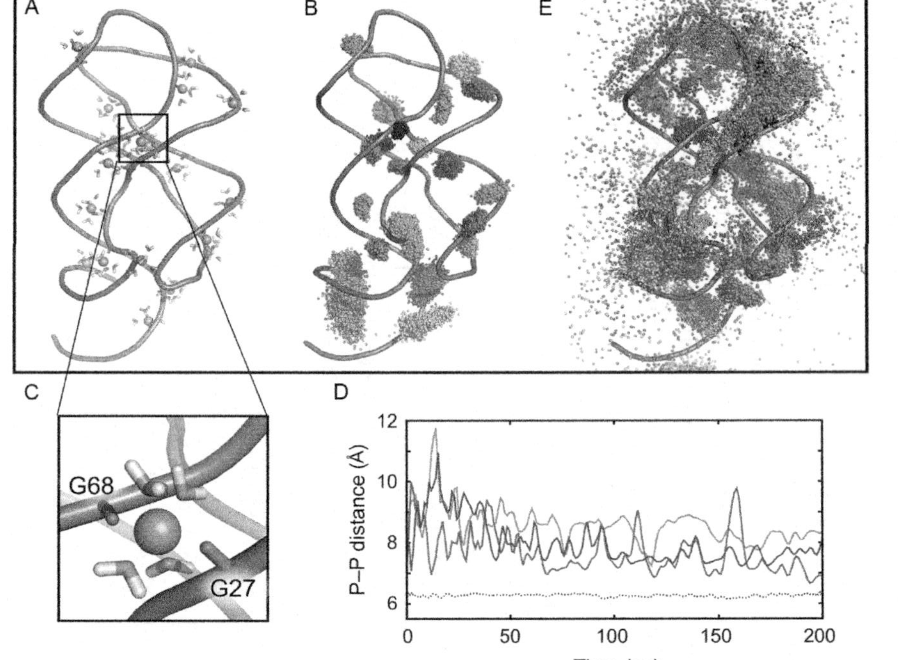

Figure 2 (A) Initial placement of 20 Mg^{2+} ions (green spheres) and surrounding water molecules (sticks) around the structure of the aptamer domain of the Gsw (with the backbone displayed as gray line). The ions were placed prior to the water molecules at electrostatically favorable locations as implemented in *leap* (Case et al., 2005) using the default procedure and are thus located very closely to the RNA backbone. (B and E) Positions of 20 Mg^{2+} ions during 100 ns of MD simulation. Different colors represent different Mg^{2+} ions. The ions were initially placed by *leap* either using the default procedure (B) or using the improved placement procedure (see Section 2) considering a first hydration shell (E). (C) Close up of a Mg^{2+} ion connecting the OP1 atoms of nucleotides G27 and G68 (red sticks connected to the RNA backbone in dark gray). Four water molecules (red and white sticks) form the hydration shell of the Mg^{2+} ion. (D) Distance calculated between the P atoms of nucleotides G27 and G68 shown in (C) during MD simulations of 200 ns length. The blue dotted line represents the distance in a simulation with the default placement of the ions; the three red solid lines, which were smoothed for better visibility, represent the distances in simulations with the improved placement procedure. (See the color plate.)

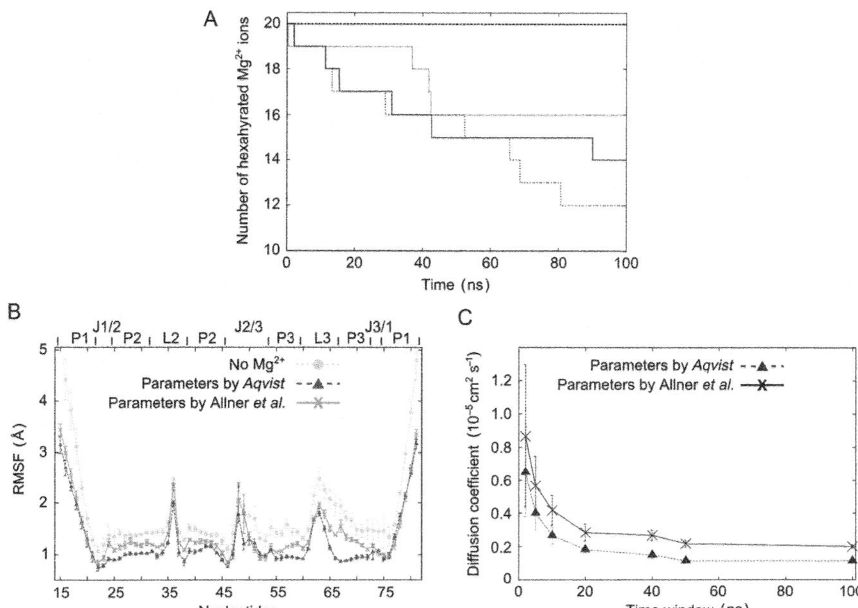

Figure 3 (A) Number of Mg^{2+} ions having six water molecules during Gsw^{apt} simulations with Mg^{2+} ions. The black dashed lines at 20 represent results for the three simulations using the Mg^{2+} parameters from Allner et al. (2012); the gray solid and dotted lines represent results for the three simulations using the Mg^{2+} parameters from Aqvist (1992). (B) Per-nucleotide RMSF values obtained from MD simulations of Gsw^{apt} without Mg^{2+} ions (green (light gray in the print version) dotted line with filled circle), with 20 Mg^{2+} ions with Mg^{2+} parameters from Aqvist (blue (dark gray in the print version) dashed line with filled triangle), or with 20 Mg^{2+} ions with Mg^{2+} parameters from Allner et al. (red (gray in the print version) solid line with "X" symbol); the mean $\pm$ SEM over three trajectories of 100 ns length is given. (C) Moving average of the diffusion coefficient ($\pm$SEM) over 20 Mg^{2+} ions calculated for varying time windows along three trajectories of 100 ns length of Gsw^{apt} simulations. The black solid line corresponds to the simulations using the Mg^{2+} parameters from Allner et al.; the gray dashed line corresponds to the simulations using the Mg^{2+} parameters from Aqvist.

molecule already there and formed an inner-sphere complex with the RNA (data not shown).

The hexahydrated Mg^{2+} ions show a dramatically different mobility during the MD simulations compared to the ions placed by the default procedure (Fig. 2B), with ions now preferentially and thoroughly sampling the space between two backbones, but also exploring almost all of the simulation box (Fig. 2E). This should allow the ions sufficient time to equilibrate prior to the formation of outer-sphere complexes with the RNA, considering fluctuation times of the bulk Mg^{2+} density of 5–8 ns, depending on the Mg^{2+} abundance, found by Hayes et al. in MD simulations of the SAM-I riboswitch (Hayes et al., 2012).

In line with the high mobility of hexahydrated Mg^{2+} ions, which show tumbling motions even when bound to RNA (Auffinger, Bielecki, & Westhof, 2003), a much less restricted structural dynamics of Gsw^{apt} is observed now in the region between G27 and G68: the distance between the two nucleotides is up to 1.5-fold larger than when restrained by inner-sphere binding of a Mg^{2+} ion and shows fluctuations up to 3 Å (Fig. 2D), resulting in noncorrelated movements of the two nucleotides (data not shown).

Thus, for all subsequent simulations, we used our optimized procedure for Mg^{2+} placement.

3.2. Parameters of Mg^{2+} ions

Describing Mg^{2+} ions by the nonbonded model, i.e., representing their interactions by Coulombic and Lennard-Jones terms, allows switching the coordination number and ligand exchange at the metal center. This and the simplicity of the model are reasons why this model is still widely used for representing Mg^{2+} ions in biomolecular simulations, despite it oversimplifying interactions between the ions and surrounding atoms and the existence of more sophisticated models (Li, Roberts, Chakravorty, & Merz, 2013). One only needs to determine Lennard-Jones parameters for the nonbonded model. Aqvist developed Lennard-Jones parameters for Mg^{2+} more than 20 years ago (Aqvist, 1992) using the hydration free energy as objective value; these parameters have been adopted for use in Amber force fields. However, these parameters lead to an ion–water exchange rate from the first solvation shell two orders of magnitude slower than the experimental rate (Allner et al., 2012). Hence, Allner et al. (2012) reparameterized the repulsive Lennard-Jones term starting from the CHARMM27 parameters for Mg^{2+} (Foloppe & MacKerell, 2000; MacKerell & Banavali, 2000) with the objective to reproduce this rate. Recently, Li et al. (2013) designed Lennard-Jones parameters for Mg^{2+} ions specifically for use in PME-MD simulations, with the objective to reproduce the hydration free energy and the ion–oxygen distance in the first hydration shell (CM set). As absolute hydration free energies for Mg^{2+} computed with the CM set deviate more strongly from the experimental value than those computed with the Amber force field adopted Aqvist parameters (Aqvist, 1992), we only tested the latter and the ones by Allner et al. (2012) with respect to their influence on the structural dynamics of Gsw^{apt}.

Atomic RMSF averaged at the nucleotide level (Fig. 3B) show that, in general, the MD simulations with 20 Mg^{2+} ions using either one of the two parameter sets are more similar to each other than to the MD simulations without Mg^{2+} ions; the latter simulations result in the highest RMSF, indicating the lack of a stabilizing effect on the RNA in the absence of Mg^{2+} ions. This result is at variance with findings by Reblova et al. (2006), which reported that a similar concentration of Mg^{2+} ions exerted surprisingly little effect on the dynamics of a long-duplex RNA, but in agreement with findings by Reblova et al. of stabilizing effects of Mg^{2+} in MD simulations of the 5*S* rRNA Loop E (Reblova et al., 2003). Comparing the two sets of MD simulations with Mg^{2+} ions, the one with parameters by Allner et al. shows slightly higher RMSF than the one with Aqvist parameters (Fig. 3B) and, hence, a less stabilizing effect by the Mg^{2+} ions. This may reflect findings by Allner et al. on *add* A-riboswitch MD simulations, where their new parameters resulted in, on average, less ions bound to the RNA via outer-sphere contacts than when using the CHARMM27 parameters, which result in similar ion–water characteristics as the Aqvist parameters (Allner et al., 2012). From a global perspective, the stabilizing influence of the Mg^{2+} ions is also reflected in the radius of gyration (R_g) as a measure of the compactness of Gsw^{apt}: $R_g = 16.5 \pm 0.2$ Å with the Mg^{2+} parameters from Allner et al. and 16.2 ± 0.1 Å with the Mg^{2+} parameters from Aqvist; for comparison, without Mg^{2+} ions, $R_g = 17.1 \pm 0.3$ Å.

Higher exchange rates have been found to be coupled to faster ion diffusion likely because the solvent structure is disrupted during the exchange (Moller, Rey, Masia, & Hynes, 2005). In line with this, our calculations of diffusion coefficients (D) over the 20 Mg^{2+} ions in the Gsw^{apt} system show 50–100% higher D values with the parameters of Allner et al. than with the ones of Aqvist for different time windows used in the moving average calculations (Fig. 3C). The computed D values for very short-time windows (2 ns; 0.86×10^{-5} cm^2 s^{-1} for the parameters of Allner et al.; 0.65×10^{-5} cm^2 s^{-1} for the parameters of Aqvist) are in very good agreement with experimental values reported for $MgSO_4$ (0.85×10^{-5} cm^2 s^{-1} at infinite dilution, Harned & Hudson, 1951) and the Mg^{2+} ion (0.71×10^{-5} cm^2 s^{-1}, Mills & Lobo, 1989). This suggests that the majority of ions move essentially freely within this time window over the course of the MD simulations of 200 ns. For time windows larger than 50 ns, D drops by $\sim$1/4 in the case of Aqvist parameters and $\sim$1/3 in the case of parameters of Allner et al. with respect to the initial values (Fig. 3C). This is a result of a reduced ion mobility over these time windows due to the formation of inner-sphere contacts with Gsw^{apt} in the case of Aqvist parameters (between 20% and 40% of all ions; Fig. 3A) and outer-sphere contacts formed by ions of both parameterizations (Fig. 3A).

While no high-resolution information on Mg^{2+} binding to the Gsw is available (Batey et al., 2004; Buck et al., 2011), it was unexpected that in neither of the three simulations with parameters of Allner et al. an inner-sphere contact was formed (Fig. 3A), taking into account that such contacts do occur in related purine-binding riboswitches (PDB ID 1Y26, Serganov et al., 2004 and 3LA5, Dixon et al., 2010). The larger drop in D in the case of parameters of Allner et al. may reflect that these parameters result in a Mg^{2+} hydration free energy that is $\sim$15 kcal mol^{-1} less favorable (-417.3 kcal mol^{-1}, Allner et al., 2012) than that computed with Aqvist parameters (-432.6 kcal mol^{-1}, Li et al., 2013), employing the PME method in both cases. We speculate that this may result in a stronger coupling between the ions, observed for outer-sphere Mg^{2+} ions recently (Hayes et al., 2012), which may not have become visible during MD simulations of only 10 ns length by Allner et al.

From the above results, we find it difficult to discern that one of the two Mg^{2+} parameterizations is advantageous over the other with respect to MD simulations of the aptamer domain. We thus decided to use the Mg^{2+} parameters by Aqvist et al. for all subsequent simulations because of the longer experience with these parameters.

3.3. Force field influence

3.3.1 Formation of ladder-like structures

The irreversible formation of ladder-like RNA structures in long-MD simulations when using ff99 and ff99 + prmbsc0 has been described for a series of systems (Banas et al., 2010; Mlynsky et al., 2010; Sklenovsky et al., 2011), with different simulation times until the structural collapse was observed. In our case of 18 MD simulations of Gsw^{apt} and Gsw^{loop} with ff99 of 200 ns length, we observed the irreversible formation of a ladder-like structure of the aptamer domain only once at the end of a simulation in the terminal P1 region. When extending some of these simulations to 500 ns (data not shown), all aptamer domains were stable but one. This is similar to MD simulations of the $preQ_1$ riboswitch where a collapse occurred after $\sim$1 μs (Banas, Sklenovsky, Wedekind, Sponer, & Otyepka, 2012). In the

18 MD simulations with ff99 + parmbsc0 of 200 ns length, two showed the sudden transition to a ladder-like structure after ~100 ns in the terminal P1 region. All trajectories with ladder-like structures were excluded from further analysis. Using ff10, which includes the parmχ_{OL3} refinement, neither of the 18 MD simulations showed a transition to a ladder-like structure.

3.3.2 Structural deviations of the aptamer domain

Data on the structural deviations of Gsw^{apt} and Gsw^{loop} with respect to the starting structures in terms of mean RMSD calculated over three independent simulations for each combination of force field (ff99, ff99 + parmbsc0, and ff10) and number of Mg^{2+} ions (0, 12, and 20) are shown in Table 2. The structural deviations are in general moderate ($\leq$3 Å for all but 4 out of the 18 cases, where the RMSD is up to 4 Å), which agrees with findings of MD simulations on related systems (Priyakumar & MacKerell,, 2010; Sharma, Bulusu, & Mitra, 2009; Villa, Wöhnert, & Stock, 2009), and experimental data according to which Gsw^{apt} with and without Mg^{2+}, and Gsw^{loop} with Mg^{2+}, have a stable tertiary structure (Buck et al., 2010). However, in simulations with ff99 and ff99 + parmbsc0, Gsw^{loop} shows in general larger RMSD values than Gsw^{apt}, and this effect is most pronounced in the absence of Mg^{2+}. This result reflects the destabilized loop–loop interactions in Gsw^{loop} (Buck et al., 2010, 2011; see also below). In contrast, in simulations with ff10, Gsw^{apt}, and Gsw^{loop} show very similar structural deviations at the respective Mg^{2+} concentrations.

3.3.3 Mobility of the aptamer domain

In order to investigate the influence of the force field on the dynamics of the RNA, we calculated atomic RMSF and averaged them per nucleotide (Fig. 4A–F). The general shapes of the curves obtained for each combination of force field and number of Mg^{2+} ions are very similar. The stem regions P2 and P3 are the least mobile, in contrast to high fluctuations observed for the aptamer terminal helix P1. Furthermore, the J2/3 region is found to be mobile, in agreement with SHAPE experiments (Stoddard et al., 2008) and the suggestion that the J2/3 region acts as an entry gate for the ligand to the binding site (Gilbert et al., 2006). Finally, pronounced mobility is observed in the L2 and L3 regions where tertiary interactions lead to the formation of two base quadruples.

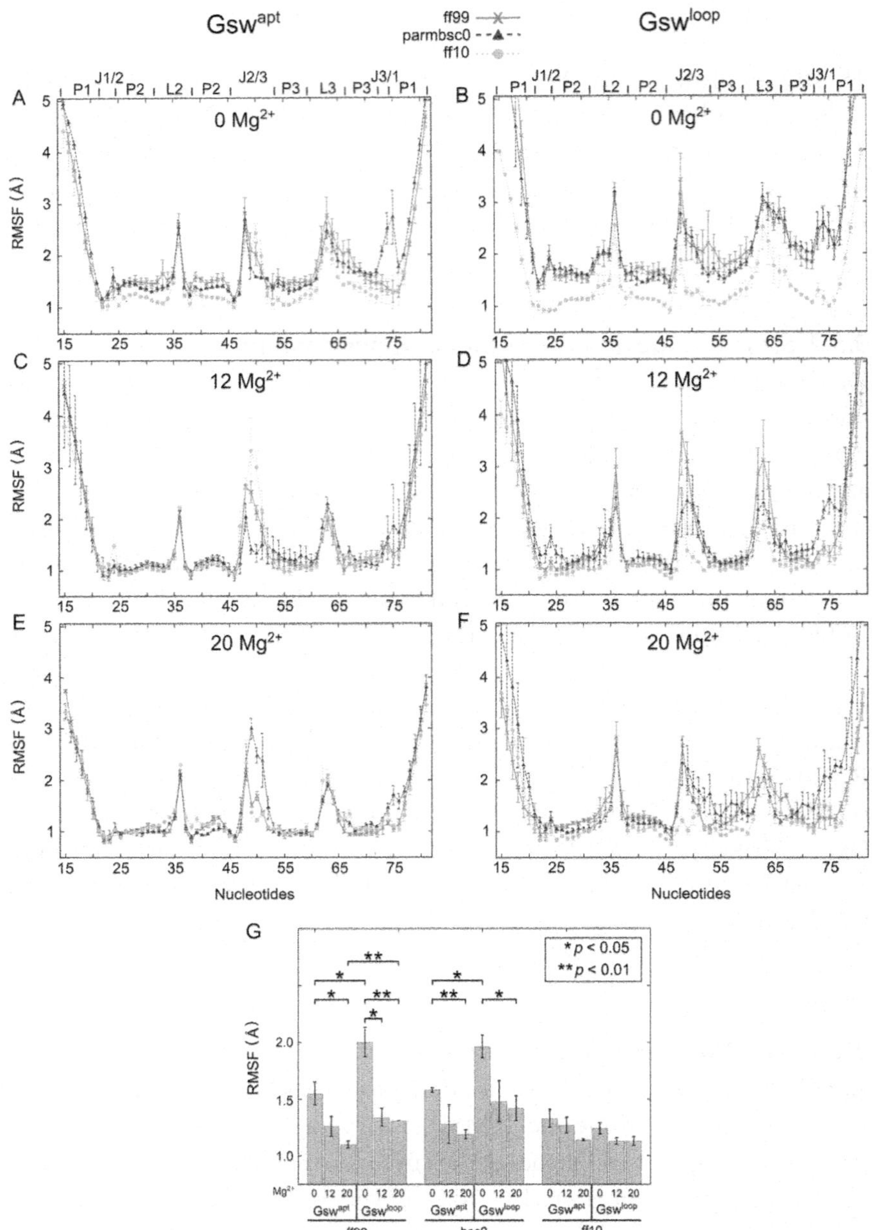

Figure 4 (A–F) Per-nucleotide RMSF obtained from MD simulations of Gsw^{apt} (A, C, and E) and Gsw^{loop} (B, D, and F) using different force fields and different numbers of Mg^{2+} ions; the mean ± SEM over three trajectories of 200 ns length is given. ff99: red (gray in the print version) solid line with "X" symbol; parmbsc0: blue (dark gray in the print version) dashed line with filled triangle; and ff10: green (light gray in the print version) dotted line with filled circle. For (A) and (B), absence of Mg^{2+} ions; (C) and (D), 12 Mg^{2+} ions per RNA molecule; and (E) and (F) 20 Mg^{2+} ions per RNA molecule. (G) Mean RMSF ± SEM of core nucleotides for the simulations shown in panels (A–F) grouped by the used force field. The stars indicate statistically significant differences between the mean values (*: $p < 0.05$; **: $p < 0.01$).

As to differences between the curves, qualitatively, MD simulations of Gsw^{apt} (Fig. 4A, C, and E) show in general lower RMSF of the aptamer domain than MD simulations of Gsw^{loop} (Fig. 4B, D, and F); within one system, the RMSF become the lower the more Mg^{2+} ions have been added. Considering mean RMSF values over the 80% of core nucleotides (see Section 2) of the aptamer domain, these are statistically significantly different in the case of ff99 and ff99 + parmbsc0 for Gsw^{apt} versus Gsw^{loop} without Mg^{2+}, as well as for Gsw^{apt} and Gsw^{loop} without versus with 20 Mg^{2+}, respectively (Fig. 4G). These findings match perfectly with experimental observations that Gsw^{loop} without Mg^{2+} does not show formation of tertiary interactions (Buck et al., 2010), with adverse effects on the structural stability of the aptamer, whereas Gsw^{apt} does; furthermore, addition of 20 Mg^{2+} results in a significant stabilization especially of the tertiary interactions in the case of Gsw^{apt} (Buck et al., 2010) or the formation of the tertiary interactions in the case of Gsw^{loop} (Buck et al., 2010).

Considering ff99 alone, two further differences are found significant (Fig. 4G): Between Gsw^{apt} and Gsw^{loop} with 20 Mg^{2+} ions and between Gsw^{loop} without and with 12 Mg^{2+} ions. The first case reflects that tertiary interactions in Gsw^{apt} are already formed without Mg^{2+} and are further stabilized upon Mg^{2+} addition (Buck et al., 2010); in contrast, in Gsw^{loop} the tertiary interactions start to form only at $[Mg^{2+}]:[RNA] > 18:1$ (Buck et al., 2010), suggesting that the Gsw^{loop} ensemble may still be more heterogeneous even in the presence of 20 Mg^{2+}. The second case agrees with the observation that addition of a few Mg^{2+} ions (ratio ~1:5–1:12, Buck et al., 2011) to Gsw^{loop} leads to only transiently formed tertiary interactions but an already ligand-binding competent aptamer ensemble (Buck et al., 2010), suggesting the beginning of structural compaction.

Likely the most unexpected observation is made for ff10 (Fig. 4G). First, MD simulations with ff10 result overall in the lowest mean RMSF when comparing to results for ff99 and ff99 + parmbsc0 and the different combinations of Gsw^{apt}/Gsw^{loop} and number of Mg^{2+} ions. Second, none of the differences between the simulated systems is significant in the case of ff10, in stark contrast to what can be expected with respect to experiment (see above). Thus, this suggests that MD simulations with ff10 lead to too strongly damped aptamer motions and/or to an overly stable aptamer domain.

In order to investigate to what extent the differences in the force fields contribute to the differences in the aptamer mobility, we computed frequency distributions of the torsion angles α, γ, and χ for Gsw^{loop} simulations in the absence of Mg^{2+}. As expected (Perez et al., 2007), ff99 + parmbsc0 and ff10 show much reduced populations of the g^+ region for α (Fig. 5A) and of the t region for γ (Fig. 5B) in comparison to ff99, and in turn higher populations of the g^- and g^+ regions, respectively. The locations of the respective distribution maxima remained largely unchanged (α: ff99: −75°; ff99 + parmbsc0: −80°; ff10: −75°), however. Likewise expected (Zgarbova et al., 2011), ff10 shows a much reduced population of the high-*anti* region of χ in comparison to ff99 and ff99 + parmbsc0 (Fig. 5C), which is a result of shifting the location of the distribution maximum by ~5° toward the *anti* region plus a narrowing of the distribution function of the main population. The population changes have a pronounced influence on the spread of the torsion angles, measured in terms of the circular variance (Eq. 1; Fig. 5D). Note that the circular variance is bounded in the interval [0,1], with 1 denoting that all torsion angle values are spread out evenly around a circle. Highly significant differences between circular variances are found for α and γ torsions between ff99 versus ff99 + parmbsc0 or ff10; similarly, highly significant differences are found for the χ torsion between ff99 or ff99 + parmbsc0 versus ff10 (Fig. 5D). These findings suggest that the modifications in the α and γ torsions are not the cause for the low-aptamer mobility observed with ff10, because the mean RMSF for Gsw^{loop} is not significantly ($p = 0.819$) different between ff99 and ff99 + parmsc0 (Fig. 4G). In contrast, a significantly reduced circular variance of the χ torsion is observed between those simulations of Gsw^{loop} that also show a significant reduction in the aptamer mobility (Fig. 4G; ff99 vs. ff10: $p = 0.005$; ff99 + parmbsc0 vs. ff10: $p = 0.003$).

Finally, we projected the per-nucleotide frequency with which a χ torsion is found in the high-*anti* region in MD simulations of Gsw^{loop} with ff99 and ff10 onto the aptamer domain (Fig. 5E and F). For some nucleotides, increased frequencies are found in both trajectories (U36, U47, U48, U61, G62, A64, and A65) with frequencies >80% for A64. Remarkably, most of the nucleotides with increased frequencies found in MD simulations with either force field also have a χ torsion in the high-*anti* region in one or both of the ligand bound Gsw^{apt} and Gsw^{loop} crystal structures (Fig. 5G; U47, G62, A64, and A65). The nucleotides with increased frequencies of their χ torsion in the high-*anti* region in MD simulations with either force field are found in those regions of the aptamer domain that are known to be the most mobile: The J2/3 region, which is suggested to act as an entry gate for the ligand to the binding site (Gilbert et al., 2006) and the loop regions (L2 and L3).

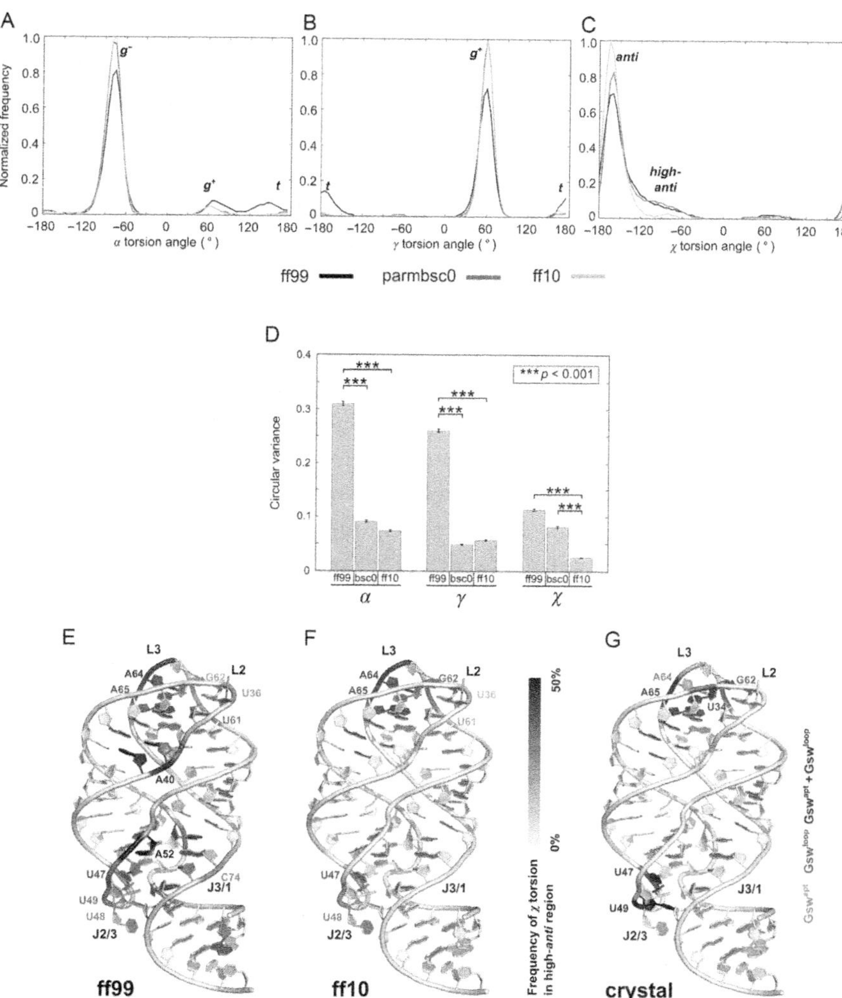

Figure 5 (A–C) Normalized histograms of torsion angles α (A), γ (B), and χ (C) in MD simulations of Gsw^{loop} in the absence of Mg^{2+} ions using different force fields. Red (gray in the print version): ff99; blue (dark gray in the print version): parmbsc0; and green (light gray in the print version): ff10. g^+, g^-, and t regions are assigned according to Perez et al. (2007), the *anti* and high-*anti* regions are assigned according to Blackburn and Gait (1996). (D) Mean circular variance (Eq. 1) ±SEM in the torsion angles α, γ, and χ for the force fields ff99, parmbsc0 (bsc0), and ff10. The stars indicate statistically significant differences between the mean values (***: $p < 0.001$). (E and F) Per-nucleotide frequency of the χ torsion angle being in the high-*anti* region (between −110° and −60°) for simulations using the ff99 (E) and the ff10 (F) force field. Frequency values between 0% (blue (dark gray in the print version)) and 50% (red (gray in the print version)) are mapped onto the structure of Gsw^{loop} according to the color scale. Nucleotides A52 and A64 (G62, A64, and A65) showed frequencies >50% and up to 90% (90%) in the case of ff99 (ff10). (G) Occupation of the high-*anti* region of the χ torsion angle in the

(Continued)

Differences for nucleotides with increased frequencies of the χ torsion in the high-*anti* region between MD simulations with ff99 and ff10 are found in the extended J2/3 region around A52, in the J3/1 region around C74, which forms one side of the binding site and gets stabilized upon ligand binding, and to a lesser extent in one strand of the P3 stem adjacent to the very mobile L3 loop. Here, simulations with ff99 show frequencies of a χ torsion in the high-*anti* region of at most 25% (Fig. 5E). None of these regions show χ torsions in the high-*anti* state in the Gsw^{apt} and Gsw^{loop} crystal structures (Fig. 5G). Note, however, that the crystal structures represent ligand bound aptamer domains with stabilized binding sites, whereas in the MD simulations analyzed here without Mg^{2+} the aptamer domains are in the least stable states. In that respect, it appears unexpected that MD simulations with ff10 do agree so well with the crystal structures in that they show no χ torsions in the high-*anti* state in these regions either (Fig. 5F). Phrased differently, together with our above results, our analyses indicate that a small propensity for visiting high-*anti* states of χ torsions is required for appropriately modeling the dynamics of known mobile regions of the Gsw aptamer domain.

3.3.4 Loop–loop interactions

The occupancy of hydrogen bonds between bases of the two base quadruples in the L2/L3 region are reported in Table 3 for MD simulations of Gsw^{apt} and Gsw^{loop} in the absence of Mg^{2+} ions and using different force fields. The values are averages over all hydrogen bonds found between two bases. The force field dependence of the results is similar to what has been found for the force field influence on the mobility of the aptamer domain. First, for Gsw^{apt}, ff10 results in the highest occupancies, with almost permanent formation of hydrogen bonds between the bases of the upper (nucleotides 34, 37, 61, and 65; Fig. 1C and D) and lower (nucleotides 33, 38, 60, and 66; Fig. 1E and F) base quadruples, respectively. ff99+parmbsc0 shows equally high occupancies for the upper

Figure 5—Cont'd ligand bound crystal structures of Gsw^{apt} (PDB ID 4FE5, Stoddard et al., 2013) and Gsw^{loop} (PDB ID 3RKF, Buck et al., 2011). Nucleotides whose χ torsion angles are not in the high-*anti* region in either of the crystal structures are colored in blue (dark gray in the print version). The other residues are colored according to the crystal structure in which their χ torsion angle is in the high-*anti* region: Gsw^{apt} only (light pink (light gray in the print version)), Gsw^{loop} only (magenta (gray in the print version)), or both (red (dark gray in the print version)).

quadruple but ~10% to ~20% lower occupancies for the lower quadruple. The lowest occupancies are found with ff99, with values <90% for the upper quadruple and <70% for the lower quadruple. Note that in this case the SEM is as high as 12%; closer inspection revealed that one out of the three independent simulations then differs with respect to the other two, suggesting that longer simulation times will be required to achieve better converged results for the more mobile system. The exception to the above are hydrogen bonds between C61 and G37 as well as between G38 and C60, which have occupancies >97% irrespective of the force field; the two Watson–Crick base pairs have been described as the most crucial components of the tertiary interaction (Gilbert, Love, Edwards, & Batey, 2007).

Second, for Gsw^{loop}, a decrease in the occupancies with respect to Gsw^{apt} is found in general for each force field used. However, for ff10, only in the case of U61–A65 an occupancy below 60% is found, whereas such low occupancies are found for four and three base pairs in the case of ff99 + parmbsc0 and ff99, respectively. Moreover, only in the case of ff99 do we find significantly lower occupancies of ~70–77% for the two base pairs U61–A37 and G38–C60 that showed very persistent hydrogen bonds in Gsw^{apt} (see above). Finally, we find a destabilization of the lower base quadruple as a result of the mutation in the upper one only in the case of ff99 and ff99 + parmbsc0; in contrast, for ff10, we find that the occupancies of the lower base quadruple are rather comparable to the ones found for Gsw^{apt}.

The G37A/C61U mutation was introduced in Gsw^{loop} to disturb the native hydrogen bond network connecting the L2 and L3 loops (Noeske, Buck, et al., 2007). Our observations of a reduced stability of tertiary interactions in Gsw^{loop} for ff99 and ff99 + parmbsc0 are in line with this. Later, NMR experiments showed that the loop–loop interactions are not formed in the absence of Mg^{2+} ions in Gsw^{loop} (Buck et al., 2010), which is at variance with our nonzero occupancies. However, one needs to consider that we start from a Gsw^{loop} crystal structure that is almost indistinguishable in the L2/L3 region from Gsw^{apt} (Fig. 1C–F) and that our simulation times of 200 ns are still short with respect to movements of an RNA on a rugged energy landscape (Sponer et al., 2014). Finally, for Gsw^{apt}, it has been reported that, even if the tertiary interactions are already present in the absence of Mg^{2+}, addition of Mg^{2+} leads to a significant stabilization especially for the loop–loop interaction (Buck et al., 2010, 2011). Considering the hydrogen bond occupancies, it is only possible to observe such a stabilization when using ff99: hydrogen bond occupancies of the base quadruples in Gsw^{apt} raise to >90% in the presence of 20 Mg^{2+} then (except for A66 and G38, where the occupancies are low throughout all simulations).

3.4. Concluding remarks

In this study, we aimed at identifying conditions for MD simulations that allow to appropriately model experimentally demonstrated differences in the structural stability of Gsw^{apt} and Gsw^{loop}, including the pronounced influence of Mg^{2+} on the structural dynamics of the aptamer domain. The most outstanding result is the strong force field dependence of the structural dynamics, with the original Amber force field ff99 yielding the best agreement with experimental observations whereas the use of ff10, implementing the parmbsc0 and parmχ_{OL} modifications, results in strongly damped aptamer motions and overly stable tertiary loop–loop interactions; ff99 + parmbsc0 yields an intermediate behavior.

We investigated the aptamer domain of the Gsw from *B. subtilis* here. The function of riboswitches inherently depends on its ability to switch between the inactive and the active states, and it requires that the aptamer domain maintains ligand-binding ability also in the inactive state in order to allow for a quick response on the presence of ligands. Especially, transcriptionally acting riboswitches such as the one investigated here have only a limited timeframe to respond to ligand binding (Stoddard et al., 2010). In order to achieve this quick response, the aptamer domain must be marginally stable, such that it can either follow the inactive route if no ligand binds, or it can quickly bind a ligand and transmit the information to the expression platform. These properties, its intricate tertiary structure, yet sufficiently small size, and the wealth of available experimental information on structure, stability, and dynamics under various conditions (Batey, 2012) make this system ideally suited for such investigations.

Our conclusions are based on detailed investigations of differences in the mobility of nucleotides and occupancies of tertiary loop–loop interactions. Ideally, our simulation results should have been compared to more direct measures of RNA dynamics, e.g., from NMR relaxation measurements (Juneja, Villa, & Nilsson, 2014). Such information is not yet available, however, on a broad per-residue level for the unbound Gsw. Still, we believe that our conclusions are well grounded. First, we performed three independent simulations for each investigated system leading to an aggregate simulation time of >11 μs, which allows us to estimate the uncertainty in the simulation results; only for occupancies of tertiary interactions in the most mobile systems (Gsw^{apt} and Gsw^{loop} in the absence of Mg^{2+} and using ff99) did we find indications that our MD simulations have not yet converged. Second, we used thorough statistical analysis for evaluating the significance of differences in mean values. Third, we compared our results to detailed experimental information on the structural stability of the aptamer domain and the formation of the loop–loop interactions (Buck et al., 2010, 2011; Stoddard et al., 2008). Finally, and most importantly, we analyze *differences* in the characteristics of two related RNA systems (Gsw^{apt} and Gsw^{loop}) and under various Mg^{2+} concentrations, rather than absolute characteristics of different RNA systems. Our approach should thus inherently profit from error cancelation.

The analysis of the spread of torsions angles sampled by the three force fields revealed a significantly reduced circular variance of the χ torsion between those simulations of Gsw^{loop} that also show a significant reduction in the aptamer mobility. Together with the population analysis of the χ torsion, the reduced variance is a result of a ~5-fold lower population of the high-*anti* region in ff10 than in ff99 and ff99 + parmbsc0. This is a consequence of an increased slope of the χ torsion profile in the high-*anti* region in ff10 compared to ff99 and ff99 + parmbsc0 (Zgarbova et al., 2011), which was successfully introduced to suppress ladder-like RNA structures (Banas et al., 2012; Sklenovsky et al., 2011).

The study by Zgarbova et al. (2011) demonstrated already that overstabilization of the χ-*anti* region has adverse effects on the geometry of A-RNA. Compared with the original effort by Zgarbova et al. (2011) to suppress ladder-like RNA structures in long MD simulations, our results suggest that even the moderate stabilization of the χ-*anti* region in ff10 can have an unwanted damping effect on the functionally relevant structural dynamics of marginally stable RNA systems. Our suggestion is supported by the finding of a high coincidence of χ torsions with high-*anti* values in the Gsw^{apt} and Gsw^{loop} crystal structures in those regions that are known to be most mobile; in contrast, MD simulations with ff10 result in very low frequencies of occurrence of χ torsions with high-*anti* values in some of the mobile regions. From the perspective of future force field development, our results suggest to include systems such as the aptamer domain of the Gsw in such studies and to extent the objectives for force field optimization beyond criteria of structural closeness to the native structure or the stabilization of signature interactions seen in experimental structures (Banas et al., 2010; Perez et al., 2007; Sklenovsky et al., 2011; Zgarbova et al., 2011) toward agreement of (differences in) the structural dynamics with experiment.

Finally, from our investigations on the influence of the set up of Mg^{2+} ions on the structural dynamics of Gsw^{apt} we strongly recommend not to use the default procedure implemented in the *leap* program of the Amber suite for placing Mg^{2+} around a nucleic acid structure but rather to use *leap* to place larger dummy ions that are later replaced by hexahydrated Mg^{2+}. Regarding the use of different Mg^{2+} ion parameters, we observed only a small influence on the structural dynamics of Gsw^{apt}. Our observation of no inner-sphere contact formation in any of the MD simulations when using the recently developed Mg^{2+} parameters of Allner et al. was unexpected. However, further investigations including other RNA systems are required to confirm this finding.

ACKNOWLEDGMENTS

We gratefully acknowledge the computing time granted by the John von Neumann Institute for Computing (NIC) and provided on the supercomputer JUROPA at Jülich Supercomputing Center (JSC) (NIC project 4722). Additional computational support was provided by the "Center for Information and Media Technology" (ZIM) at the Heinrich-Heine-Universität Düsseldorf (Germany).

REFERENCES

Allner, O., Nilsson, L., & Villa, A. (2012). Magnesium ion–water coordination and exchange in biomolecular simulations. *Journal of Chemical Theory and Computation, 8*(4), 1493–1502.

Aqvist, J. (1990). Ion water interaction potentials derived from free-energy perturbation simulations. *Journal of Physical Chemistry, 94*(21), 8021–8024.

Aqvist, J. (1992). Modeling of ion ligand interactions in solutions and biomolecules. *Journal of Molecular Structure: Theochem, 88*, 135–152.

Auffinger, P. (2012). Ions in molecular dynamics simulations of RNA systems. In N. Leontis, & E. Westhof (Eds.), *RNA 3D structure analysis and prediction* (pp. 299–318). Berlin, Heidelberg: Springer.

Auffinger, P., Bielecki, L., & Westhof, E. (2003). The Mg^{2+} binding sites of the 5S rRNA loop E motif as investigated by molecular dynamics simulations. *Chemistry & Biology, 10*(6), 551–561.

Banas, P., Hollas, D., Zgarbova, M., Jurecka, P., Orozco, M., Cheatham, T. E., III, et al. (2010). Performance of molecular mechanics force fields for RNA simulations: Stability of UUCG and GNRA hairpins. *Journal of Chemical Theory and Computation, 6*(12), 3836–3849.

Banas, P., Sklenovsky, P., Wedekind, J. E., Sponer, J., & Otyepka, M. (2012). Molecular mechanism of $preQ_1$ riboswitch action: A molecular dynamics study. *Journal of Physical Chemistry B, 116*(42), 12721–12734.

Batey, R. T. (2012). Structure and mechanism of purine-binding riboswitches. *Quarterly Reviews of Biophysics, 45*(3), 345–381.

Batey, R. T., Gilbert, S. D., & Montange, R. K. (2004). Structure of a natural guanine-responsive riboswitch complexed with the metabolite hypoxanthine. *Nature, 432*(7015), 411–415.

Besseova, I., Otyepka, M., Reblova, K., & Sponer, J. (2009). Dependence of A-RNA simulations on the choice of the force field and salt strength. *Physical Chemistry Chemical Physics, 11*(45), 10701–10711.

Blackburn, G. M., & Gait, M. J. (1996). *Nucleic acids in chemistry and biology*. Oxford: Oxford University Press.

Buck, J., Noeske, J., Wöhnert, J., & Schwalbe, H. (2010). Dissecting the influence of Mg^{2+} on 3D architecture and ligand-binding of the guanine-sensing riboswitch aptamer domain. *Nucleic Acids Research, 38*(12), 4143–4153.

Buck, J., Wacker, A., Warkentin, E., Wöhnert, J., Wirmer-Bartoschek, J., & Schwalbe, H. (2011). Influence of ground-state structure and Mg^{2+} binding on folding kinetics of the guanine-sensing riboswitch aptamer domain. *Nucleic Acids Research, 39*(22), 9768–9778.

Case, D. A., Cheatham, T. E., III, Darden, T., Gohlke, H., Luo, R., Merz, K. M., et al. (2005). The Amber biomolecular simulation programs. *Journal of Computational Chemistry, 26*(16), 1668–1688.

Case, D. A., Darden, T. A., Cheatham, T. E., III, Simmerling, C. L., Wang, J., Duke, R. E., et al. (2010). *AMBER 11*. San Francisco, CA: University of California.

Case, D. A., Darden, T. A., Cheatham, T. E., III, Simmerling, C. L., Wang, J., Duke, R. E., et al. (2012). *AMBER 13*. San Francisco, CA: University of California.

Cheatham, T. E., III, & Case, D. A. (2013). Twenty-five years of nucleic acid simulations. *Biopolymers, 99*(12), 969–977.

Darden, T., York, D., & Pedersen, L. (1993). Particle mesh Ewald: An N log(N) method for Ewald sums in large systems. *Journal of Chemical Physics, 98*(12), 10089–10092.

Dixon, N., Duncan, J. N., Geerlings, T., Dunstan, M. S., McCarthy, J. E. G., Leys, D., et al. (2010). Reengineering orthogonally selective riboswitches. *Proceedings of the National Academy of Sciences of the United States of America, 107*(7), 2830–2835.

Draper, D. E., Grilley, D., & Soto, A. M. (2005). Ions and RNA folding. *Annual Review of Biophysics and Biomolecular Structure, 34*, 221–243.

Foloppe, N., & MacKerell, A. D., Jr. (2000). All-atom empirical force field for nucleic acids: I. Parameter optimization based on small molecule and condensed phase macromolecular target data. *Journal of Computational Chemistry, 21*(2), 86–104.

Gilbert, S. D., Love, C. E., Edwards, A. L., & Batey, R. T. (2007). Mutational analysis of the purine riboswitch aptamer domain. *Biochemistry, 46*(46), 13297–13309.

Gilbert, S. D., Stoddard, C. D., Wise, S. J., & Batey, R. T. (2006). Thermodynamic and kinetic characterization of ligand binding to the purine riboswitch aptamer domain. *Journal of Molecular Biology, 359*(3), 754–768.

Gohlke, H., Kuhn, L. A., & Case, D. A. (2004). Change in protein flexibility upon complex formation: Analysis of Ras-Raf using molecular dynamics and a molecular framework approach. *Proteins: Structure, Function, and Bioinformatics, 56*(2), 322–337.

Harned, H. S., & Hudson, R. M. (1951). The diffusion coefficient of magnesium sulfate in dilute aqueous solution at 25°. *Journal of the American Chemical Society, 73*(12), 5880–5882.

Hayes, R. L., Noel, J. K., Mohanty, U., Whitford, P. C., Hennelly, S. P., Onuchic, J. N., et al. (2012). Magnesium fluctuations modulate RNA dynamics in the SAM-I riboswitch. *Journal of the American Chemical Society, 134*(29), 12043–12053.

Jorgensen, W. L., Chandrasekhar, J., Madura, J. D., Impey, R. W., & Klein, M. L. (1983). Comparison of simple potential functions for simulating liquid water. *Journal of Chemical Physics, 79*(2), 926–935.

Joung, I. S., & Cheatham, T. E., III. (2008). Determination of alkali and halide monovalent ion parameters for use in explicitly solvated biomolecular simulations. *Journal of Physical Chemistry B, 112*(30), 9020–9041.

Juneja, A., Villa, A., & Nilsson, L. (2014). Elucidating the relation between internal motions and dihedral angles in an RNA hairpin using molecular dynamics. *Journal of Chemical Theory and Computation, 10*(8), 3532–3540.

Li, P. F., Roberts, B. P., Chakravorty, D. K., & Merz, K. M. (2013). Rational design of particle mesh Ewald compatible Lennard-Jones parameters for +2 metal cations in explicit solvent. *Journal of Chemical Theory and Computation, 9*(6), 2733–2748.

MacArthur, M. W., & Thornton, J. M. (1993). Conformational analysis of protein structures derived from NMR data. *Proteins, 17*(3), 232–251.

MacKerell, A. D., Jr., & Banavali, N. K. (2000). All-atom empirical force field for nucleic acids: II. Application to molecular dynamics simulations of DNA and RNA in solution. *Journal of Computational Chemistry, 21*(2), 105–120.

Mills, R., & Lobo, V. M. M. (1989). *Self-diffusion in electrolyte solutions: A critical examination of data compiled from the literature*. Amsterdam: Elsevier.

Mlynsky, V., Banas, P., Hollas, D., Reblova, K., Walter, N. G., Sponer, J., et al. (2010). Extensive molecular dynamics simulations showing that canonical G8 and protonated $A38H^+$ forms are most consistent with crystal structures of hairpin ribozyme. *Journal of Physical Chemistry B, 114*(19), 6642–6652.

Moller, K. B., Rey, R., Masia, M., & Hynes, J. T. (2005). On the coupling between molecular diffusion and solvation shell exchange. *Journal of Chemical Physics, 122*(11), 114508.

Noeske, J., Buck, J., Fürtig, B., Nasiri, H. R., Schwalbe, H., & Wöhnert, J. (2007). Interplay of 'induced fit' and preorganization in the ligand induced folding of the aptamer domain of the guanine binding riboswitch. *Nucleic Acids Research, 35*(2), 572–583.

Noeske, J., Schwalbe, H., & Wöhnert, J. (2007). Metal-ion binding and metal-ion induced folding of the adenine-sensing riboswitch aptamer domain. *Nucleic Acids Research, 35*(15), 5262–5273.

Ohtaki, H., & Radnai, T. (1993). Structure and dynamics of hydrated ions. *Chemical Reviews, 93*(3), 1157–1204.

Perez, A., Marchan, I., Svozil, D., Sponer, J., Cheatham, T. E., III, Laughton, C. A., et al. (2007). Refinement of the AMBER force field for nucleic acids: Improving the description of α/γ conformers. *Biophysical Journal, 92*(11), 3817–3829.

Priyakumar, U., & MacKerell, A. D., Jr. (2010). Role of the adenine ligand on the stabilization of the secondary and tertiary interactions in the adenine riboswitch. *Journal of Molecular Biology, 396*(5), 1422–1438.

Pyle, A. M. (2002). Metal ions in the structure and function of RNA. *Journal of Biological Inorganic Chemistry, 7*(7–8), 679–690.

Reblova, K., Lankas, F., Razga, F., Krasovska, M. V., Koca, J., & Sponer, J. (2006). Structure, dynamics, and elasticity of free 16S rRNA helix 44 studied by molecular dynamics simulations. *Biopolymers, 82*(5), 504–520.

Reblova, K., Spackova, N., Stefl, R., Csaszar, K., Koca, J., Leontis, N. B., et al. (2003). Non-Watson–Crick base pairing and hydration in RNA motifs: Molecular dynamics of 5S rRNA loop E. *Biophysical Journal, 84*(6), 3564–3582.

Robinson, H., Gao, Y. G., Sanishvili, R., Joachimiak, A., & Wang, A. H. J. (2000). Hexahydrated magnesium ions bind in the deep major groove and at the outer mouth of A-form nucleic acid duplexes. *Nucleic Acids Research, 28*(8), 1760–1766.

Roe, D. R., & Cheatham, T. E., III. (2013). PTRAJ and CPPTRAJ: Software for processing and analysis of molecular dynamics trajectory data. *Journal of Chemical Theory and Computation, 9*(7), 3084–3095.

Ryckaert, J. P., Ciccotti, G., & Berendsen, H. J. C. (1977). Numerical integration of cartesian equations of motion of a system with constraints: Molecular dynamics of n-alkanes. *Journal of Computational Physics, 23*(3), 327–341.

Saini, J. S., Homeyer, N., Fulle, S., & Gohlke, H. (2013). Determinants of the species selectivity of oxazolidinone antibiotics targeting the large ribosomal subunit. *Biological Chemistry, 394*(11), 1529–1541.

Salomon-Ferrer, R., Gotz, A. W., Poole, D., Le Grand, S., & Walker, R. C. (2013). Routine microsecond molecular dynamics simulations with AMBER on GPUs. 2. Explicit solvent particle mesh Ewald. *Journal of Chemical Theory and Computation, 9*(9), 3878–3888.

Schafmeister, C. E. A. F., Ross, W. S., & Romanovski, V. (1995). *Leap*. San Francisco, CA: University of California.

Serganov, A., & Nudler, E. (2013). A decade of riboswitches. *Cell, 152*(1–2), 17–24.

Serganov, A., Yuan, Y. R., Pikovskaya, O., Polonskaia, A., Malinina, L., Phan, A. T., et al. (2004). Structural basis for discriminative regulation of gene expression by adenine- and guanine-sensing mRNAs. *Chemistry & Biology, 11*(12), 1729–1741.

Sharma, M., Bulusu, G., & Mitra, A. (2009). MD simulations of ligand-bound and ligand-free aptamer: Molecular level insights into the binding and switching mechanism of the *add* A-riboswitch. *RNA, 15*(9), 1673–1692.

Sklenovsky, P., Florova, P., Banas, P., Reblova, K., Lankas, F., Otyepka, M., et al. (2011). Understanding RNA flexibility using explicit solvent simulations: The ribosomal and group I intron reverse kink-turn motifs. *Journal of Chemical Theory and Computation, 7*(9), 2963–2980.

Snedecor, G. W., & Cochran, W. G. (1989). *Statistical methods* (8th ed.). Ames: Iowa State University Press.

Sponer, J., Banas, P., Jurecka, P., Zgarbova, M., Kuhrova, P., Havrila, M., et al. (2014). Molecular dynamics simulations of nucleic acids. From tetranucleotides to the ribosome. *Journal of Physical Chemistry Letters, 5*(10), 1771–1782.

Sponer, J., Cang, X. H., & Cheatham, T. E., III. (2012). Molecular dynamics simulations of G-DNA and perspectives on the simulation of nucleic acid structures. *Methods, 57*(1), 25–39.

Stoddard, C. D., Gilbert, S. D., & Batey, R. T. (2008). Ligand-dependent folding of the three-way junction in the purine riboswitch. *RNA, 14*(4), 675–684.

Stoddard, C. D., Montange, R. K., Hennelly, S. P., Rambo, R. P., Sanbonmatsu, K. Y., & Batey, R. T. (2010). Free state conformational sampling of the SAM-I riboswitch aptamer domain. *Structure, 18*(7), 787–797.

Stoddard, C. D., Widmann, J., Trausch, J. J., Marcano-Velazquez, J. G., Knight, R., & Batey, R. T. (2013). Nucleotides adjacent to the ligand-binding pocket are linked to activity tuning in the purine riboswitch. *Journal of Molecular Biology, 425*(10), 1596–1611.

Tucker, B. J., & Breaker, R. R. (2005). Riboswitches as versatile gene control elements. *Current Opinion in Structural Biology, 15*(3), 342–348.

Villa, A., Wöhnert, J., & Stock, G. (2009). Molecular dynamics simulation study of the binding of purine bases to the aptamer domain of the guanine sensing riboswitch. *Nucleic Acids Research, 37*(14), 4774–4786.

Wang, J. M., Cieplak, P., & Kollman, P. A. (2000). How well does a restrained electrostatic potential (RESP) model perform in calculating conformational energies of organic and biological molecules? *Journal of Computational Chemistry, 21*(12), 1049–1074.

Welch, B. L. (1947). The generalization of students problem when several different population variances are involved. *Biometrika, 34*(1–2), 28–35.

Woodson, S. A. (2005). Metal ions and RNA folding: A highly charged topic with a dynamic future. *Current Opinion in Chemical Biology, 9*(2), 104–109.

Zgarbova, M., Otyepka, M., Sponer, J., Mladek, A., Banas, P., Cheatham, T. E., III, et al. (2011). Refinement of the Cornell et al. nucleic acids force field based on reference quantum chemical calculations of glycosidic torsion profiles. *Journal of Chemical Theory and Computation, 7*(9), 2886–2902.

CHAPTER EIGHT

Thermodynamic and Kinetic Folding of Riboswitches

Stefan Badelt*, Stefan Hammer*,†, Christoph Flamm[1],*, Ivo L. Hofacker*,†

*Institute for Theoretical Chemistry, University of Vienna, Vienna, Austria
†Research Group Bioinformatics and Computational Biology, University of Vienna, Vienna, Austria
[1]Corresponding author: e-mail address: xtof@tbi.univie.ac.at

Contents

Abstract

Riboswitches are structured RNA regulatory elements located in the 5′-UTRs of mRNAs. Ligand-binding induces a structural rearrangement in these RNA elements, effecting events in downstream located coding sequences. Since they do not require proteins for their functions, they are ideally suited for computational analysis using the toolbox of RNA structure prediction methods. By their very definition riboswitch function depends on structural change. Methods that consider only the thermodynamic equilibrium of an RNA are therefore of limited use. Instead, one needs to employ computationally more expensive methods that consider the energy landscape and the folding dynamics on that landscape. Moreover, for the important class of kinetic riboswitches, the mechanism of riboswitch function can only be understood in the context of co-transcriptional folding. We present a computational approach to simulate the dynamic behavior of riboswitches during co-transcriptional folding in the presence and absence of a ligand. Our investigations show that the abstraction level of RNA secondary

Methods in Enzymology, Volume 553
ISSN 0076-6879
http://dx.doi.org/10.1016/bs.mie.2014.10.060

structure in combination with a dynamic folding landscape approach is expressive enough to understand how riboswitches perform their function. We apply our approach to a experimentally validated theophylline-binding riboswitch.

1. INTRODUCTION

The past decades witnessed a dramatic expansion of our knowledge on RNA as a regulatory molecule. A myriad of functional small RNAs influencing a diverse set of cellular processes have been described for bacteria and eukaryotes. Among them are riboswitches (Serganov & Nudler, 2013), structured RNA elements located in the 5′-UTR of mRNAs (Nudler & Mironov, 2004), that are capable of regulating gene expression. Regulation works either on the transcriptional or on the translational level. Translational riboswitches regulate the formation of the translation initiation complex enabling them to switch between an on- and off-state. In contrast, transcriptional riboswitches induce early termination of the whole transcription process and are therefore not reversible.

Usually, riboswitches are composed of two parts: (i) a relatively conserved aptamer domain responsible for ligand binding and (ii) a variable sequence region termed expression platform for regulating the downstream located coding sequences. The ligand recognition sites vary greatly in size and complexity of their secondary and tertiary structures. Environmental stimuli like temperature changes or the binding of ligands such as ions, enzyme cofactors, RNA, or DNA trigger switching due to changes in the expression platform which are then translated into a modulation of downstream events.

Riboswitches can furthermore be classified into thermodynamic and kinetic switches. Thermodynamic switches are found in energetic equilibrium between their on- and off-state, i.e. if switching is triggered, the equilibrium distribution shifts towards the new energetically best conformation. This implies that thermodynamic switches can reversibly and repeatedly toggle between on- and off-states. In contrast, kinetic switches are trapped in one state, depending on whether the trigger was present at the time of folding. The functional states correspond to local minima of the energy landscape that cannot be escaped during the lifetime of the molecule. Therefore, kinetic switches are not reversible without addition of extrinsic energy, RNA turnover through degradation and synthesis is responsible for changing the state of a cell.

The high modularity in the structural architecture of riboswitches allows for a high degree of functional portability to other contexts making riboswitches an highly attractive design target to achieve context-dependent gene regulation in synthetic biology (Dawid, Cayrol, & Isambert, 2009; Isaacs et al., 2004; Qi, Lucks, Liu, Mutalik, & Arkin, 2012; Rodrigo, Landrain, Majer, Daros, & Jaramillo, 2013).

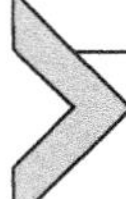

2. CHARACTERIZATION AND PREDICTION OF RIBOSWITCHES

Riboswitches implement a particularly direct mechanism of gene expression, as they effect the expression of an mRNA via the structure of the RNA molecule itself without requiring any protein cofactors. Therefore these mechanisms can be easily modeled *in-silico* on the level of secondary structure, using well-established and computationally efficient methods. Nevertheless a number of caveats make the application of such programs to riboswitches less than straightforward. (i) The effect of ligand binding to the aptamer is not included in energy models for RNA secondary. (ii) Many aptamers form pseudoknots or complex tertiary structures ignored by secondary structure prediction and (iii) the commonly used methods for describing RNA molecules in thermodynamic equilibrium are insufficient for modeling riboswitches whose mechanisms depend on RNA folding kinetics.

The computational effort to characterize riboswitches is therefore dependent on the type of the riboswitch. Temperature-dependent riboswitches can be modeled with standard free energy parameters, while modeling riboswitches that bind ligands need empirical data on the binding free energy. Also, thermodynamic switches can be characterized by methods predicting equilibrium properties, while kinetic switches require the much harder computation of folding kinetics.

Figure 1 shows a designed and experimentally tested example (Waldminghaus, Kortmann, Gesing, & Narberhaus, 2008) of a temperature-dependent, thermodynamic switch. Computing the specific heat using `RNAheat` (Hofacker et al., 1994) readily identifies a structural transition at around 34°C.

To model switches responsive to small RNA molecules, it is sufficient to use methods able to predict RNA–RNA interactions, such as `RNAcofold` (Bernhart et al., 2006) or `RNAup` (Mückstein et al., 2008), given one assumes high concentrations of the small RNA. Ligand-binding riboswitches are

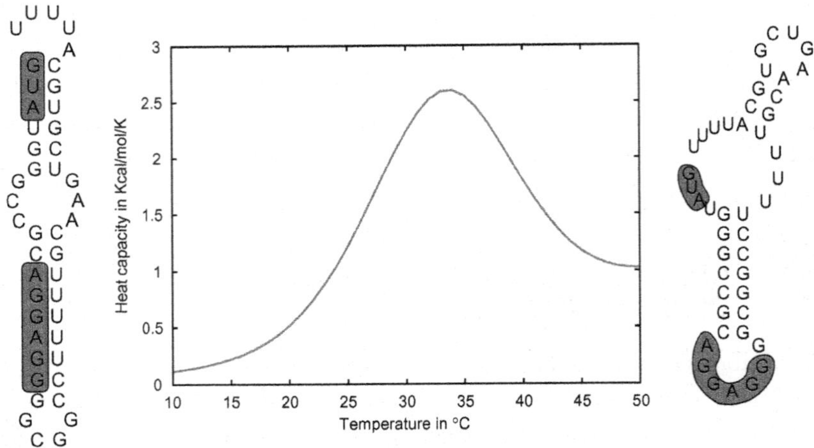

Figure 1 *RNAthermometer*- specific heat of a thermosensitive RNA switch. The peak at around 34°C marks the structural transition between the low-temperature (left) and high-temperature (right) structures. The Shine–Dalgarno sequence and the start codon (highlighted in red (gray in the print version)) are inaccessible at the low-temperature structure, but accessible at high temperatures.

more difficult to analyze, as the mechanism for binding such small molecules is not captured by standard RNA energy models. However, experimentalists have measured binding free energies of ligands interacting with particular RNA motifs, e.g., Jenison, Gill, Pardi, and Polisky (1994), Jucker, Phillips, McCallum, and Pardi (2003), and Gouda, Kuntz, Case, and Kollman (2003) for the theophylline aptamer. These free energies can be included in energy landscape predictions as a energy correction on binding competent structures.

Section 3 describes fast approaches for thermodynamic riboswitches. For switches that are trapped in a kinetically favored structure, we describe methods based on RNA landscape computations in section 4. The most complex case are switches depending on co-transcriptional folding, as they have a dynamic energy landscape that changes with every newly transcribed nucleotide. Section 5 shows how a co-transcriptional theophylline riboswitch (Wachsmuth, Findeiß, Weissheimer, Stadler, & Mörl, 2013) can be modeled by computing RNA folding kinetics on such dynamic energy landscapes.

3. THERMODYNAMIC RNA FOLDING

If we define the set of RNA secondary structures Ω as all structures that (i) are formed from nested, isosteric base pairs (GC, CG, AU, UA,

GU, UG), (ii) have hairpins with at least 3 unpaired nucleotides, and (iii) have interior loops of at most 30 unpaired nucleotides, then this includes the vast majority of known pseudoknot free secondary structures. There are experimentally determined energy parameters (Mathews et al., 2004; Mathews, Sabina, Zuker, & Turner, 1999) that enable to compute a free energy for any RNA secondary structure $S \in \Omega$. In particular, these energy parameters assign energies to every loop (hairpin, interior, exterior, and multiloop). They mostly depend on the loop type and size, with some sequence dependence. Most beneficial are stacking energies, i.e., base pairs that close an interior loop with no unpaired bases in between, but there are also tabulated energy values, e.g., for common interior loops that are known for stable non-canonical interactions. This energy model is known as the Nearest Neighbor energy model (Turner & Mathews, 2010). The total energy of an RNA structure can be computed as the sum of all loops

$$E(S) = \sum_{L \in S} E(L) \tag{1}$$

3.1. RNA structure prediction

Based on the described energy model several methods exist to efficiently predict the minimum free energy (MFE) structure as well as various equilibrium properties of the RNA. These methods solve the problem by dynamic programming and typically require $\mathcal{O}(n^2)$ space and $\mathcal{O}(n^3)$ time. They can thus be used routinely even for very long RNA molecules. In this contribution, we will focus on methods available in the `ViennaRNA package` (Lorenz et al., 2011), which provides an especially large selection of prediction methods. Other popular methods include, e.g., `RNAstructure` (Mathews, 2014) and `mfold / UNAfold` (Markham & Zuker, 2008).

The most common mode of structure prediction, e.g., of programs such as `RNAfold`, will return a single structure corresponding to the lowest free energy state of the RNA. Since riboswitch function depends on the presence of at least two functional conformations, MFE folding is clearly insufficient.

A more complete picture of the thermodynamic folding can be gained by computing the partition function Z of an RNA molecule. From the partition function

$$Z = \sum_{S \in \Omega} e^{\frac{-E(S)}{RT}} \tag{2}$$

various equilibrium properties can be derived. In particular, we can compute the probability P of observing a structure S

$$P(S) = \frac{1}{Z} e^{\frac{-E(S)}{RT}} \tag{3}$$

and the ensemble free energy G

$$G = -RT \ln(Z) \tag{4}$$

The partition function can be computed with the same $\mathcal{O}(n^3)$ effort as computing the MFE structure. In addition, the algorithm allows to compute the equilibrium probability p_{ij} for every possible base pair (i,j). Pair probabilities provide a compact representation of the complete Boltzmann ensemble of structures of an RNA molecule.

Most folding programs allow to specify constraints, such as base pairs that have to be present or positions that are not allowed to pair. For a riboswitch with a known aptamer structure, this can be used to compute the partition function only over those structures which form the aptamer, i.e., binding competent structures. The ratio of the constrained and unconstrained partition function yields the equilibrium probability that the aptamer structure is formed

$$\mathrm{P(aptamer)} = \frac{Z^{\mathrm{constrained}}}{Z}. \tag{5}$$

If we have information on how strong the aptamer structure is stabilized by ligand binding, e.g., from measurements of the dissociation constant K_d, we can even compute the fraction of ligand-bound RNAs as a function of the concentrations.

Another approach to gain a more complete picture than only a single MFE structure is to compute suboptimal structures in addition to the MFE structure. At least three commonly used strategies for this exist. `mfold` (Zuker, 1989) first introduced an algorithm to compute all suboptimals detectable by picking one base pair and asking for the optimal structure containing this pair. This approach yields a small, but generally incomplete list of alternative structures. `RNAsubopt` (Wuchty, Fontana, Hofacker, & Schuster, 1999) will produce all suboptimal structures in a defined energy range, resulting in a number of structures that grows exponentially with sequence length. Finally, it is possible to directly sample structures from the Boltzmann ensemble after computing the partition function Z.

As riboswitches possess at least two functionally important conformations, it seems natural to use the prediction of suboptimal structures to search for novel riboswitches. One of the first methods to attempt this was paRNAss (Giegerich, Haase, & Rehmsmeier, 1999; Voss, Meyer, & Giegerich, 2004). This program generates a sample of suboptimal structures, computes pairwise distances between those structures using two different distance measures and performs a clustering. RNAs which exhibit two well-separated clusters of structures are classified as RNA switches. This procedure works well for a number of known switching RNA molecules, such as attenuator sequences, but is less successful for ligand-binding aptamers. The reason simply is that the aptamer binding conformation typically is only stable in the presence of the ligand. As the structure predictions do not take ligand binding into account, they fail to recognize the aptamer conformation as a low energy state. In practice, computational efforts for riboswitch discovery have therefore focused on the detection of known aptamer structures using structural homology search.

3.2. RNA2Dfold

The paRNAss method, mentioned above, introduced a so-called validation plot as visualization of the clustering result. Once the procedure has identified two clusters and their representative structure, it computes for every suboptimal structure the distances d_1,d_2 to these two reference structures. The resulting distance pairs plotted as points in a 2D coordinate system.

The idea of classifying each structure by its distance to two reference structures is pursued in a more principled way in RNA2Dfold (Lorenz, Flamm, & Hofacker, 2009). Rather than working with a sample of suboptimal structures, RNA2Dfold considers all possible secondary structures and performs a classified dynamic programming. In short, we define a distance class (κ,λ) to comprise all structures with distance κ to the first reference structure and λ to the second. An extension of classical RNA folding algorithms then computes the MFE structure (or partition function) for every distance class. In effect, RNA2Dfold computes a projection of the high-dimensional conformation space into two dimensions spanned by the distance to the reference structures. The result is ideal for visualizing the folding landscape by plotting the folding energy as a function of κ and λ, see Fig. 2.

The additional bookkeeping makes RNA2Dfold much more expensive than normal RNA folding, requiring $\mathcal{O}(n^7)$ time and $\mathcal{O}(n^4)$ space. Nevertheless, the approach is readily applicable to sequences of up to about 400 nt,

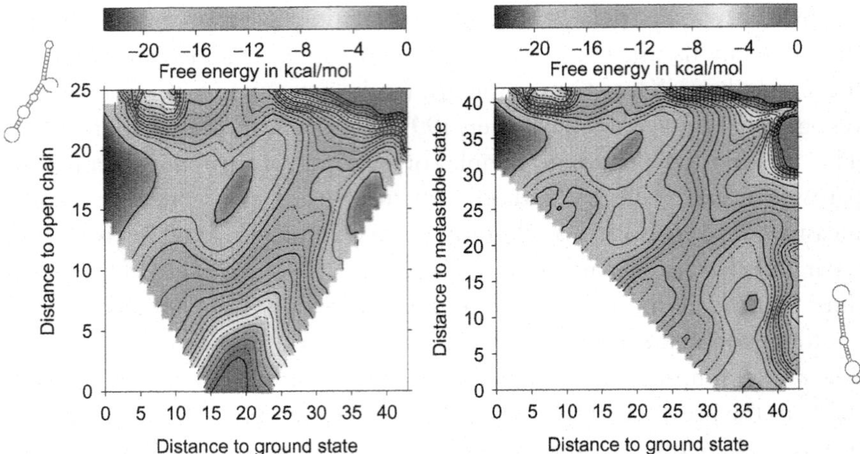

Figure 2 RNA2Dfold computed projection of the energy landscape for the leader sequence of the *E. coli* $tRNA^{phe}$ synthetase operon. Left panel: Projection using the MFE structure (terminator hairpin, far left) and the open chain as references. Right panel: Projection using the MFE structure and the metastable structure found at position (36,17) of the first projection. The two structures are shown on the left and right of the landscapes. (See the color plate.)

easily exceeding the length of typical riboswitches. A remaining problem is to choose the two reference structures. In general, one can choose the MFE structure as the first and a metastable structure as the second reference. A common work flow is to first perform a run of RNA2Dfold with the MFE and the open chain conformation as references, and choose a suitable metastable structure from the results of this first run. The metastable structure is then used as reference in a subsequent second RNA2Dfold computation.

This procedure is illustrated in Fig. 2 using the leader sequence of the *E. coli* $tRNA^{phe}$ synthetase operon. The ground state structure of the sequence forms a terminator hairpin that switches the transcription of the downstream operon off. The 2D landscape clearly indicates the presence of a metastable state with 17 base pairs and a base pair distance of 36 to the MFE structure. Using the metastable structure as the second reference shows the clearly separated conformational states of the leader sequence even better.

It is worth noting that 2D landscapes immediately provide a lower bound on the energy barrier between the two structures and thus an estimate how quickly the RNA can switch conformations. They can also be used as a starting point for more sophisticated path-finding heuristics. By computing a series of 2D landscapes for successively longer sequences, one can obtain a qualitative impression of co-transcriptional folding in order to study kinetic switches.

The caveat about the effect of ligand binding applies here as well, as we only obtain a landscapes for the unbound riboswitch. An upcoming version of the Vienna RNA package will allow to specify flexible soft constraints, such as energy bonuses for particular structural motifs. Given suitable experimental binding energies, this should allow us to compute 2D landscapes for the riboswitch in the presence of the ligand.

3.3. `RNAsubopt`, `barriers`

`RNA2Dfold` as described above will generally not find all local minima, i.e., metastable states, of an RNA. For a more complete characterization of riboswitches, we need to consider the whole energy landscape $\mathcal{L}$ of an RNA molecule and identify all stable alternative conformations. In general, RNA molecules can adopt multiple conformations and also non-riboswitches might have alternative structures that are kinetically favored. Moreover, the lifespan of an RNA molecule can be simply too short to reach the MFE structure at all. Whether this is the case for a particular RNA can be determined by analyzing the energy landscape.

More formally, denote the energy landscape as $\mathcal{L} = (\Omega, \mathcal{M}, E)$, with Ω being the previously introduced set of RNA conformations, $\mathcal{M}$ being a move-set to define a neighborhood relation, and E being an energy function to assign a fitness value to each conformation. For an ergodic move-set $\mathcal{M}$, we chose the most elementary modification of an RNA secondary structure, the formation or opening of a single base pair.

`RNAsubopt` (Wuchty et al., 1999) computes all conformations $S \in \Omega$ that are within a certain energy range above the MFE. As RNA energy landscapes grow exponentially with sequence length, this results in a massive amount of secondary structures even for very short sequences. The program `barriers` (Flamm, Hofacker, Stadler, & Wolfinger, 2002) can then process such an energetically sorted list of suboptimal structures with a flooding algorithm to find all local minima and the according saddle points connecting them. In particular, every structure is either a local minimum, a saddle point connecting at least two local minima, or it belongs to the *basin* of one local minimum. This allows for computing the partition functions (see Eq. (2)) for every basin. The level of coarse graining can be adjusted to the inspected landscape by specifying the minimal depth of a local minimum or the total number of energetically best local minima. The results can be visualized in form of a barrier-tree.

For temperature sensitive RNA switches, as shown in Fig. 1, one can compute the suboptimal structures for two temperatures and compare the energy landscapes. See Fig. 3 for barrier trees depicting the landscapes at temperatures 30 and 40°C.

As mentioned previously, modeling ligand-binding riboswitches requires to take into account the stabilizing effect of the bound ligand to the aptamer structure. For some aptamers, binding affinities and thus binding free energies have been experimentally determined and in addition, the structural requirements for ligand binding are often known. The energy landscape of the riboswitch in presence of the ligand can then be analyzed by adding the binding free energy to all conformations that are binding competent, i.e., contain an intact aptamer structure.

In the following, we use an artificially designed theophylline-dependent riboswitch termed RS10 (Wachsmuth et al., 2013). It is positioned at the 5′UTR of its target gene (*bgaB*) and leads to the formation of an early terminator hairpin in the absence of theophylline. As soon as theophylline is present, a co-transcriptionally formed aptamer structure is stabilized, the terminator cannot be formed and the mRNA is transcribed in its full length. The binding energy of theophylline to the aptamer was estimated from the dissociation constant of $K_d = 0.32\mu M$ at 25°C (Jenison et al., 1994) as $\Delta G = -\mathrm{RT} \ln K_d = -8.86$kcal/mol. As RS10 regulates at the transcription level, it necessarily falls into the category of kinetic switches. The terminator hairpin can be effective only if it forms quickly enough, i.e., before the polymerase has continued into the coding region. Transcription speed and therefore the choice of nucleotides and the length of the spacer region play a crucial role for a proper functionality.

Figure 4 shows two barrier trees of the RS10 riboswitch, representing the energy landscape with and without the ligand. Structures A and B contain the terminator hairpin, while structure I does not and therefore represents the on-state. Note that even in the presence of theophylline, structure A remains the ground state. The terminator free structure I is only metastable, but separated by an energy barrier of $\approx$ 12 kcal/mol from the ground state. The static landscape picture is, however, insufficient to decide whether the on-state structure I will indeed be reached by the co-transcriptional folding process.

The limitation of the `RNAsubopt/barriers` approach lies in the lengths of inspected molecules. The `barriers` program has to read and store all low

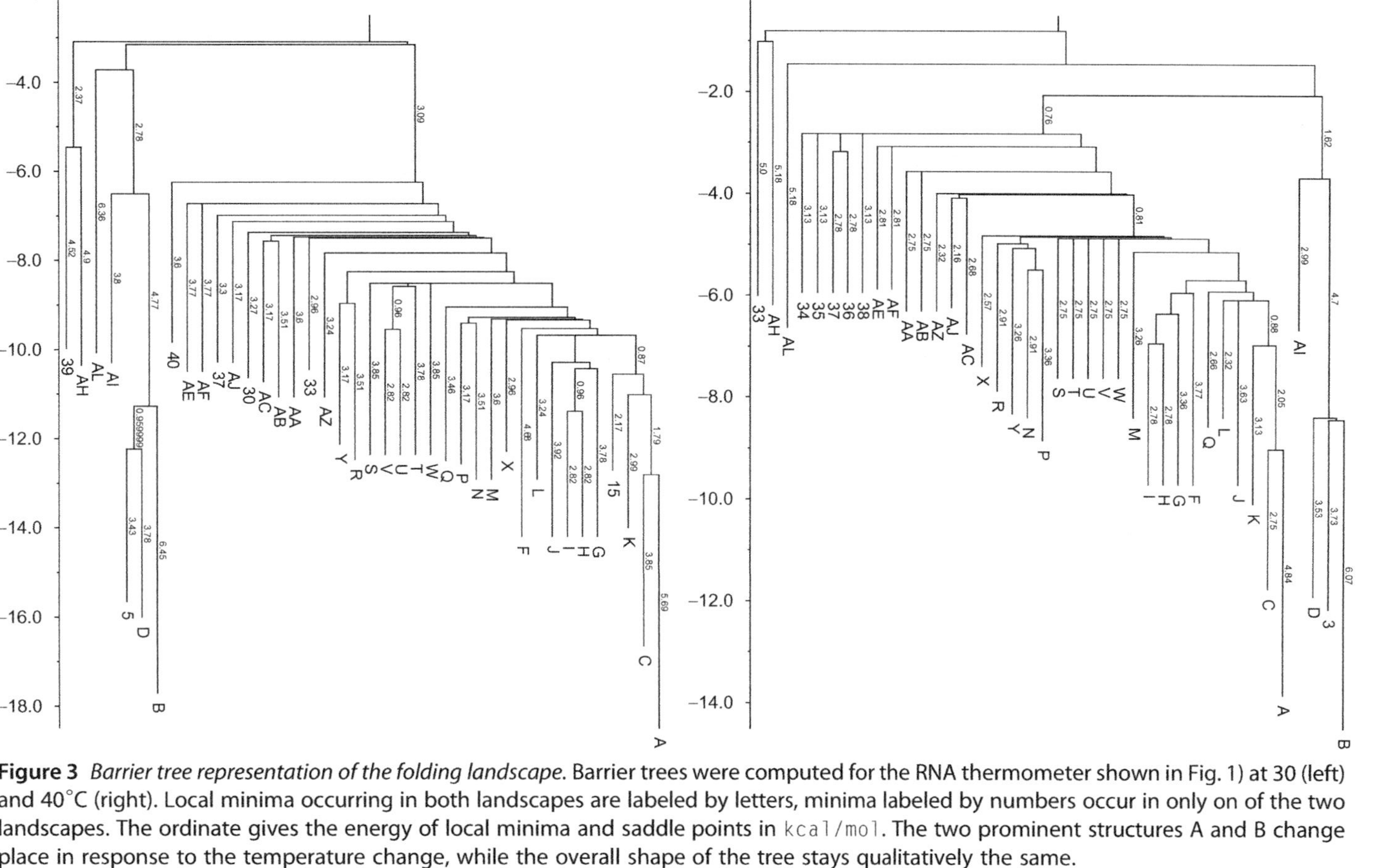

Figure 3 *Barrier tree representation of the folding landscape.* Barrier trees were computed for the RNA thermometer shown in Fig. 1) at 30 (left) and 40°C (right). Local minima occurring in both landscapes are labeled by letters, minima labeled by numbers occur in only on of the two landscapes. The ordinate gives the energy of local minima and saddle points in `kcal/mol`. The two prominent structures A and B change place in response to the temperature change, while the overall shape of the tree stays qualitatively the same.

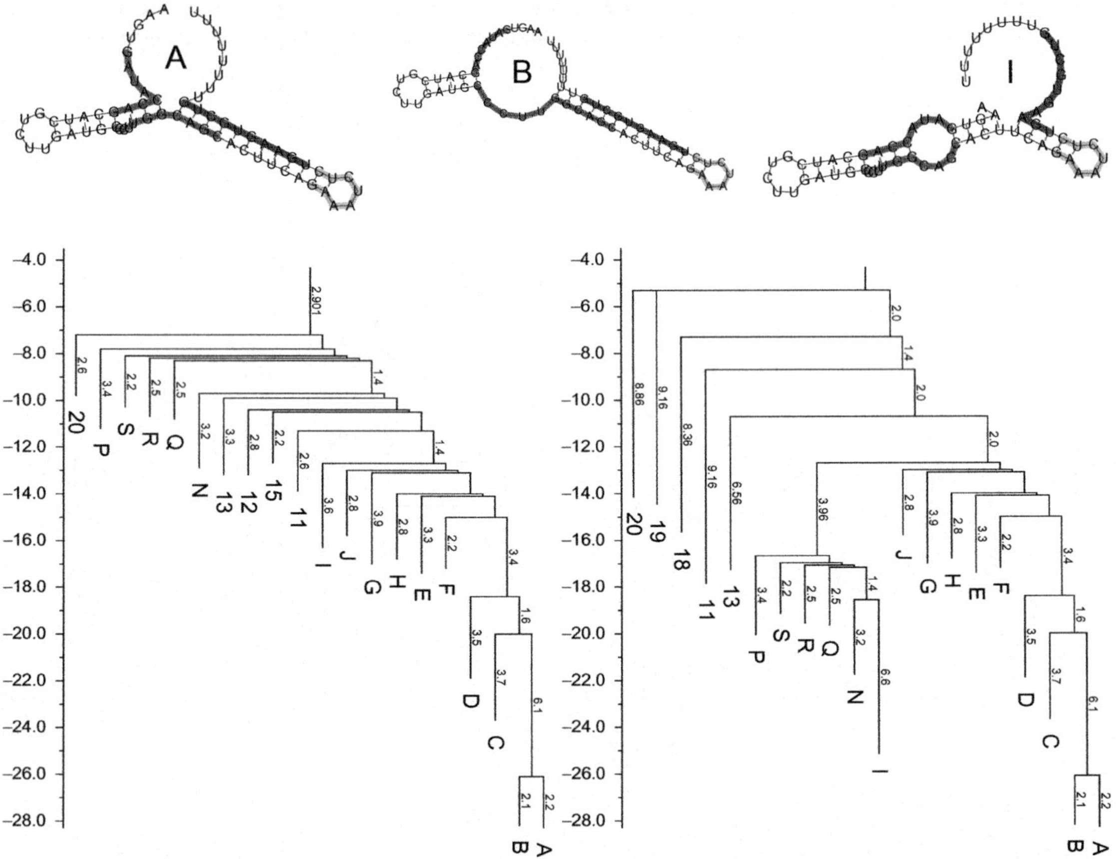

Figure 4 See legend on opposite page.

energy structures in memory. This limits the approach to RNA molecules where the relevant low energy part of the landscape comprises less than, say, 10^8 conformations which is often reached by molecules of about 100 nt. For longer RNAs, it might be still possible to identify the most important local minima, but not the saddle points connecting them.

Recent work has aimed to overcome the length limits of `barriers` by using heuristics for sampling low lying local minima as well as for estimating barrier heights between local minima. The Basin hopping graph approach of Kucharik, Hofacker, Stadler, and Qin (2014), for example, can handle RNAs of several 100 nt at the expense of loosing exact barrier heights.

4. RNA FOLDING KINETICS ON STATIC LANDSCAPES

The physical process of RNA folding is usually modeled as a stochastic process on an RNA energy landscape specifying (i) the structure space, (ii) the neighborhood relation, and (iii) the transition rate model. With these three key concepts at hand the folding process can be described as a continuous time Markov process

$$\frac{\mathrm{d}P_x(t)}{\mathrm{d}t} = \sum_{y \neq x} [P_y(t)k_{xy} - P_x(t)k_{yx}] \tag{6}$$

where $P_x(t)$ gives the probability to observe the folding RNA chain in conformation x at time t, and k_{xy} is the transition rate from conformation y to conformation x. Clearly, $k_{xy} > 0$ only if conformation x is reachable from conformation y via the neighborhood relation. Most existing approaches for kinetic RNA folding are based on the master equation model above and mainly differ in the set of allowed states (e.g., with or without

Figure 4 *Folding landscape picture* (top): The three main structures of the RS10 riboswitch, using the color code as in (Wachsmuth et al., 2013): red (gray in the print version) indicates the aptamer region, blue (dark gray in the print version) is the 3′-part of the terminator hairpin and cyan (light gray in the print version) is a spacer region. Structures A and B contain the terminator hairpin and therefore correspond to the off-state of the switch. Structure I is the theophylline-binding competent structure corresponding to the on-state of the switch. This structure is stabilized by −8.86 kcal/mol upon ligand binding. The barrier tree on the left and the right correspond to the undistorted (theophylline free) and distorted folding landscape, respectively. Note the gain in ruggedness for the distorted folding landscape and the emergence of a distinct subtree containing structure I.

pseudoknots), the neighborhood relation, as well as in the energy rules and the resulting rate model. However, the existing approaches can be partitioned into two major classes according to the method on how the master equation is solved. The first class of approaches apply Gillespie-type simulation algorithms (Gillespie, 1977) to generate statistically correct trajectories as possible solutions. The second class of approaches solve the master equation directly.

4.1. Stochastic simulation of folding kinetics

The program `Kinfold` (Flamm, Fontana, Hofacker, & Schuster, 2000) implements a rejection-less Monte–Carlo method together with the most elementary neighborhood relation, the insertion or deletion of a single base pair. While this combination allows for a very detailed simulation of folding pathways, the elementary step resolution leads necessarily to long simulation runs. Many approaches therefore choose to allow larger structural changes by using the formation or destruction of an entire helix as the basic step (Danilova, Pervouchine, Favorov, & Mironov, 2006; Huang & Voß, 2014; Isambert & Siggia, 2000; Mironov & Lebedev, 1993). Using helix insertion/deletion as basic transformation strongly restricts the space of allowed conformations. This reduction allows to explore the conformation space in a much smaller number of steps. Consequently, simulations of larger RNAs become feasible. However, due to the larger structural changes during a simulation step, the quality of the rate model becomes extremely important. The extension of these approaches to gain folding during transcription, or the incorporation of pseudoknotted structures is straight forward. For a recent review on the advantages and problems of kinetic folding approaches, see Flamm and Hofacker (2008).

On the downside, stochastic simulation approaches require a fairly large number of trajectories in order to give statistically robust results. In general, they also require sophisticated post-processing in order to interpret the trajectories in a meaningful way.

4.2. Barriers/treekin

Formally, the master equation (see Eq. (6)) is solved by

$$P(t) = \mathrm{e}^{t \cdot \mathbf{K}} \cdot P(0) \tag{7}$$

where $P(0)$ is the vector of initially populated conformations for $t = 0$, and $\mathbf{K} = (k_{xy})$ is the matrix of transition rates between individual conformations of the conformation space. Integrating the master equation thus involves computing matrix exponentials, usually by first diagonalizing the matrix **K**. This limits the dimension of the number of **K** to a few thousand. As the number of conformations grows exponentially with sequence length, the Eq. (7) is applicable only for short toy examples.

In order to treat RNAs of biological interest, we need a coarse graining that reduces the number of conformations. The program `barriers` (Flamm et al., 2002) performs such a coarse graining of the conformation space into macrostates, by partitioning the folding landscape into gradient basins and their connecting saddle points. The resulting hierarchical structure, called barrier tree (see Fig. 3), offers a compact representation of the entire folding landscape, where leaf nodes of the tree correspond to local minima and internal tree nodes to the energetically lowest saddle points connecting two local minima. During the construction of the barrier tree, the program `barriers` identifies these "gradient basins" and calculates the partition function of each macrostate as well as effective transition rates between any two macrostates α,β as

$$k(\alpha \to \beta) \approx \sum_{x \in \alpha} \sum_{y \in \beta} k(x \to y) \mathrm{e}^{-E(x)/\mathrm{RT}} / Z_{\alpha}.$$

The approximation assumes a local equilibrium between the conformations within each macrostate such that the partition function Z_{α} can be used to calculate the probability of being in conformation x in macrostate α. The Metropolis rule is used to assign the microstate transition probabilities $k(x \to y)$. The macrostate transition matrix and a vector of initial populations is then handed to the program `treekin` (Wolfinger, Svrcek-Seiler, Flamm, Hofacker, & Stadler, 2004), which numerically integrates the master equation for arbitrary long times t by computing the matrix exponential. The time evolution of the population density is returned as a result (see Fig. 5). The folding dynamics of RNA molecules up to the size of tRNAs can therefore easily be computed for arbitrary long time scales using the `barriers/treekin` approach.

For illustration, we again use the RS10 riboswitch and compute folding kinetics starting at the aptamer conformation on either the undisturbed energy landscape or the landscape corrected for ligand-binding energies. We predict that the unbound RS10 riboswitch refolds to the off-state in about one hundredth of a second, while the theophylline-bound off-state remains stable for more than 15 min, see Fig. 5.

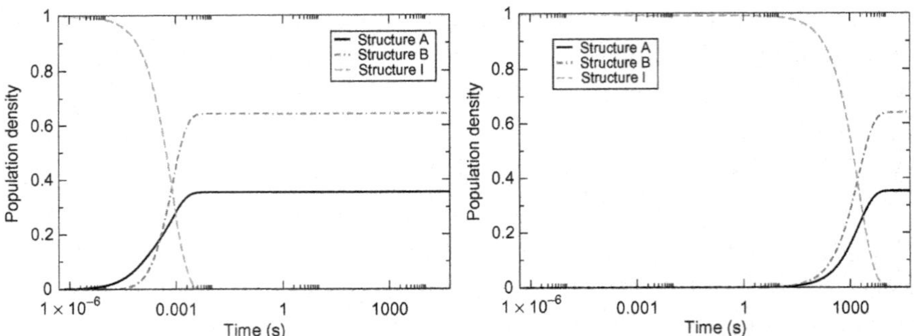

Figure 5 `treekin` simulation of the unbound RS10 riboswitch (left) and the theophylline-bound switch (right). We set the start conditions of the simulation to structure I being the only population. While the refolding to the MFE structure and a close neighbor happens really fast in the unbound condition, the molecule is trapped for a long time (> 10^3 s) in the ligand-bound state.

A general problem with such simulations is that the computation uses an internal time scale whose relation to actual wall clock time is unknown. Recently, Sauerwine and Widom (2013) performed `Kinfold` simulations on short RNAs and compared `Kinfold` time to refolding times determined from NMR experiments and determined that 1 `Kinfold` time step corresponds to roughly $5\mu s$. We performed `Kinfold` and `treekin` simulations on the same molecule in order to verify that `treekin` and `Kinfold` time units are approximately equivalent and used this number to convert simulation time to seconds.

We note, that the computation of barrier trees and `treekin` trajectories can be performed using the Vienna RNA web services (Gruber, Lorenz, Bernhart, Neuböck, & Hofacker, 2008) at http://www.tbi.univie.ac.at/. The web version, however, does not support the inclusion of ligand-binding energies.

5. RNA FOLDING KINETICS ON DYNAMIC LANDSCAPES

In a cellular context the nascent RNA molecule starts folding before the transcription process is completed (Lai, Proctor, & Meyer, 2013) and the folded structure may therefore depend on the speed of elongation, on site-specific pausing of the RNA polymerase (Wong, Sosnick, & Pan, 2007), and interactions of the nascent RNA molecule with proteins or small-molecule metabolites (Pan & Sosnick, 2006). Many riboswitches are thought to co-transcriptionally fold into their on- or off-state, depending on the presence of their trigger, and will then stay trapped in that conformation even if the trigger is removed.

The hybrid-simulation framework `BarMap` (Hofacker et al., 2010) enables to study the interplay between the kinetic folding process and time-dependent changes of the folding landscape. The main idea is to compute a mapping between macrostates of successive folding landscapes and use this information to determine the initial population densities for successive kinetic simulations. In the case of co-transcriptional folding, an energy landscape for each RNA elongation step (adding a single nucleotide) is computed using `barriers`.

`BarMap` then constructs a mapping between the energy landscapes $\mathcal{L}_n \rightarrow \mathcal{L}_{n+1}$. Since a newly transcribed nucleotide cannot initially interact with the previously transcribed part, every minimum in $\mathcal{L}_n$ is appended by an unpaired base. In the easiest case, this new structure is a minimum in landscape $\mathcal{L}_{n+1}$, then it can be directly mapped. Alternatively, a heuristic is used to compute the next best local minimum conformation which, if still not found in $\mathcal{L}_{n+1}$, is mapped to the state with the least base pair distance. The three possible cases that result from this mapping are illustrated in Fig. 6.

Folding kinetics can now be simulated using `treekin` starting with the first landscape that has more than one macrostate. The amount of time should correspond to the elongation time of the polymerase. The distribution of populated minima after the simulation is then transfered to the successive landscape according to the mapping computed by `BarMap`. Again, a folding simulation is performed starting from these conditions. This interleaving sequence of kinetic folding and transfer of the population density to the successive landscape is done until the folding landscape of the full length sequence is reached. The amount of time the folding chain spends

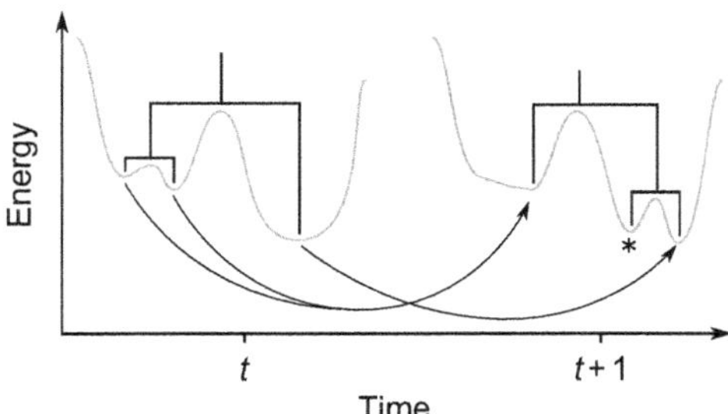

Figure 6 Schematic representation of the mapping process between two consecutive landscapes at time t and $t + 1$. Three types of events need to be distinguished: (i) A simple one-to-one correspondence between two local minima (right), (ii) two minima are merged into one (left), and (iii) a new minimum appears in $t + 1$ (*).

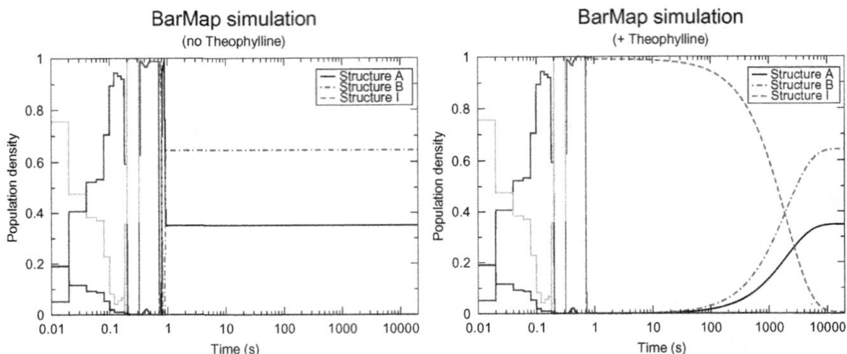

Figure 7 `BarMap` simulation of co-transcriptional folding with a transcription speed of approximately 50 nt/s. Structures A, B, and I correspond to the ones shown in Fig. 4. Both simulations start after transcribing the first 16 nt. The simulation on the landscape without theophylline reaches the equilibrium (with the riboswitch in the off-state) as soon as the last nucleotide is added. With theophylline, almost 100% of the RNA is in the on-state (structure I) at the end of the elongation period. The molecule then needs on the order of 1000 s to refold into the equilibrium off-state.

on a particular landscape (in the series) before being remapped allows to implement any type of coupling between the dynamics of the folding chain and the dynamics of the changing landscape.

The RS10 riboswitch introduced above is a good example for a system whose function can only be understood in view of its co-transcriptional folding behavior. A terminator hairpin can only be effective if it is formed almost immediately after the transcript reaches the poly-U tract adjacent to the hairpin. The interplay between the height of energy barriers and the speed of transcription is therefore crucial for riboswitch function. This fact also makes the design of such switches especially challenging. Wachsmuth et al. (2013) thus designed a series of candidates among which RS10 was the most effective.

Following Bremer and Dennis (1996), the transcription rate of *E. coli* polymerase is around 50 nt/s. Hence, we used the `BarMap` framework with an elongation time of 4000 in `treekin` units to compute co-transcriptional folding dynamics of RS10 in presence/absence of theophylline. The resulting population density for the different conformational states as a function of time can be seen in Fig. 7.

6. CONCLUSION

The well-known RNA structure prediction methods assume an RNA in thermodynamic equilibrium and are therefore of limited use for studying the conformational switching at the heart of riboswitch function. A number of approaches exist that characterize the folding landscape and the resulting dynamics of RNA molecules. While these are well suited to study mechanisms of riboswitch function, they are both computationally more demanding as well as more challenging for the user.

In particular, many riboswitches can only be understood in the context of co-transcriptional folding. The `BarMap` approach treats co-transcriptional folding as a process on a time-varying landscape. Using the example of a recently designed theophylline riboswitch, we show that this approach predicts riboswitch behavior in good agreement with experimental observations.

Several limitations remain in the computational approaches. (i) The energetics of RNA–ligand interactions cannot be predicted within the secondary structure model, although binding energies from experiments can be incorporated. (ii) In general, the structural prerequisites for ligand binding are not precisely known, making it difficult to judge which conformations along a folding pathway are binding competent. (iii) Secondary structure prediction ignores pseudoknots and tertiary interactions which can be essential for aptamer function. (iv) The most accurate computational methods are expensive and limited to moderate sequence lengths.

Nevertheless, the examples presented here illustrate that the secondary structure model captures enough detail of the molecular mechanism to provide a realistic picture of riboswitch function.

ACKNOWLEDGMENTS

This work was supported in part by the FWF International Programme I670, the DK RNA program FG748004, the EU-FET grant RiboNets 323987, and the COST Action CM1304 "Emergence and Evolution of Complex Chemical Systems."

REFERENCES

Bernhart, S., Tafer, H., Mückstein, U., Flamm, C., Stadler, P., & Hofacker, I. (2006). Partition function and base pairing probabilities of RNA heterodimers. *Algorithms for Molecular Biology*, *1*, 3.

Bremer, H., & Dennis, P. P. (1996). Modulation of chemical composition and other parameters of the cell by growth rate. *Escherichia coli and Salmonella: Cellular and Molecular Biology*, *2*, 1553–1569.

Danilova, L. V., Pervouchine, D. D., Favorov, A. V., & Mironov, A. A. (2006). RNAKinetics: A web server that models secondary structure kinetics of an elongating RNA. *Journal of Bioinformatics and Computational Biology*, *4*(2), 589–596.

Dawid, A., Cayrol, B., & Isambert, H. (2009, June). RNA synthetic biology inspired from bacteria: Construction of transcription attenuators under antisense regulation. *Physical Biology*, *6*(2), 025007.

Flamm, C., Fontana, W., Hofacker, I. L., & Schuster, P. (2000). RNA folding at elementary step resolution. *RNA*, *6*, 325–338.

Flamm, C., & Hofacker, I. L. (2008). Beyond energy minimization: Approaches to the kinetic folding of RNA. *Monatshefte für Chemie*, *139*(4), 447–457.

Flamm, C., Hofacker, I. L., Stadler, P. F., & Wolfinger, M. T. (2002). Barrier trees of degenerate landscapes. *Zeitschrift für physikalische Chemie*, *216*, 155–173.

Giegerich, R., Haase, D., & Rehmsmeier, M. (1999). Prediction and visualization of structural switches in RNA. *Pacific Symposium on Biocomputing*, 126–137.

Gillespie, D. T. (1977). Exact stochastic simulation of coupled chemical reactions. *The Journal of Physical Chemistry*, *81*, 2340–2361.

Gouda, H., Kuntz, I. D., Case, D. A., & Kollman, P. A. (2003). Free energy calculations for theophylline binding to an RNA aptamer: Comparison of MM-PBSA and thermodynamic integration methods. *Biopolymers*, *68*(1), 16–34.

Gruber, A. R., Lorenz, R., Bernhart, S. H., Neuböck, R., & Hofacker, I. L. (2008). The Vienna RNA websuite. *Nucleic Acids Research*, *36*, W70–W74.

Hofacker, I. L., Flamm, C., Heine, C., Wolfinger, M. T., Scheuermann, G., & Stadler, P. F. (2010). BarMap: RNA folding on dynamic energy landscapes. *RNA*, *16*, 1308–1316.

Hofacker, I. L., Fontana, W., Stadler, P. F., Bonhoeffer, S., Tacker, M., & Schuster, I. P. (1994). Fast folding and comparison of RNA secondary structures (the Vienna RNA Package). *Monatshefte für Chemie*, *125*(2), 167–188.

Huang, J., & Voß, B. (2014). Analysing RNA-kinetics based on folding space abstraction. *BMC Bioinformatics*, *15*, 60.

Isaacs, F. J., Dwyer, D. J., Ding, C., Pervouchine, D. D., Cantor, C. R., & Collins, J. J. (2004, July). Engineered riboregulators enable post-transcriptional control of gene expression. *Nature Biotechnology*, *22*(7), 841–847.

Isambert, H., & Siggia, E. D. (2000). Modeling RNA folding paths with pseudoknots: Application to hepatitis delta virus ribozyme. *Proceedings of the National Academy of Sciences USA*, *97*(12), 6515–6520.

Jenison, R. D., Gill, S. C., Pardi, A., & Polisky, B. (1994, March). High-resolution molecular discrimination by RNA. *Science*, *263*(5152), 1425–1429.

Jucker, F. M., Phillips, R. M., McCallum, S. A., & Pardi, A. (2003). Role of a heterogeneous free state in the formation of a specific RNA-theophylline complex. *Biochemistry*, *42*(9), 2560–2567.

Kucharik, M., Hofacker, I. L., Stadler, P. F., & Qin, J. (2014, July). Basin hopping graph: A computational framework to characterize RNA folding landscapes. *Bioinformatics*, *30*(14), 2009–2017.

Lai, D., Proctor, J. R., & Meyer, I. M. (2013). On the importance of cotranscriptional RNA structure formation. *RNA*, *19*(11), 1461–1473.

Lorenz, R., Bernhart, S. H., Siederdissen, C., Höner zu, Tafer, H., Flamm, C., Stadler, P. F., et al. (2011). ViennaRNA package 2.0. *Algorithms for Molecular Biology*, *6*, 26.

Lorenz, R., Flamm, C., & Hofacker, I. L. (2009). 2D projections of RNA folding landscapes. In I. Grosse, S. Neumann, S. Posch, F. Schreiber, & P. Stadler (Eds.), *German conference on bioinformatics 2009 (Vol. 157, pp. 11–20). Bonn: Gesellschaft f.* Informatik.

Markham, N. R., & Zuker, M. (2008). Unafold: Software for nucleic acid folding and hybridization. *Methods in Molecular Biology*, *453*, 3–31.

Mathews, D. H. (2014). RNA secondary structure analysis using RNA structure. *Current Protocols in Bioinformatics*, *46*, 12.6.1–12.6.25.

Mathews, D. H., Disney, M. D., Childs, J. L., Schroeder, S. J., Zuker, M., & Turner, D. H. (2004). Incorporating chemical modification constraints into a dynamic programming algorithm for prediction of RNA secondary structure. *Proceedings of the National Academy of Sciences USA*, *101*(19), 7287–7292.

Mathews, D. H., Sabina, J., Zuker, M., & Turner, D. H. (1999). Expanded sequence dependence of thermodynamic parameters improves prediction of RNA secondary structure. *Journal of Molecular Biology*, *288*(5), 911–940.

Mironov, A. A., & Lebedev, V. F. (1993). A kinetic modle of RNA folding. *BioSystems*, *30*, 49–56.
Mückstein, U., Tafer, H., Bernhart, S. H., Hernandez-Rosales, M., Vogel, J., Stadler, P. F., et al. (2008). Translational control by RNA–RNA interaction: Improved computation of RNA–RNA binding thermodynamics. In M. Elloumi, J. Küng, M. Linial, R. Murphy, K. Schneider, & C. Toma (Eds.), *Bioinformatics research and development (Vol. 13, pp. 114–127)*: Springer.
Nudler, E., & Mironov, A. S. (2004). The riboswitch control of bacterial metabolism. *Trends in Biochemical Sciences*, *29*(1), 11–17.
Pan, T., & Sosnick, T. (2006). RNA folding during transcription. *Annual Review of Biophysics and Biomolecular Structure*, *35*, 161–175.
Qi, L., Lucks, J. B., Liu, C. C., Mutalik, V. K., & Arkin, A. P. (2012, July). Engineering naturally occurring trans-acting non-coding RNAs to sense molecular signals. *Nucleic Acids Research*, *40*(12), 5775–5786.
Rodrigo, G., Landrain, T. E., Majer, E., Daros, J.-A., & Jaramillo, A. (2013, August). Full design automation of multi-state RNA devices to program gene expression using energy-based optimization. *PLoS Computational Biology*, *9*(8), e1003172.
Sauerwine, B., & Widom, M. (2013). Folding kinetics of riboswitch transcriptional terminators and sequesterers. *Entropy*, *15*(8), 3088–3099.
Serganov, A., & Nudler, E. (2013). A decade of riboswitches. *Cell*, *152*(1–2), 17–24.
Turner, D. H., & Mathews, D. H. (2010, January). NNDB: The nearest neighbor parameter database for predicting stability of nucleic acid secondary structure. *Nucleic Acids Research*, *38*(suppl. 1), D280–D282.
Voss, B., Meyer, C., & Giegerich, R. (2004). Evaluating the predictability of conformational switching in RNA. *Bioinformatics*, *20*(10), 1573–1582.
Wachsmuth, M., Findeiß, S., Weissheimer, N., Stadler, P. F., & Mörl, M. (2013). De novo design of a synthetic riboswitch that regulates transcription termination. *Nucleic Acids Research*, *41*(4), 2541–2551.
Waldminghaus, T., Kortmann, J., Gesing, S., & Narberhaus, F. (2008). Generation of synthetic RNA-based thermosensors. *Biological Chemistry*, *389*, 1319–1326. http://dx.doi.org/10.1515/BC.2008.150.
Wolfinger, M. T., Svrcek-Seiler, W. A., Flamm, C., Hofacker, I. L., & Stadler, P. F. (2004). Efficient computation of RNA folding dynamics. *Journal of Physics A: Mathematical and General*, *37*, 4731–4741.
Wong, T. N., Sosnick, T. R., & Pan, T. (2011). Folding of non-coding RNAs during transcription facilitated by pausing-induced non-native structures. *Proceedings of the National Academy of Sciences USA*, *104*, 17995–18000.
Wuchty, S., Fontana, W., Hofacker, I. L., & Schuster, P. (1999). Complete suboptimal folding of RNA and the stability of secondary structures. *Biopolymers*, *49*(2), 145–165.
Zuker, M. (1989). On finding all suboptimal foldings of an RNA molecule. *Science*, *244*, 48–52.

CHAPTER NINE

Integrating Molecular Dynamics Simulations with Chemical Probing Experiments Using SHAPE-FIT

Serdal Kirmizialtin[*,†,1], Scott P. Hennelly[*,†], Alexander Schug[‡], Jose N. Onuchic[§,¶,||,#,], Karissa Y. Sanbonmatsu[*,†,1]**

[*]New Mexico Consortium, Los Alamos, New Mexico, USA
[†]Theoretical Biology and Biophysics, Theoretical Division, Los Alamos National Laboratory, Los Alamos, New Mexico, USA
[‡]Steinbuch Centre for Computing, Karlsruhe Institute of Technology, Karlsruhe, Germany
[§]Center for Theoretical Biological Physics, Rice University, Houston, Texas, USA
[¶]Department of Physics and Astronomy, Rice University, Houston, Texas, USA
[||]Department of Chemistry, Rice University, Houston, Texas, USA
[#]Department of Biosciences, Rice University, Houston, Texas, USA
[**]Department of Biochemistry and Cell Biology, Rice University, Houston, Texas, USA
[1]Corresponding authors: e-mail address: serdal@lanl.gov; kys@lanl.gov

Contents

Abstract

Integration and calibration of molecular dynamics simulations with experimental data remain a challenging endeavor. We have developed a novel method to integrate chemical probing experiments with molecular simulations of RNA molecules by using a native structure-based model. Selective 2′-hydroxyl acylation by primer extension (SHAPE)

Methods in Enzymology, Volume 553
ISSN 0076-6879
http://dx.doi.org/10.1016/bs.mie.2014.10.061

characterizes the mobility of each residue in the RNA. Our method, SHAPE-FIT, automatically optimizes the potential parameters of the force field according to measured reactivities from SHAPE. The optimized parameter set allows simulations of dynamics highly consistent with SHAPE probing experiments. Such atomistic simulations, thoroughly grounded in experiment, can open a new window on RNA structure–function relations.

1. INTRODUCTION

Molecular dynamics simulations enable studies of biomolecules in atomic resolution. Over the past few decades, the predictive capability of this method has improved significantly due to the advances in hardware technologies (Shaw et al., 2008) and novel computational methods (Adcock & McCammon, 2006; Kirmizialtin & Elber, 2011; Laio & Parrinello, 2002; Sugita & Okomoto, 1999). These advances invite the development of more accurate force fields for biomolecular simulations. However, the development of highly accurate force field potential functions remains a challenge for molecular simulation. Many studies have been successful in producing dynamics consistent with NMR spectroscopy studies (Lange et al., 2008; Lipari, Szabo, & Levy, 1982; Maragakis et al., 2008; Zagrovic & Gunsteren, 2006). In addition to NMR studies, the development of nucleotide resolution chemical probing assays in the RNA community presents a new source of experimental data that can be used to benchmark and improve molecular simulation force fields. (Merino, Wilkinson, Coughlan, & Weeks, 2005; Soukup & Breaker, 1999).

From a biochemical perspective, RNA has the advantage over proteins in being amenable to reverse transcription readout assays, yielding information at nucleotide resolution. These assays were used extensively in ribosome studies to determine the ribosome secondary structure, binding sites, and conformational changes (Moazed & Noller, 1986, 1989; Woese et al., 1980). The development of in-line probing in the riboswitch community by Breaker and coworkers enabled readout of backbone mobility (Soukup & Breaker, 1999). Selective 2′-hydroxyl acylation by primer extension (SHAPE) was developed by Weeks and coworkers (Merino et al., 2005). This method is a rapid assay capable of backbone mobility readout at nucleotide resolution for a variety of environmental conditions (e.g., magnesium titration). While NMR spectroscopy studies produce superb data sets monitoring RNA mobility (Blanchard & Puglisi, 2001; Chen, Zuo, Wang, & Dayie, 2012; Clore & Kuszewski, 2003; Davis, Foster, Tonelli, & Butcher, 2007; Eichhorn et al., 2012; Fourmy, Recht,

Blanchard, & Puglisi, 1996; Gherghe, Shajani, Wilkinson, Varani, & Weeks, 2008; Hall, 2008; Proctor et al., 2004; Showalter & Hall, 2002; Zhang, Kang, Peterson, & Feigon, 2011), SHAPE allows one to obtain mobility information in experiments over the course of a few days and also for very large RNA systems (Fig. 1). This technique has opened the door to studies using a wide variety of environmental conditions, mutation sequences, and system sizes (Hennelly & Sanbonmatsu, 2011). This technique is a powerful, widespread method in the RNA community that has produced important experimental data sets for comparison with molecular simulations. Weeks and coworkers have used SHAPE probing to generate three-dimensional structural models of the tRNA based on a three-bead model. Here, we investigated dynamics and calibrate dynamics with chemical probing reactivity measurements (Gherghe, Leonard, Ding, Dokholyan, & Weeks, 2009).

From the perspective of RNA molecular simulations, important advances have been made in recent years regarding force field parameters for all-atom explicit solvent molecular dynamics simulations (Hart et al., 2012; Zgarbova et al., 2013). Few studies have compared RNA simulation with experiment in a detailed manner including a recent PreQ riboswitch study (Eichhorn et al., 2012; Feng, Walter, & Brooks, 2011; Sarkar, 2009; Sarkar, Nguyen, & Gruebele, 2010) and studies of small-angle X-ray scattering (SAXS) (Kirmizialtin, Pabit, Meisburger, Pollack, & Elber, 2012; Meisburger et al., 2013). While these studies are essential for improving force fields, their high computational costs limit their sampling capability and therefore affect the accuracy of the entropic component of the free energy. Specifically, the functional dynamics of many RNA systems occurs on the timescale of hundreds of milliseconds to seconds (Al-Hashimi & Walter, 2008; Blanchard, 2009). While large-scale simulations have produced millisecond simulations of small proteins (Shaw et al., 2010) and microsecond simulations of large systems (Whitford, Blanchard, Cate, & Sanbonmatsu, 2013), current computing capabilities prevent all-atom explicit solvent molecular dynamics simulations from accessing the physiological timescales of 100 ms–1 s.

To improve molecular simulation sampling, structure-based potentials have been used (Lutz, Faber, Verma, Klumpp, & Schug, 2014; Lutz, Sinner, Heuermann, Verma, & Schug, 2013; Noel, Whitford, Sanbonmatsu, & Onuchic, 2010; Ratje et al., 2010; Whitford, Geggier, et al., 2010; Whitford, Noel, et al., 2009; Whitford, Onuchic, & Sanbonmatsu, 2010; Whitford, Schug, et al., 2009). This potential is defined by the crystallographic structure and has the advantage of preserving stereochemistry in the crystallographic structure while sampling hundreds of

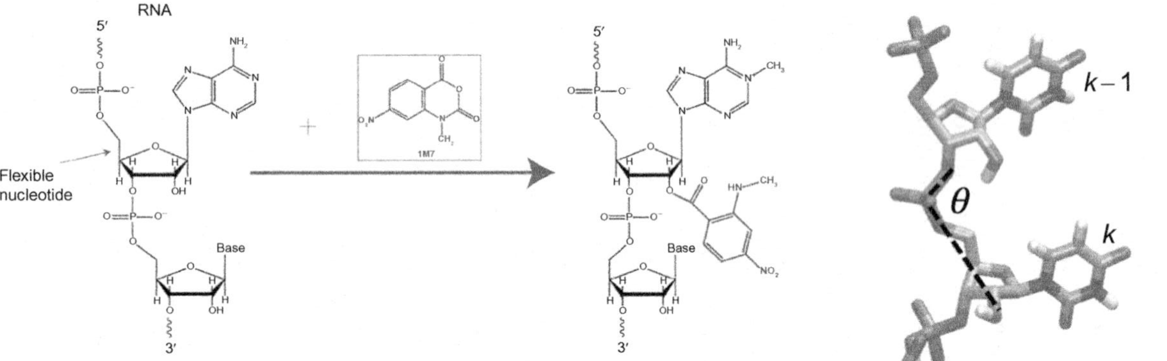

Figure 1 Detecting nucleotide mobility experimentally and computationally. (A) Schematic for the acylation reaction and the 2′-hydroxyl group of an RNA nucleotide with the SHAPE reagent (1M7). The acylation reaction is more probable when backbone is mobile and base is unpaired. (B) Mobility of the 2′-hydroxyl group is characterized in molecular dynamics simulations using the RMS fluctuations of the angle between the 2′-hydroxyl group, phosphate group, and the 5′-oxygen.

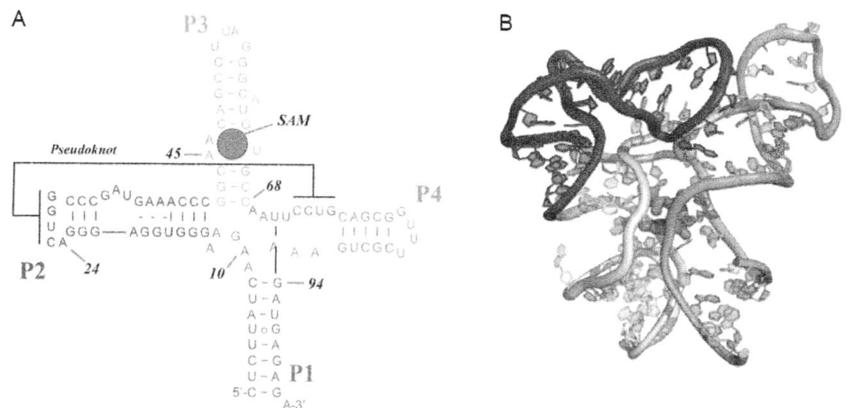

Figure 2 The *T. tengcongensis metF* SAM-I riboswitch aptamer domain in the off state. (A) Secondary structure of the aptamer domain with different colors representing secondary structure elements. (B) Tertiary structure of the sequence in the presence of *S*-adenosylmethionine (SAM) ligand. (See the color plate.)

milliseconds. The method allows reproducibly folding and unfolding small to medium size protein and nucleic acid structures, hence dramatically improving sampling and therefore the accuracy of the entropic component of the free energy. An additional advantage is that the potential is robust to changes in parameters, enabling calibration to experimental data while leaving the stereochemistry intact.

In this chapter, we present SHAPE-FIT, a novel technique to automatically calibrate molecular simulations to RNA chemical probing experiments. We demonstrate this method on the *Thermoanaerobacter tengcongensis metF* *S*-adenosylmethionine (SAM-I) riboswitch aptamer domain (Fig. 2), a useful test system that has previously been studied using a variety of experimental and computational techniques. Our method is easily extendable to large RNA systems. The approach can also be combined with explicit water all-atom simulations. SHAPE data integrated with molecular simulations improve the force field and produce mechanistic studies of RNA systems grounded in experimental data.

2. MATERIALS AND METHODS

2.1. Computation of SHAPE reactivity

SHAPE reactivity is inversely correlated with the base stability. The acylation reaction rate is higher if the nucleotide is mobile or easily accessible to the probing molecule (Fig. 1). However, the exact relationship between the SHAPE reactivity and stability is not yet known. To account for the relationship between the base stability and SHAPE reactivity Deigan, Li, Mathews, and Weeks (2009) proposed a pseudo-energy function. Here, the stability of nucleotide k is given by the relation $\Delta G(k) = m\ln(a(k)+1)+n$, where $a(k)$ is the normalized SHAPE reactivity of nucleotide k, and m and n are fitting parameters fitted to $m = 2.6$ kcal/mol and $n = 0.8$ kcal/mol using *Escherichia coli* 16S ribosomal RNA SHAPE data. Here in our study, we will use the same pseudo-energy term and the coefficients.

To account for the base stability and backbone mobility, we use the fluctuations in the angle between the 2′-hydroxyl group, the phosphate, and the adjacent O5′ group (O2′-P-O5′) (see Fig. 1). This angle characterizes the mobility of the 2′-hydroxyl atom and is relevant to both SHAPE and in-line probing reaction geometry. Effective stability is characterized by the fluctuations as

$$\Delta G_{\mathrm{SIM}}(k) = k_B T \ln \frac{\langle \theta_k^2 \rangle - \langle \theta_k \rangle^2}{\sum_{i=1}^{N} \left(\langle \theta_i^2 \rangle - \langle \theta_i \rangle^2 \right)/N} \tag{1}$$

Here, the stability of nucleotide k is computed from the fluctuation of this angle where $\langle \ldots \rangle$ represents the ensemble average computed from the time series of the simulation trajectory. The value is normalized with the average fluctuation of the RNA chain of length N nucleotides.

Combining the pseudo-energy term above with Eq. (1), we obtain the computed reactivity as

$$a_{\mathrm{SIM}}(k) = \exp\left(\frac{\Delta G_{\mathrm{SIM}}(k) - n}{m} \right) - 1 \tag{2}$$

Note that neither the choice of the order parameter nor the formulation of the stability is unique.

2.2. Optimization of potential energy function

To integrate SHAPE reactivity into the structure-based potential, we optimize the native structure-based potential (SBM) (Lutz et al., 2014, 2013; Noel et al., 2010; Ratje et al., 2010; Whitford, Geggier, et al., 2010; Whitford, Noel, et al., 2009; Whitford, Onuchic, et al., 2010; Whitford, Schug, et al., 2009) by steepest-descent search in parameter space. The basic function of the previously defined SBM potential (Eq. 3) is a summation of harmonic potentials restraining the bond lengths, bond angles, and dihedral angles to the native state structure that is given *a priori* by X-ray or NMR studies:

$$E = \sum_{\text{bonds}} K_r(r - r_0)^2 + \sum_{\text{angles}} K_\theta(\theta - \theta_0)^2 + \sum_{\text{dihedrals}} K_\phi^{(n)}[1 - \cos(n \times (\phi - \phi_0))] + \sum_{\text{impropers/planars}} (\chi_i - \chi_0)^2 + \sum_{i<j-3} \left\{ \varepsilon(i,j)\left[\left(\frac{\sigma_{ij}}{r_{ij}}\right)^{12} - 2\left(\frac{\sigma_{ij}}{r_{ij}}\right)^{6}\right] + \varepsilon_2(i,j)\left(\frac{\sigma_{ij}}{r_{ij}}\right)^{12} \right\} \tag{3}$$

The interactions between atoms that are not bonded are represented by a repulsive term that accounts for excluded volume of the polymer and an attractive term is used to account for the native interactions dictated by the structure. Native interactions are defined as contact pairs with a simple cutoff distance of $r \leq r_c = 4$ Å for nucleic acids and 6 Å for proteins. We emphasize that while more elaborate definitions for contacts exist (Noel, Whitford, & Onuchic, 2012), the exact contact definition will not significantly affect local dynamics in this study. This choice is more critical for large-scale conformational transitions such as those found in protein folding. The functional form for native contacts is a Lennard–Jones potential where the minimum is set to reproduce the native structure (Whitford, Noel, et al., 2009). The barrier height of the dihedral potential $K_\phi^{(n)}$ and the strength of the nonbonding native interactions between atom pairs $\varepsilon(i,j) = (\varepsilon_i \varepsilon_j)^{1/2}$ assume a uniform weight in such a way that the ratio of the total nonbonded native interaction to the sum of all torsional angle contributions $\sum \varepsilon(i,j) / \sum K_\varphi^{(n)}$ is set to 4.

The results of simulations using the SBM potential (Eq. 3) depend on the choice of the set of parameters $\{\{K_r\}, \{K_\theta\}, \{K_\phi^{(n)}\}, \{\chi_i\}, \{\varepsilon\}, \{\varepsilon_2\}\}$. Here, the most sensitive parameters for dynamics and at the same time the least-known parameters to us are torsional angle parameters and nonbonded native interaction terms. For that purpose, we search for the parameter space for these set of parameters. Hence, our sequence-dependent parameter space is defined as $\pi = \{\{K_\phi^{(n)}\}, \{\varepsilon\}\}$.

Following Di Pierro and Elber (2013), we search the parameter space to minimize the difference between the experimental and simulation result of the same observable. The target function is defined as the distance between the two data sets:

$$\Psi(\pi) = (1/N) \sum_{k=1}^{N} \left[\ln a_{\text{SIM}}{}^{\pi}(k) - \ln a_{\text{EXP}}(k) \right]^2 \tag{4}$$

Here, $a^{\pi}_{\mathrm{SIM}}(k)$ denotes the computed SHAPE reactivity of nucleotide k with respect to the parameter set π, while $a_{\mathrm{EXP}}(k)$ is the reactivity of the same nucleotide in experiment. We aim to minimize the target function. Typically achieving this goal can be difficult. In some cases, oversimplification of the physical interactions by the potential makes it difficult to find a parameter set that reproduces the experimental data. In other cases, the sought-after parameter space becomes too large for an exhaustive search. Here, we tested the functional form of the SBM potential against SHAPE experiments. We reduce the parameter space by focusing only the torsional and native interactions. To effectively search this reduced space, we introduced the steepest-descent minimization algorithm by iteratively solving the following differential equation (see Eq. 5):

$$\pi_i = \pi_{i-1} - \alpha \nabla_{\pi} \Psi(\pi_{i-1}) \tag{5}$$

where π_i is the parameter set in iteration i and α is a scalar that determines the gradient step length. We note that the choice of α can be optimized to reduce the computational cost (Kirk, 2004). We use constant α at values of $0.01 \leq \alpha \leq 0.1$. The iterative procedure is terminated when both $\Psi(\pi_i) < \gamma_1$ and $|\Psi(\pi_{i-1}) - \Psi(\pi_i)| < \gamma_2$ are satisfied, where γ_1 and γ_2 are preselected positive numbers chosen as 0.2 and 0.05, respectively, in our study.

2.3. Molecular simulations

We studied wild-type sequences of the *T. tengcongensis metF* riboswitch aptamer domain (residues 1–101) in two different solution conditions: (i) riboswitch in the presence of cognate ligand SAM-I and (ii) in the absence of the ligand. Because crystallographic structures of these sequences were not available, homology models were constructed using our previously published RNA homology modeling techniques used to model the ribosome (Korostelev, Trakhanov, Laurberg, & Noller, 2006; Tung, Joseph, & Sanbonmatsu, 2002; Tung & Sanbonmatsu, 2004). Our model of the SAM-bound aptamer was derived directly from the crystallographic structure published by Batey and coworkers (Lu et al., 2010; Montange & Batey, 2006; Stoddard et al., 2010). To obtain the atomic model of the SAM-I free configuration, we used the Kratky plots published in earlier studies (Stoddard et al., 2010).

We performed all-atom simulations of the SAM-I riboswitch aptamer by using structure-based potential (Ratje et al., 2010; Whitford, Geggier, et al.,

2010; Whitford, Noel, et al., 2009; Whitford, Schug, et al., 2009). Simulations were performed using GROMACS 4.6 suit of programs (Hess, Kutzner, VanderSpoel, & Lindahl, 2008). We used leap-frog stochastic integrator with an inverse friction coefficient of 1 ps with a reference temperature of 77 K and a time step of 1 fs throughout our simulations. VdW interactions are computed with twin range cutoffs with a cutoff distance of 12 Å. To compute the SHAPE reactivity, we simulate each system and compute the ensemble average from the time trace of trajectories. Ideally, statistical errors decrease as the length of the simulation increases; however, the computational cost also increases. Here, we determine the optimal length of the simulation by computing the average effective stability (Eq. 1) of the RNA chain as a function of simulation length L, $C(L) = (1/LN)\int_0^L \sum_{k=1}^{N} \Delta G_k^{\mathrm{SIM}}(t)\,\mathrm{d}t$ and compute the standard error for the observable C. Our analysis indicated that $L = 5 \times 10^6$ steps gives less than %1 error in the C estimate. Therefore, 5×10^6 steps is used to obtain a statistically converged ensemble averages in our study. SHAPE reactivity from simulations is computed using Eqs. (1) and (2).

2.4. SHAPE probing experiments

We use SHAPE probing and in-line probing to follow changes in secondary and tertiary structure occurring upon ligand binding (Hennelly & Sanbonmatsu, 2011). We study these changes for *T. tengcongensis metF* aptamers. In-line probing measures the ability of more mobile bases to spontaneously cleave (Regulski & Breaker, 2008). SHAPE probing measures the ability of more mobile bases to react with the 1-methyl-7-nitroisatoic anhydride (1M7) reagent (Fig. 1). While the methods measure the same basic properties, we have found in-line probing to be less ambiguous in some structures. 1M7 is relatively insensitive to temperature and Mg^{2+} concentration.

2.5. Preparation of RNA systems

Synthetic DNA oligonucleotide (IDT, Coralville, IA) templates were PCR amplified for use as *in vitro* transcription templates. Transcription reactions were performed using Ampliscribe (EPICENTRE Biotechnologies, Madison, WI) high-yield T7 RNA polymerase transcription kits as per instructions.

2.6. SHAPE chemical probing

The aptamer domain RNA was folded at a concentration of 5 nM in 1 × HMK buffer (50 mM HEPES–KOH pH 8.0, 2 mM $MgCl_2$, 100 mM

KCl) and various concentrations of SAM. Thirty microliters of 60 m*M* 1M7 in DMSO was added to a 300 μl volume of RNA. The reaction proceeded for 5 min at 25 °C and was then precipitated by the addition of 0.1 volumes 3 *M* Na:Acetate pH 6.5, 75 μg glycogen, and 3 volumes EtOH. The recovered RNA was then subjected to reverse transcription analysis as described above. Capillary electrophoresis data were integrated by simultaneously fitting Gaussian curves to the whole trace using in-house scripts. The traces were normalized using residues whose reactivity does not change in response to SAM.

2.7. In-line chemical probing

All chemical probing reactions were performed a minimum of three times with and without SAM. SAM (NEB, Ipswich, MA) was added to a final concentration of 10 μ*M*. In-line probing was performed as previously described (Hennelly & Sanbonmatsu, 2011) in in-line probing buffer (50 m*M* Tris–HCl pH 8.3, 20 m*M* $MgCl_2$, 100 m*M* KCl) with the following modifications: RNAs were unfolded by heating to 90 °C in water for 2 min and then crash cooled on ice for 2 min followed by addition of buffer with or without SAM. In-line probing reactions were performed at RNA concentrations of 0.3 μ*M* for 40 h at 25 °C. Reactions were purified by precipitation with 3 volumes EtOH and 50 μg RNase-free glycogen (Ambion) and then analyzed by capillary electrophoresis.

2.8. Analysis of chemical probing reactions

In-line probing and SHAPE probing reactions were analyzed by fluorescently labeling of the RNA on the 3′-terminus. The labeled RNA was then diluted 1–5 μl into 20 μl of formamide, depending on recovery and labeling efficiency, followed by heating to 90 °C for 2 min prior to loading on an ABI Prism 310 genetic analyzer (Applied Biosystems) capillary electrophoresis system equipped with a laser-induced fluorescence detector. The reactions were electrokinetically injected at 12.5 kV for 30 s and run in POP-6 polymer (ABI) at 70 °C for 1 h. Sequencing reactions were performed by transcribing the RNA constructs in the presences of a low level of α-phosphorothioate nucleotides followed by iodine cleavage and capillary electrophoresis.

3. RESULTS

We performed molecular dynamics simulations of the SAM-I riboswitch aptamer wild-type sequence in the presence of SAM (Fig. 2A

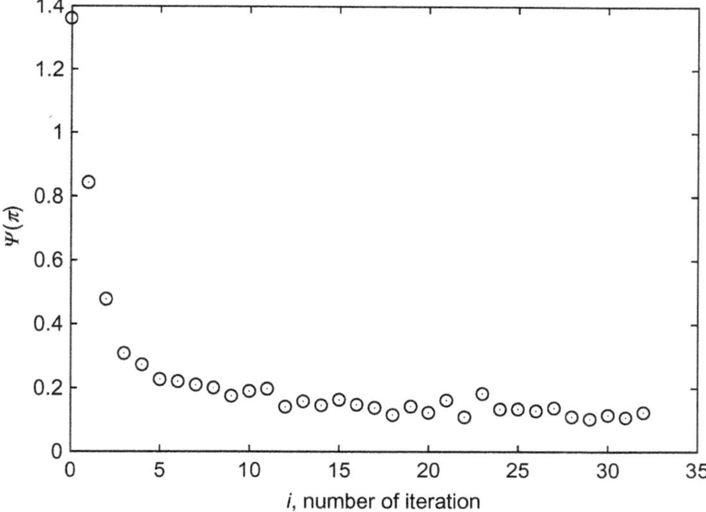

Figure 3 Evolution of the distance function $\Psi(\pi)$, quantifying agreement between simulated and measured SHAPE reactivities, as a function of the iteration number.

and B) with the unmodified potential energy function. Computed SHAPE reactivity is then used to calculate the distance function. We modify the potential as detailed in Section 2 to minimize the difference between the experimental and computed SHAPE data. Our iterative procedure effectively decreases the target function in each iteration. Figure 3 shows the change in the target function as a function of the number of iterations. The procedure is stopped when a convergence in the distance function is achieved, as there was no more improvement in the target function.

Figure 4A displays the reactivities based on fluctuations from the simulations with the optimized potential and normalized reactivity based on in-line probing experiments. We achieve excellent agreement between simulation and experiment. Comparison of the reactivities with the unmodified potential (see Fig. 5A) on the other hand significantly differs from the experiment, showing the success of our automatic procedure. We repeated the optimization procedure for the riboswitch in the absence of SAM-I. Our results are again in agreement (Fig. 4B) with the experiment and the data are quite different with the unmodified simulation results (see Fig. 5B). Despite the excellent agreement, there are still remaining questions: What is the reason for a nucleotide to show high reactivity while the other do not? What is the difference between liganded and unliganded states that leads to a significant change in the SHAPE spectra? What is the mode of dynamics in the region of high reactivities?

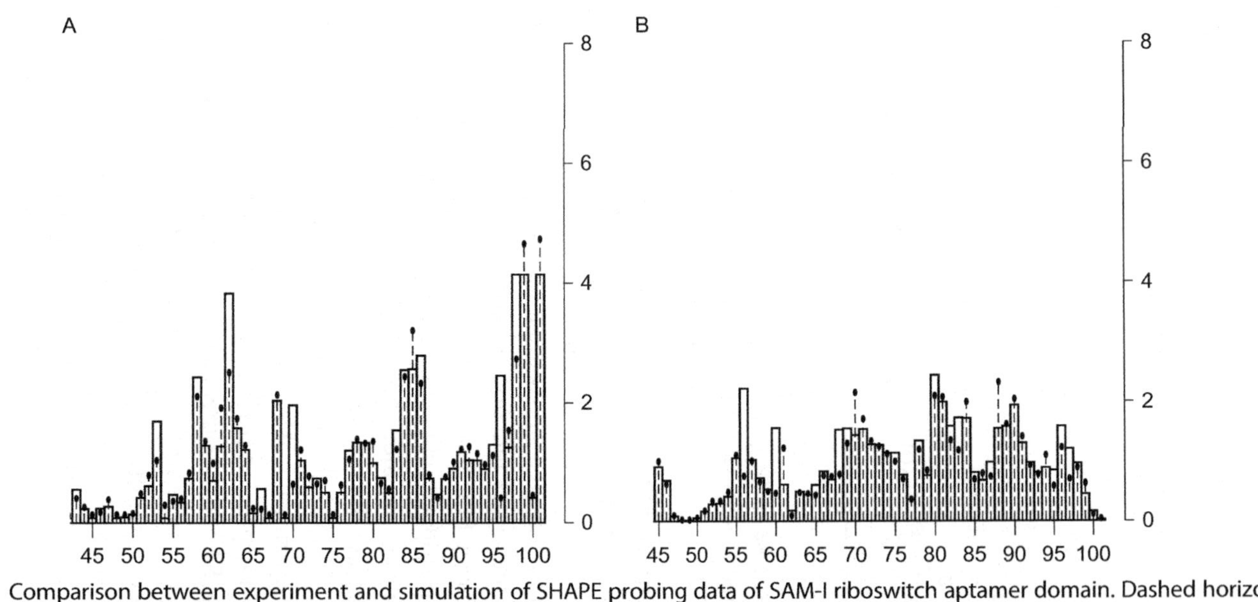

Figure 4 Comparison between experiment and simulation of SHAPE probing data of SAM-I riboswitch aptamer domain. Dashed horizontal lines with points are normalized activity from SHAPE reactivity measurements. Bars are activities computed from structure-based molecular dynamics simulations *after* optimization of the potential. Close agreement is achieved after the optimization of the potential (A) in the presence of SAM and (B) in the absence of it.

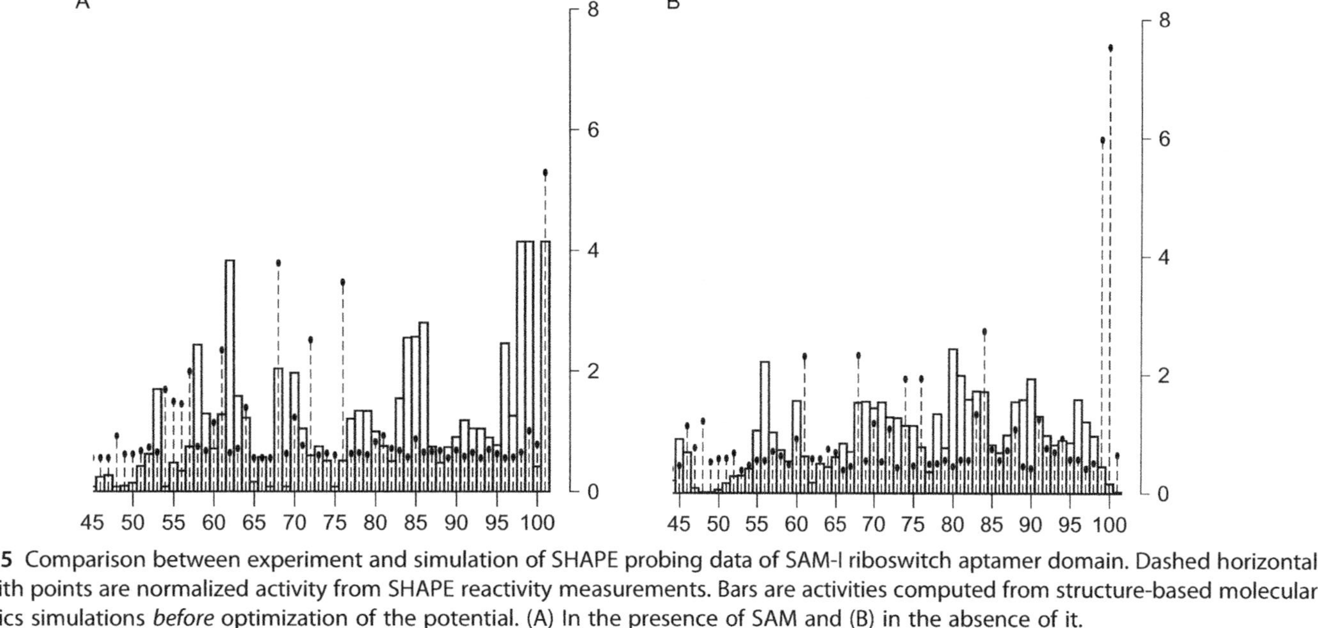

Figure 5 Comparison between experiment and simulation of SHAPE probing data of SAM-I riboswitch aptamer domain. Dashed horizontal lines with points are normalized activity from SHAPE reactivity measurements. Bars are activities computed from structure-based molecular dynamics simulations *before* optimization of the potential. (A) In the presence of SAM and (B) in the absence of it.

To accurately address the differences in the reactivities and to better understand the changes in mobility between the liganded and unliganded states, we used the trajectories from our optimized potentials. Figure 6A and B shows a conformation with bound SAM-I ligand. We report here the most populated conformation in the ensemble color coded according to the SHAPE reactivities. The most populated structure is obtained by clustering the trajectory with RMSD as distance measure. The cluster with the highest weight in the ensemble is the most populated state. We choose the cluster center configuration to be representative for the respective state. We observe that SHAPE reactivity strongly correlates with riboswitch topology. Unpaired regions are, in general, significantly more reactive relative to base paired regions. Regions on the exterior of the molecule tend to be more reactive than those in the core. Simulations that incorporate SHAPE data

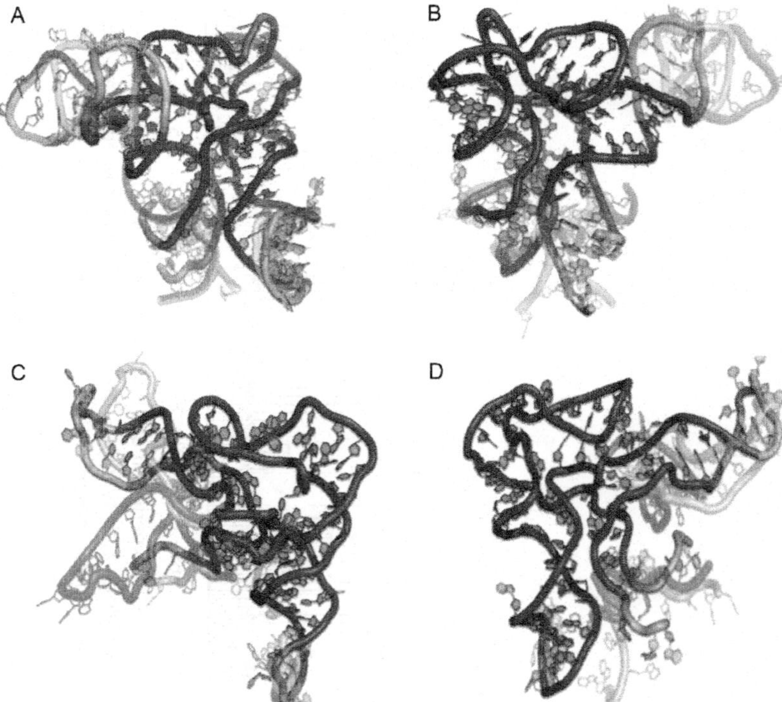

Figure 6 Comparison of SHAPE reactivities of SAM-I riboswitch aptamer domain with and without SAM-I. Residues are colored according to their SHAPE reactivity, blue (gray in the print version) is unreactive and red (dark gray in the print version) is highly reactive. The type of motion in reactive regions is depicted by overlying conformations (partially transparent). (A) The dynamics of riboswitch with bound ligand and (B) the same molecule rotated by 180°. (C and D) The dynamics of the riboswitch unliganded.

show the dynamics in highly reactive regions. For example, P1, which is close to the SAM-I ligand, remains formed during our simulations, yet its high mobility is due to the rotational motion of the domain in solution (see Fig. 6A and B). In contrast, P4, another highly reactive domain, goes back and forth toward P2 making contacts with it which results in exposing the buried residues to the solvent in the off state and hence increases the SHAPE reactivity.

A detailed comparison of the structure and dynamics in the *absence* of the ligand on the other hand depicts that the P1 domain is now unfolded (Fig. 6C and D). The high reactivity in this region is due to the resulting high mobility of the backbone chain as well as due to unpaired nucleotides exposed to the solution. The P4 domain, which also shows high reactivity, is now an extended helix. The structure that is dominated in our simulation with optimized potential is similar to that constructed by Lilley and coworkers (Lemay, Penedo, Tremblay, Lilley, & Lafontaine, 2006). Our structure also reveals a lack of pseudoknot tertiary interactions in the absence of the ligand. The lack of pseudoknot contacts leads to more extended conformations in the conformational ensemble increasing the radius of gyration by ~10%. This observation is also in accord with previous SAXS studies (Stoddard et al., 2010). Our simulation results are consistent with nucleotide analog interference modification studies and suggest that ligand promotes a global collapse of the aptamer domain by stabilizing tertiary contacts (Stoddard et al., 2010).

4. DISCUSSIONS

We have developed a systematic method to integrate molecular dynamics simulations with chemical probing experiments. The method produces simulated RNA fluctuations consistent with experiment at each nucleotide. The method called SHAPE-FIT hence offers a solution to fine-tune an energy function toward better agreement with experiment with applications in RNA structure and dynamics. While SHAPE probing has been extremely successful in predicting secondary structures and evaluating the mobility of RNA molecules, the exact relation between SHAPE reactivity, nucleotide mobility, and solvent exposure is not entirely understood. The order parameter we choose to monitor has successfully recapitulated the experimental data. Improvements may result from a clearer understanding of the SHAPE reaction. This may be achieved by, for example, via quantum chemical calculations of acylation reaction or understanding of RNA polymerase transcription mechanism.

Here in our study, we use the aptamer domain of the SAM-I riboswitch that is designed to fold to a single RNA native state in solution. This may not be the general case for the complete sequence that switches between on and off states. It may be also difficult to justify a single native state in many RNA systems as significant populations of alternative folds known to present in solution. Still, native interactions have been found to dominate RNA-folding (Behrouzi, Roh, Kilburn, Briber, & Woodson, 2012). A future direction would be to account for multiple native states in our potential. Multiple basin Go models have been successfully applied in protein folding (Okazaki, Koga, Takada, Onuchic, & Wolynes, 2006; Schug, Whitford, Levy, & Onuchic, 2007). Application of this model to SHAPE-FIT will be a future direction. Extensions to the method will also include nonnative interactions, explicit electrostatic interactions, and additional calibration of our model with thermodynamic melting studies.

While our method achieved excellent agreement with experimental data, our use of a local minimization procedure leads to a single solution of the parameter set. One could easily imagine that multiple solutions are possible. These different solutions may lead to different interpretations of the same data. Modeling multiple simultaneous solutions is an opportunity for future research. Another direction is to add multiple experiments to the target function in addition to SHAPE such as SAXS and melting temperatures. This would reduce the parameter space and result in a more robust potential across experimental data sets.

Overall, our method calibrates the structure-based potential to our experimental data, producing simulations consistent with chemical probing experiments. We note that our method is easily scalable to large systems. Thus, mechanistic studies of the folding and function of RNAs using these potentials will be consistent with SHAPE measurements, yielding conclusions grounded in experiment. Accordingly, predictions from such studies have the potential to be more precise. At the same time, these simulations would help interpret SHAPE measurements. Dynamics at each residue can be used for mechanistic interpretation of the SHAPE data as well as providing an atomic level interpretation of the reactivities based on fluctuations. The method has the potential to provide new mechanistic insights into the interplay between structure and dynamics, tightly integrating experiment and atomistic simulation to obtain an atomic level picture of RNA function.

ACKNOWLEDGMENTS

This work was supported by the Center for Theoretical Biological Physics sponsored by the National Science Foundation (Grant PHY-1427654), by the NSF (Grant MCB-1214457), and by the Human Frontiers Science Program. J. N. O. is a CPRIT Scholar in Cancer Research sponsored by the Cancer Prevention and Research Institute of Texas.

REFERENCES

Adcock, S. A., & McCammon, J. A. (2006). Molecular dynamics: Survey of methods for simulating the activity of proteins. *Chemical Reviews*, *106*(5), 1589–1615.

Al-Hashimi, H. M., & Walter, N. G. (2008). RNA dynamics: It is about time. *Current Opinion in Structural Biology*, *18*(3), 321–329.

Behrouzi, R., Roh, J. H., Kilburn, D., Briber, R. M., & Woodson, S. A. (2012). Cooperative tertiary interaction network guides RNA folding. *Cell*, *149*(2), 348–357.

Blanchard, S. C. (2009). Single-molecule observations of ribosome function. *Current Opinion in Structural Biology*, *19*(1), 103–109.

Blanchard, S. C., & Puglisi, J. D. (2001). Solution structure of the A loop of 23S ribosomal RNA. *Proceedings of the National Academy of Sciences of the United States of America*, *98*(7), 3720–3725.

Chen, B., Zuo, X., Wang, Y. X., & Dayie, T. K. (2012). Multiple conformations of SAM-II riboswitch detected with SAXS and NMR spectroscopy. *Nucleic Acids Research*, *40*(7), 3117–3130.

Clore, G., & Kuszewski, J. (2003). Improving the accuracy of NMR structures of RNA by means of conformational database potentials of mean force as assessed by complete dipolar coupling cross-validation. *Journal of the American Chemical Society*, *125*(6), 1518–1525.

Davis, J. H., Foster, T. R., Tonelli, M., & Butcher, S. E. (2007). Role of metal ions in the tetraloop-receptor complex as analyzed by NMR. *RNA*, *13*(1), 76–86.

Deigan, K. E., Li, T. W., Mathews, D. H., & Weeks, K. M. (2009). Accurate SHAPE-directed RNA structure determination. *Proceedings of the National Academy of Sciences of the United States of America*, *106*(1), 97–102.

Di Pierro, M., & Elber, R. (2013). Automated optimization of potential parameters. *Journal of Chemical Theory and Computation*, *9*(8), 3311–3320.

Eichhorn, C. D., Feng, J., Suddala, K. C., Walter, N. G., Brooks, C. L., 3rd., & Al-Hashimi, H. M. (2012). Unraveling the structural complexity in a single-stranded RNA tail: Implications for efficient ligand binding in the prequeuosine riboswitch. *Nucleic Acids Research*, *40*(3), 1345–1355.

Feng, J., Walter, N. G., & Brooks, C. L., 3rd. (2011). Cooperative and directional folding of the preQ1 riboswitch aptamer domain. *Journal of the American Chemical Society*, *133*(12), 4196–4199.

Fourmy, D., Recht, M. I., Blanchard, S. C., & Puglisi, J. D. (1996). Structure of the A site of Escherichia coli 16S ribosomal RNA complexed with an aminoglycoside antibiotic. *Science*, *274*(5291), 1367–1371.

Gherghe, C. M., Leonard, C. W., Ding, F., Dokholyan, N. V., & Weeks, K. M. (2009). Native-like RNA tertiary structures using a sequence-encoded cleavage agent and refinement by discrete molecular dynamics. *Journal of the American Chemical Society*, *131*(7), 2541–2546.

Gherghe, C. M., Shajani, Z., Wilkinson, K. A., Varani, G., & Weeks, K. M. (2008). Strong correlation between SHAPE chemistry and the generalized NMR order parameter (S2) in RNA. *Journal of the American Chemical Society*, *130*(37), 12244–12245.

Hall, K. B. (2008). RNA in motion. *Current Opinion in Chemical Biology*, *12*(6), 612–618.

Hart, K., Foloppe, N., Baker, C. M., Denning, E. J., Nilsson, L., & Mackerell, A. D., Jr. (2012). Optimization of the CHARMM additive force field for DNA: Improved treatment of the BI/BII conformational equilibrium. *Journal of Chemical Theory and Computation, 8*(1), 348–362.

Hennelly, S. P., & Sanbonmatsu, K. Y. (2011). Tertiary contacts control switching of the SAM-I riboswitch. *Nucleic Acids Research, 39*(6), 2416–2431.

Hess, B., Kutzner, C., VanderSpoel, D., & Lindahl, E. (2008). GROMACS 4: Algorithms for highly efficient, load-balanced, and scalable molecular simulation. *Journal of Chemical Theory and Computation, 4*(3), 435–447.

Kirk, D. E. (2004). Optimal control theory: An introduction. In *Dover books on electrical engineering*. Englewood Cliffs, N.J.: Prentice-Hall.

Kirmizialtin, S., & Elber, R. (2011). Revisiting and computing reaction coordinates with directional milestoning. *Journal of Physical Chemistry A, 115*(23), 6137–6148.

Kirmizialtin, S., Pabit, S. A., Meisburger, S. P., Pollack, L., & Elber, R. (2012). RNA and its ionic cloud: Solution scattering experiments and atomically detailed simulations. *Biophysical Journal, 102*(4), 819–828.

Korostelev, A., Trakhanov, S., Laurberg, M., & Noller, H. F. (2006). Crystal structure of a 70S ribosome–tRNA complex reveals functional interactions and rearrangements. *Cell, 126*(6), 1065–1077.

Laio, A., & Parrinello, M. (2002). Escaping free-energy minima. *Proceedings of the National Academy of Sciences of the United States of America, 99*(20), 12562–12566.

Lange, O. F., Lakomek, N. A., Farès, C., Schröder, G. F., Walter, K. F., Becker, S., et al. (2008). Recognition dynamics up to microseconds revealed from an RDC-derived ubiquitin ensemble in solution. *Science, 320*(5882), 1471–1475.

Lemay, J. F., Penedo, J. C., Tremblay, R., Lilley, D. M., & Lafontaine, D. A. (2006). Folding of the adenine riboswitch. *Chemistry & Biology, 13*(8), 857–868.

Lipari, G., Szabo, A., & Levy, R. M. (1982). Protein dynamics and NMR relaxation: Comparison of simulations with experiment. *Nature, 300*(5888), 197–198.

Lu, C., Ding, F., Chowdhury, A., Pradhan, V., Tomsic, J., Holmes, W. M., et al. (2010). SAM recognition and conformational switching mechanism in the *Bacillus subtilis* yitJ S box/SAM-I riboswitch. *Journal of Molecular Biology, 404*(5), 803–818.

Lutz, B., Faber, M., Verma, A., Klumpp, S., & Schug, A. (2014). Differences between cotranscriptional and free riboswitch folding. *Nucleic Acids Research, 42*(4), 2687–2696.

Lutz, B., Sinner, C., Heuermann, G., Verma, A., & Schug, A. (2013). eSBMTools 1.0: Enhanced native structure-based modeling tools. *Bioinformatics, 29*(21), 2795–2796.

Maragakis, P., Lindorff-Larsen, K., Eastwood, M. P., Dror, R. O., Klepeis, J. L., Arkin, I. T., et al. (2008). Microsecond molecular dynamics simulation shows effect of slow loop dynamics on backbone amide order parameters of proteins. *The Journal of Physical Chemistry B, 112*(19), 6155–6158.

Meisburger, S. P., Sutton, J. L., Chen, H., Pabit, S. A., Kirmizialtin, S., Elber, R., et al. (2013). Polyelectrolyte properties of single stranded DNA measured using SAXS and single-molecule FRET: Beyond the wormlike chain model. *Biopolymers, 99*(12), 1032–1045.

Merino, E. J., Wilkinson, K. A., Coughlan, J. L., & Weeks, K. M. (2005). RNA structure analysis at single nucleotide resolution by selective 2′-hydroxyl acylation and primer extension (SHAPE). *Journal of the American Chemical Society, 127*(12), 4223–4231.

Moazed, D., & Noller, H. F. (1986). Transfer RNA shields specific nucleotides in 16S ribosomal RNA from attack by chemical probes. *Cell, 47*(6), 985–994.

Moazed, D., & Noller, H. F. (1989). Intermediate states in the movement of transfer RNA in the ribosome. *Nature, 342*(6246), 142–148.

Montange, R. K., & Batey, R. T. (2006). Structure of the S-adenosylmethionine riboswitch regulatory mRNA element. *Nature, 441*(7097), 1172–1175.

Noel, J. K., Whitford, P. C., & Onuchic, J. N. (2012). The shadow map: A general contact definition for capturing the dynamics of biomolecular folding and function. *Journal of Physical Chemistry B, 116*(29), 8692–8702.

Noel, J. K., Whitford, P. C., Sanbonmatsu, K. Y., & Onuchic, J. N. (2010). SMOG@ctbp: Simplified deployment of structure-based models in GROMACS. *Nucleic Acids Research, 38*(Web Server issue), W657–W661.

Okazaki, K., Koga, N., Takada, S., Onuchic, J. N., & Wolynes, P. G. (2006). Multiple-basin energy landscapes for large-amplitude conformational motions of proteins: Structure-based molecular dynamics simulations. *Proceedings of the National Academy of Sciences of the United States of America, 103*(32), 11844–11849.

Proctor, D. J., Ma, H., Kierzek, E., Kierzek, R., Gruebele, M., & Bevilacqua, P. C. (2004). Folding thermodynamics and kinetics of YNMG RNA hairpins: Specific incorporation of 8-bromoguanosine leads to stabilization by enhancement of the folding rate. *Biochemistry, 43*(44), 14004–14014.

Ratje, A. H., Loerke, J., Mikolajka, A., Brünner, M., Hildebrand, P. W., Starosta, A. L., et al. (2010). Head swivel on the ribosome facilitates translocation by means of intra-subunit tRNA hybrid sites. *Nature, 468*(7324), 713–716.

Regulski, E. E., & Breaker, R. R. (2008). In-line probing analysis of riboswitches. *Methods in Molecular Biology, 419*, 53–67.

Sarkar, K., Meister, K., Sethi, A., & Gruebele, M. (2009). Fast folding of an RNA tetraloop on a rugged energy landscape detected by a stacking-sensitive probe. *Biophysical Journal, 97*(5), 1418–1427.

Sarkar, K., Nguyen, D. A., & Gruebele, M. (2010). Loop and stem dynamics during RNA hairpin folding and unfolding. *RNA, 16*(12), 2427–2434.

Schug, A., Whitford, P. C., Levy, Y., & Onuchic, J. N. (2007). Mutations as trapdoors to two competing native conformations of the Rop-dimer. *Proceedings of the National Academy of Sciences of the United States of America, 104*(45), 17674–17679.

Shaw, D. E., Deneroff, M. M., Dror, R. O., Kuskin, J. S., Larson, R. H., Salmon, J. K., et al. (2008). Anton, a special-purpose machine for molecular dynamics simulation. *Communications of the ACM, 51*(7), 91–97.

Shaw, D. E., Maragakis, P., Lindorff-Larsen, K., Piana, S., Dror, R. O., Eastwood, M. P., et al. (2010). Atomic-level characterization of the structural dynamics of proteins. *Science, 330*(6002), 341–346.

Showalter, S. A., & Hall, K. B. (2002). A functional role for correlated motion in the N-terminal RNA-binding domain of human U1A protein. *Journal of Molecular Biology, 322*(3), 533–542.

Soukup, G. A., & Breaker, R. R. (1999). Relationship between internucleotide linkage geometry and the stability of RNA. *RNA, 5*(10), 1308–1325.

Stoddard, C. D., Montange, R. K., Hennelly, S. P., Rambo, R. P., Sanbonmatsu, K. Y., & Batey, R. T. (2010). Free state conformational sampling of the SAM-I riboswitch aptamer domain. *Structure, 18*(7), 787–797.

Sugita, Y., & Okomoto, Y. (1999). Replica-exchange molecular dynamics method for protein folding. *Chemical Physics Letters, 314*(1–2), 141–151.

Tung, C. S., Joseph, S., & Sanbonmatsu, K. Y. (2002). All-atom homology model of the *Escherichia coli* 30S ribosomal subunit. *Nature Structural Biology, 9*(10), 750–755.

Tung, C. S., & Sanbonmatsu, K. Y. (2004). Atomic model of the *Thermus thermophilus* 70S ribosome developed in silico. *Biophysical Journal, 87*(4), 2714–2722.

Whitford, P. C., Blanchard, S. C., Cate, J. H., & Sanbonmatsu, K. Y. (2013). Connecting the kinetics and energy landscape of tRNA translocation on the ribosome. *PLoS Computational Biology, 9*(3), e1003003.

Whitford, P. C., Geggier, P., Altman, R. B., Blanchard, S. C., Onuchic, J. N., & Sanbonmatsu, K. Y. (2010). Accommodation of aminoacyl-tRNA into the ribosome involves reversible excursions along multiple pathways. *RNA, 16*(6), 1196–1204.

Whitford, P. C., Noel, J. K., Gosavi, S., Schug, A., Sanbonmatsu, K. Y., & Onuchic, J. N. (2009). An all-atom structure-based potential for proteins: Bridging minimal models with all-atom empirical forcefields. *Proteins, 75*(2), 430–441.

Whitford, P. C., Onuchic, J. N., & Sanbonmatsu, K. Y. (2010). Connecting energy landscapes with experimental rates for aminoacyl-tRNA accommodation in the ribosome. *Journal of the American Chemical Society, 132*(38), 13170–13171.

Whitford, P. C., Schug, A., Saunders, J., Hennelly, S. P., Onuchic, J. N., & Sanbonmatsu, K. Y. (2009). Nonlocal helix formation is key to understanding S-adenosylmethionine-1 riboswitch function. *Biophysical Journal, 96*(2), L7–L9.

Woese, C. R., Magrum, L. J., Gupta, R., Siegel, R. B., Stahl, D. A., Kop, J., et al. (1980). Secondary structure model for bacterial 16S ribosomal RNA: Phylogenetic, enzymatic and chemical evidence. *Nucleic Acids Research, 8*(10), 2275–2293.

Zagrovic, B., & Gunsteren, F. v. W. (2006). Comparing atomistic simulation data with the NMR experiment: How much Can NOEs actually tell us? *Proteins, 63*, 210–218.

Zgarbova, M., Luque, F. J., Sponer, J., Cheatham, T. E., 3rd., Otyepka, M., & Jurečka, P. (2013). Toward improved description of DNA backbone: Revisiting epsilon and zeta torsion force field parameters. *Journal of Chemical Theory and Computation, 9*(5), 2339–2354.

Zhang, Q., Kang, M., Peterson, R. D., & Feigon, J. (2011). Comparison of solution and crystal structures of preQ1 riboswitch reveals calcium-induced changes in conformation and dynamics. *Journal of the American Chemical Society, 133*(14), 5190–5193.

CHAPTER TEN

Using Simulations and Kinetic Network Models to Reveal the Dynamics and Functions of Riboswitches

Jong-Chin Lin[*,†], **Jeseong Yoon**[‡], **Changbong Hyeon**[‡,1], **D. Thirumalai**[*,†]

[*]Institute for Physical Science and Technology, University of Maryland, College Park, MD, USA
[†]Department of Chemistry and Biochemistry, University of Maryland, College Park, MD, USA
[‡]School of Computational Sciences, Korea Institute for Advanced Study, Seoul Republic of Korea
[1]Corresponding author: e-mail address: hyeoncb@kias.re.kr

Contents

Abstract

Riboswitches, RNA elements found in the untranslated region, regulate gene expression by binding to target metaboloites with exquisite specificity. Binding of metabolites to the conserved aptamer domain allosterically alters the conformation in the downstream expression platform. The fate of gene expression is determined by the changes in the downstream RNA sequence. As the metabolite-dependent cotranscriptional folding and unfolding dynamics of riboswitches are the key determinant of gene expression, it is important to investigate both the thermodynamics and kinetics of riboswitches both in the presence and absence of metabolite. Single molecule force experiments that decipher the free energy landscape of riboswitches from their mechanical responses, theoretical and computational studies have recently shed light on the

Methods in Enzymology, Volume 553
ISSN 0076-6879
http://dx.doi.org/10.1016/bs.mie.2014.10.062

distinct mechanism of folding dynamics in different classes of riboswitches. Here, we first discuss the dynamics of water around riboswitch, highlighting that water dynamics can enhance the fluctuation of nucleic acid structure. To go beyond native state fluctuations, we used the Self-Organized Polymer model to predict the dynamics of add adenine riboswitch under mechanical forces. In addition to quantitatively predicting the folding landscape of add-riboswitch, our simulations also explain the difference in the dynamics between pbuE adenine- and add adenine-riboswitches. In order to probe the function *in vivo*, we use the folding landscape to propose a system level kinetic network model to quantitatively predict how gene expression is regulated for riboswitches that are under kinetic control.

1. INTRODUCTION AND SCOPE OF THE REVIEW

The remarkable discovery just over a decade ago that riboswitches, which are RNA elements in the untranslated regions in mRNA, control gene expression by sensing and binding target metabolites with exquisite sensitivity is another example of the versatility of RNA in controlling crucial cellular functions (Serganov & Nudler, 2013). In the intervening time, considerable insights into their functions have come from a variety of pioneering biochemical and biophysical experiments. In addition, determination of structures of a number of riboswitches has greatly aided in molecular understanding of their functions and in the design of synthetic riboswitches. Typically, riboswitches contain a conserved aptamer domain to which a metabolite binds, producing a substantial conformational change in the downstream expression platform leading to control of gene expression. The variability in the functions of structurally similar aptamer domain is remarkable. The *add* adenine (A) riboswitch activates translation upon binding the metabolite (purine) whereas the structurally similar *pbuE* adenine (A) riboswitch controls transcription (Mandal, Boese, Barrick, Winkler, & Breaker, 2003). We can classify both these as ON riboswitches, which means that translation or transcription is activated only upon binding of the metabolite. In contrast, OFF riboswitches (for example, flavin mononucleotide (FMN) binding aptamer) shut down gene expression when the metabolite binds to the aptamer domain. Finally, some of the riboswitches are under kinetic control (FMN riboswitch (Wickiser, Winkler, Breaker, & Crothers, 2005) and *pbuE* adenine riboswitch (Frieda & Block, 2012)) whereas others (for example, *add* adenine riboswitch and SAM-III riboswitch) may be under thermodynamic control.

The functions of riboswitches are vastly more complicated than indicated by *in vitro* studies, which typically focus on limited aspects of their activities. Under cellular conditions the metabolite which binds to the aptamer domain itself is a product of gene expression. Thus, regulation of gene expression involves negative or positive (or a combination) feedback. This implies that a complete understanding of riboswitch function must involve a system level description, which should minimally include the machinery of gene expression, rate of transcription or translation, degradation rates of mRNA, and activation rate of the synthesized metabolite as well as rate of binding (through feedback loop) to the aptamer domain. Many *in vitro* studies have dissected these multisteps into various components in order to quantify them as fully as possible. In this context, single molecule studies in which response of riboswitches to mechanical force are probed have been particularly insightful (Frieda & Block, 2012; Greenleaf, Frieda, Foster, Woodside, & Block, 2008).

More recently, computational and theoretical studies have been initiated to develop a quantitative description of the folding landscape and dynamics at the single molecule level (Allner, Nilsson, & Villa, 2013; Feng, Walter, & Brooks, 2011; J. Lin & Thirumalai, 2008; Quarta, Sin, & Schlick, 2012; Whitford et al., 2009). The findings have been combined to produce a framework for describing the function of riboswitches at the system level. Because atomically detailed simulations can only provide limited information of the dynamics in the folded aptamer states it is necessary to develop suitable coarse grained (CG) model for more detailed exploration of the dynamics. The CG model-based simulations of nucleic acids, first introduced by Hyeon and Thirumalai 2005, are particularly efficacious to deal with riboswitches that undergo large scale conformational fluctuations for functional purposes. The goal of this article is limited to a brief review of the insights molecular simulations have brought to the understanding of the folding landscape of riboswitches using purine-binding and *S*-adenosylmethionine (SAM) riboswitches as examples. We begin with the description of the fluctuations in the folded state of the small preQ1 riboswitch, which exhibits rich dynamics in the native state that can be probed using atomic detailed simulations in explicit water. The hydration dynamics could have a functional role, which can be resolved by spectroscopic experiments. We then describe the response of three classes of riboswitches to forces and map the entire folding landscape from which we have made testable predictions. The data from these free energy landscapes are used to construct a network model, which provides system level description of the OFF riboswitches (Fig. 1).

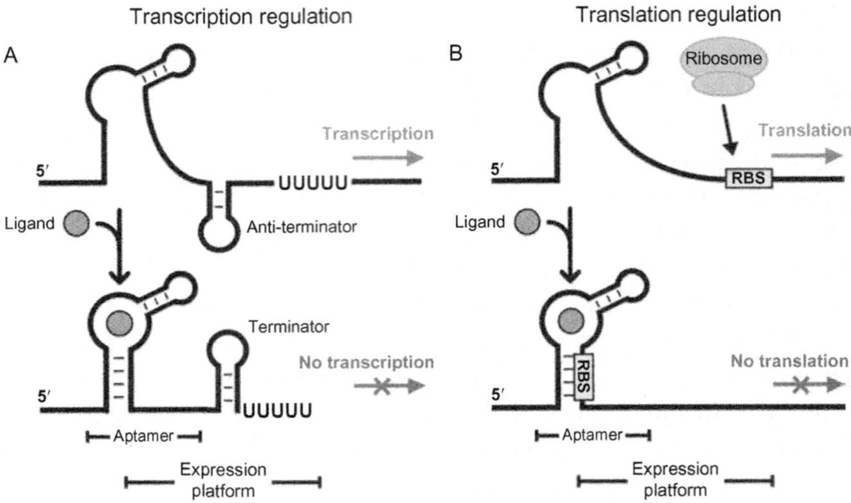

Figure 1 Principles of transcription and translation regulation by OFF-riboswitches. (A) Metabolite binding to the aptamer domain results in the formation of the terminator hairpin exposing a stretch of Uracil nucleotides. The polymerase disengages from the gene, thus terminating transcription. (B) Binding of the metabolite encrypts the Ribosome Binding Site (RBS) preventing the ribosome to initiate translation. *The figure adapted from Kim and Breaker (2008).*

2. HYDRATION DYNAMICS AROUND THE FOLDED STATE: ALL ATOM SIMULATIONS

It is known that unlike proteins there are many low free energy excitations (alternate structures) that a folded RNA can access. Consequently, dynamical fluctuations of the folded states are critical for RNA to execute their biological functions. There is growing evidence that hydration plays a key role in triggering conformational fluctuations in RNA. First, RNA can access low-lying excitation states via local melting of bases (Dethoff, Chugh, Mustoe, & Al-Hashimi, 2012; Jacob, Santer, & Dahlberg, 1987). Recent NMR studies suggest that a potential premelting of the hydration shell is required for the base pair disruption in response to elevation in temperature (Nikolova & Al-Hashimi, 2010; Rinnenthal, Klinkert, Narberhaus, & Schwalbe, 2010). Second, the versatile functional capacity of RNAs can be attributed to their ability to access alternative conformations (Zhang, Sun, Watt, & Al-Hashimi, 2006). Local conformational fluctuations from a few nanosecond dynamics enable RNA to explore a heterogeneous conformational ensemble, giving them the capacity to recognize and bind a diverse set of ligands. Third, binding of metabolites to riboswitches to control gene expression (Montange & Batey, 2008) may also be linked to local fluctuations in specific regions of cotranscriptional folded UTR regions of mRNA. In all of these examples, hydration of RNA is likely to play important role.

In order to illustrate the importance of hydration, we performed atomically detailed simulations of $PreQ_1$ riboswitch. We showed that water dynamics is spatially heterogeneous with metastable functionally relevant states whose dynamics spans many orders of magnitude. This behavior is reminiscent of glassy behavior (Thirumalai, Mountain, & Kirkpatrick, 1989). The glassy behavior of water molecules may indicate that RNA molecule is able to access low-lying free energy states around the putative folded state. We identified distinct classes of water molecules near the RNA surface, which can be classified as "bulk," "surface," "cleft," and "buried" water in the order of increasing water hydrogen bond relaxation time (Yoon, Lin, Hyeon, & Thirumalai, 2014). In this section, we review the molecular details of hydration around various regions of RNA and discuss how water dynamics gives rise to local structural fluctuations by using atomic simulations of $PreQ_1$-riboswitches (Yoon et al., 2014; Yoon, Thirumalai, & Hyeon, 2013).

2.1 Water hydrogen bond kinetics around nucleotides

The time- and ensemble-averaged autocorrelation function $c(t)$ (see definition in Luzar & Chandler, 1996; Yoon et al., 2014) are used to quantify the structure and dynamics of water molecules near the surface of RNA. The relaxation kinetics of the water HB at T = 310 K around three different nucleotide groups (B: base, R: ribose, P: phosphate) is fitted to a multi-exponential function $c_\xi(t)=\sum_{i=1}^{N}\phi_i e^{-t/\tau_i}$ where $\sum_{i=1}^{N}\phi_i=1$ with $N=4$ and ξ denotes B, P, or R (Fig. 2A, left). The time constant τ_i ranges from $\mathcal{O}(1)$ ps to $\mathcal{O}(10^4)$ ps, but 90% of kinetics is described by the dynamics of $\lesssim\mathcal{O}(10^2)$ ps (Fig. 2A, left). The average lifetime of the water HB at each nucleotide group ($\langle\tau_\xi\rangle=\int_0^\infty dt c_\xi(t)$) reveals that water molecules exhibit the slowest relaxation dynamics near bases instead of phosphate groups as $\langle\tau_P^w\rangle=$ 193 ps, $\langle\tau_R^w\rangle=$ 81 ps, $\langle\tau_B^w\rangle$ = 289 ps, and hence $\langle\tau_R^w\rangle<\langle\tau_P^w\rangle<\langle\tau_B^w\rangle$. It is worth pointing out that these time scales are much longer than found in proteins, and far exceed by a few orders of magnitude hydrogen bond dynamics in bulk water.

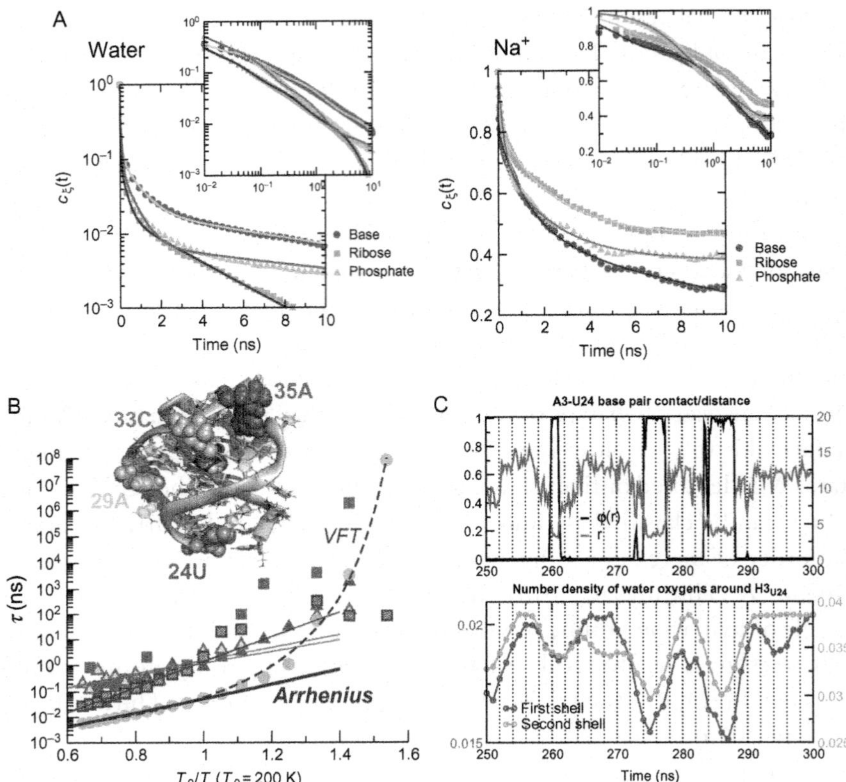

Figure 2 (A) The data for the dynamics of water HBs (left) and Na^+ (right) with base, ribose, and base (left) are described by the appropriate autocorrelation functions. (B) Temperature dependence of water HB relaxation time of bulk water and around four nucleotides, 24U, 29A, 33C, and 35A or $PreQ_1$. Arrhenius fits are made to $T > 200$ K for the four nucleotide and bulk water. Alternatively, the relaxation dynamics of bulk water HB can be fit using the Vogel–Fulcher–Tamman (VFT) equation over the full range of temperatures (dashed line) (Yoon et al., 2014). (C) Water dynamics induced fluctuations of a base pair. The status of base pair ($N1_{A3}$–$N3_{U24}$), quantified by calculating logistic function $\varphi(r) = (1 + e^{(r-r_0)/\sigma})^{-1}$ as well as distance r, shows apparent fluctuations, whose time scale is ~10 ns. *Figures adapted from Yoon et al. (2014, 2013).* (See the color plate.)

Excess monovalent counter-ions are distributed around RNA to neutralize the negative charges on the phosphodiester backbone of nucleic acids. The autocorrelation functions computed for Na^+ ions bound to P, R, and B (Fig. 2A, right) show that the time constant of ion relaxation is a few orders of magnitude greater than the water hydrogen bonds, $\langle \tau_P^{Na^+} \rangle = 294$ ns, $\langle \tau_R^{Na^+} \rangle = 63$ ns, $\langle \tau_B^{Na^+} \rangle = 9.2$ ns. The order of lifetime differs from that of water HB as $\langle\tau_B^{\mathrm{Na}^+}\rangle < \langle\tau_R^{\mathrm{Na}^+}\rangle < \langle\tau_P^{\mathrm{Na}^+}\rangle$. In contrast to water HB, monovalent counterions have the slowest dynamics near the phosphate group. Most importantly, while binding or release of a Na^+ ion to or from the surface of RNA certainly perturbs the water environment (Song, Franck, Pincus, Kim, & Han, 2014), the time scale separation between water and counterion dynamics ensures that the hydration dynamics around RNA occurs essentially in a static ionic environment.

2.2 Heterogeneity of Water Dynamics on the RNA Surface

The time scale of hydrated water varies many orders of magnitude depending on its location on the surface of RNA. Calculations of electrostatic potential on the solvent accessible surface confirm (Yoon et al., 2014, 2013) that the charge distribution on RNA surface is indeed not uniform but heterogeneous. Multiexponential function $c(t) = \sum_{i=1}^{N} \phi_i e^{-t/\tau_i}$ with different weights (ϕ_i) and well-separated time constants (τ_i) are needed to quantitate the relaxation dynamics of water molecules around four selected nucleotides of preQ$_1$-riboswitch, 24U, 29A, 33C, and 35A (Fig. 2C). The rich dynamics reflects the heterogeneity and justifies the interpretation that there are distinct class of water molecules, which can be divided into multiple classes such as "bulk," "surface," "cleft," and "buried" water (Yoon et al., 2014). At high temperatures, the population of fast, bulk water-like dynamics is dominant, but as the temperature decreases, the population of slow dynamics grows. The average lifetime of water molecules near RNA is at least 1–2 orders of magnitude slower than that of bulk water over the broad range of temperatures (Fig. 2C).

2.3 Water-induced fluctuations of base-pair dynamics

Dynamic feature of water that induce local conformational fluctuation of RNA is captured by probing the base pair dynamics along with surface water (Yoon et al., 2013). The space made of base stacks and base pairings is generally dry and hydrophobic, and thus devoid of any water molecules. However, in base pairs located at the end of stacks, it is possible to observe an enhanced fluctuation of base pair. Figure 2C shows the dynamics of base pairs A3-U24 located in the 5′- and 3′-end in preQ$_1$ riboswitch in aqueous solution. Remarkably, when the time series of water density around H3 of U24 and breathing dynamics of the base pair are compared, the change in water density always precedes the change in base-pair distance. The water densities calculated in the first and second solvation shell around H3 of

AU24 show that water population starts to increase before the base pair disruption; the decrease of water population always precedes the event of base-pair formation. Thus, we conclude that the dynamics of water hydration and dehydration induces the breathing dynamics of base pairs. The spontaneous fluctuations in base pair opening induced by water are important in protein–DNA interactions as well and may be responsible for transcription initiation by RNA polymerase.

3. STABILITY OF ISOLATED HELICES CONTROL THE FOLDING LANDSCAPES OF PURINE RIBOSWITCHES

A key event in the function of riboswitches is the conformational change in the aptamer domain leading to the formation of the terminator with the downstream expression platform (Fig. 1A) or sequestration of the ribosome binding site upon ligand binding (Fig. 1B). In order to assess the time scale in which such conformational change takes place and how it competes with ligand binding, it is first important to quantitatively map the folding landscapes of riboswitches. From such landscapes, the time scales for the conformational change in the switching region in the aptamer can be estimated (Hyeon, Morrison, & Thirumalai, 2008; J. Lin & Thirumalai, 2008).

In a pioneering experiment, Block and coworkers used single molecule pulling experiments to map the folding landscape as a function of the extension of the RNA. Purine (guanine and adenine) riboswitches are remarkably selective in their affinity for ligands and carry out markedly different functions despite the structural similarity of their aptamers. For the pbuE adenine (A) riboswitch, whose response to force was first probed in the LOT experiments, ligand binding activates the gene expression when an antiterminator is formed. In the absence of adenine, part of the aptamer region is involved in the formation of a terminator stem with the expression platform resulting in transcription termination. The add A-riboswitch activates the gene expression by forming a translational activator upon ligand binding. In the absence of adenine, the riboswitch adopts the structure with a translational repressor stem in the downstream region. At the heels of the first single molecule studies, we reported the entire folding landscape and calculated the time scale for switching of helix that engages in hairpin formation with the downstream sequence using the self-organized polymer (SOP) model (Hyeon, Dima, & Thirumalai, 2006; Hyeon & Thirumalai, 2007; J. Lin &

Thirumalai, 2008). As we show below comparison of the landscapes of these two riboswitches underscores the importance of the stability of the isolated helices in the assembly and rupture of the folded strauctureure.

Structures of purine riboswitch aptamers are characterized by a three-way junction consisting of P1, P2 and P3 helices, which are further stabilized by tertiary interactions in the folded state (Fig. 3A). For pbuE A-riboswitch, binding of metabolite (adenine) activates the gene expression by enabling the riboswitch to form an antiterminator. Without adenine, the molecule forms a terminator stem with the expression platform, resulting in transcription termination. On the other hand, the add A-riboswitch uses adenine to regulate the process of translation. Recent single molecule experiments (Greenleaf et al., 2008; Neupane, Daniel, Foster, Wang, & Woodside, 2011) and our simulation studies (J. Lin & Thirumalai, 2008; J.-C. Lin, Hyeon, & Thirumalai, 2014) have shown that, despite the marked structural

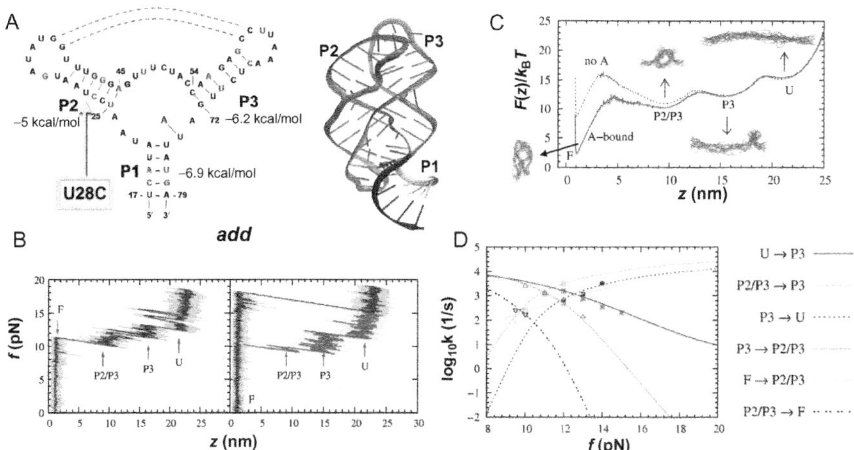

Figure 3 Force-induced dynamics of add A-riboswitch (RS). (A) Structure of the conserved domain of purine riboswitch containg a three way junction. On the left is the secondary structure map and on the right the three dimensional structure is shown. (B) Force-extension curves (FECs) obtained by pulling the RS at loading rate of 960 pN/s in the presence (left) and absence (right) of metabolite. The FEC in red on the right panel was obtained during the refolding of the RS while the exerted force is reduced. (C) Free energy profile *F*(*z*) with (red) and without (blue) the metabolite. (D) Force-dependent transition rates. The data points are directly from simulation; the lines were obtained by calculating mean first passage time using *F*(*z*) with a force-independent diffusion constant, which was calibrated by equating the theoretical and simulation rates. *Figure adapted from J. Lin and Thirumalai (2008).* (See the color plate.)

similarity, these two aptamers have different folding landscapes, thus providing a fingerprint of their function.

Single molecule optical tweezer experiments have been used to directly observe the hierarchical folding of both pbuE A- and add A-riboswitch aptamers (Greenleaf et al., 2008; Neupane et al., 2011). Here, we summarize force (f)-triggered unfolding and refolding of the A-riboswitch aptamer theoretically using Brownian dynamics simulations of the SOP model (Hyeon et al., 2006; Hyeon & Thirumalai, 2007). The crystal structure of add A-riboswitch (PDB id: 1Y26 (U17 to A79)) is available while that of pbuE A-riboswitch is not. However, since the sequence similarity between add-A and pbuE A-riboswitch is unusually high, we modeled the atomic structure of pbuE A-riboswitch by substituting the sequences of pbuE A-riboswitch into the crystal structure of add A-riboswitch and produced an ensemble of pbuE A-riboswitch structures via conformational sampling with molecular dynamics simulations (J.-C. Lin et al., 2014).

In the absence of adenine, our simulations show that force-induced unfoldings of both pbuE-A and add A-riboswitches occur in three distinct steps. Force extension curves of riboswitch generated under constant loading condition ($r_f = 960$ pN/s) reveal three distinct steps for both RS. Investigating the loss of secondary and tertiary contacts during the unfolding process, we found that the order of unfolding events differs qualitatively in add A-riboswitch and pbuE A-riboswitch. In add A-riboswitch, the unfolding occurred in the order of ΔP1$\rightarrow$ ΔP2/P3$\rightarrow$ ΔP3$\rightarrow$U. The order of forced unfolding of pbuE A-riboswitch is ΔP1$\rightarrow$ ΔP2/P3$\rightarrow$ ΔP2$\rightarrow$U, where ΔP2/P3 denotes the disruption of kissing loop interaction between P2 and P3 due to force. In the absence of adenine thermal fluctuations transiently disrupt this kissing-loop interaction, which is consistent with the observation that stable P2/P3 tertiary interactions require adenine.

The presence of adenine in the binding pocket in the triple-helix junction of add A-riboswitch changes the force-response of RS completely: (i) The unfolding force increases from $\sim$ 10 pN to $\sim$ 18 pN, the value of which is comparable to the one found in experiments for the pbuE A-riboswitch aptamer (Greenleaf et al., 2008); and (ii) the unfolding of RS occurs in all-or-none fashion without intermediate unfolding steps. After the complete unfolding, when refolding of the add A-riboswitch is initiated by reducing the force, we find that the refolding pathway follows the reverse order of unfolding pathway as U$\rightarrow$P3$\rightarrow$P2$\rightarrow$P2/P3$\rightarrow$P1. Refolding of P3 preceding that of P2 implies that P3 is more stable than P2, which is consistent with the implication from the stability of each helix

($\Delta G_{P2}^{add} = -5$ kcal/mol > $\Delta G_{P3}^{add} = -6.2$ kcal/mol) calculated using the Vienna RNA package (Hofacker, 2003).

Remarkably, despite the structural similarity between pbuE-A and add A-riboswitch aptamers, experiments show that P2 in pbuE unfolds at the last moment, which implies that P2 is the first structural element to refold upon force quench (or reduction). In agreement with the experiments, our results also imply that P2 ought to be more stable than P3 in the pbuE A-riboswitch aptamer, and Vienna RNA package indeed predicts that the stability of P2 is lower than that of P3 by 2 kcal/mol ($\Delta G_{P2}^{pbuE} = -7.3$ kcal/mol < $\Delta G_{P3}^{pbuE} = -5.3$ kcal/mol) (Fig. 3). The difference of the stability in the two RS aptamers explains the reversed order of the folding of P2 and P3 in the pbuE A-riboswitch aptamer. The order of unfolding of the helices, which is in accord with single molecule pulling experiments, is determined by the relative stabilities of the individual helices. Our results show that the stability of isolated helices determines the order of assembly and response to force in these noncoding regions. Thus, the folding landscape is determined by the local stability of the structural elements, a finding that also holds good in the thermal refolding of a number of RNA pseudoknots (Cho, Pincus, & Thirumalai, 2009).

Based on the stability hypothesis as the determining factor of the RNA folding landscape, we make an interesting prediction for pulling experiments in a mutant of the add A-riboswitch. One of the major differences that contributes the different stability of P2 helix in the two purine riboswitches is that the P2 of add A-riboswitch has one G-U and two G-C base pairs, whereas P2 in the pbuE A-riboswitch has three G-C base pairs (see Figs. 3A and 4A). At the level of stability of secondary structure, a point mutation of U28C in the add A-riboswitch, which leads to three G-C base pairs in P2, would increase the secondary structure stability of P2 to $\sim$7.3 kcal/mol. Thus, the U28C mutation stabilizes the add A-riboswitch P2 by 1.1 kcal/mol lower than P3. The folding landscape of the U28C add A-riboswitch would be qualitatively similar to the WT pbuE A-riboswitch. As a consequence, we predict that U28C would reverse the order of unfolding of the add A-riboswitch.

It is noteworthy that the number of contacts between P2 and P3 hairpin loops with and without adenine is almost identical, which suggests that adenine binding does not affect the interactions between P2 and P3 hairpin loops. Rather, the adenine binding stabilizes the triple-helix junction and makes unfolding of P1 more difficult. Hence, the rate limiting step in the fully folded aptamer is the formation of P1.

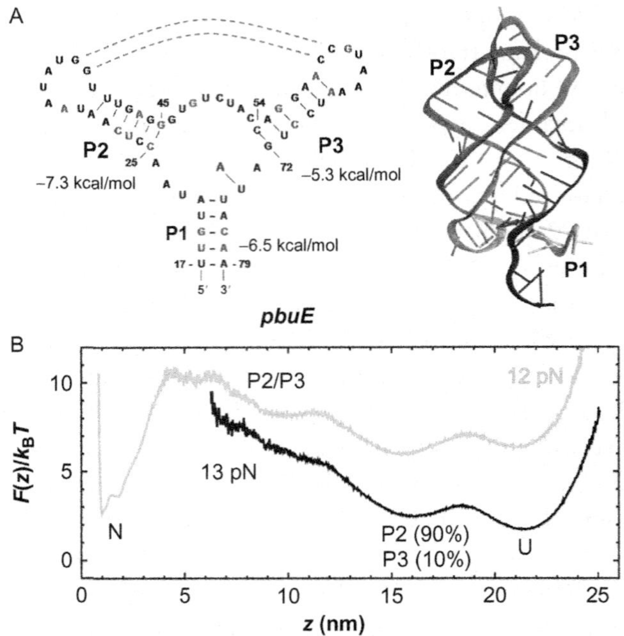

Figure 4 pbuE adenine-riboswitch. (A) Secondary structure map (on the left) and the three-dimensional structure on the right. (B) Free energy profile calculated from simulations at $f = 12$, 13 pN. *Figure adapted from J.-C. Lin et al. (2014).*

The free energy profiles ($G(z)$), obtained from the distribution of the molecular extension at force f ($P(z; f)$) using $G(z;f) = -k_BT \log P(z;f)$, make explicit the hierarchical characteristics of RNA assembly and disassembly. In the absence of adenine, the position of first transition barrier from the folded state ($z = 1$ nm) is $\sim$ 2.5 nm. Thus, the unfolding transition over the first barrier amounts to the unzipping of three base pairs in P1 helix. In the presence of adenine, the position of the first barrier shifts to $\sim$ 4 nm, implying that five base pairs of P1 in direct interaction with adenine should be disrupted at the first TS. Thus, adenine binding makes the first unfolding barrier the rate limiting step. We also find that, at $f = 10$ pN, binding of adenine stabilizes the folded state by ~ 6 k_BT and increases the energy barrier for leaving the folded state by 2 k_BT. These results support the hypothesis that the ligand binding stabilizes P1, a feature that is common to most riboswitches.

The binding of adenine stabilizes the folded basin of attraction, which slows down the unfolding transition by two orders of magnitude. The slow unfolding rate, which makes the conformational sampling difficult, is in agreement with that observed for FMN riboswitches (Wickiser et al., 2005), suggesting that functions of riboswitches are kinetically, rather than thermodynamically, controlled (see below for further discussion). The result is crucial for transcription of the complete riboswitch.

4. FOLDING LANDSCAPES OF SAM RIBOSWITCH

Upon binding SAM the riboswitch undergoes an allosteric transition to control translation. There are at least five distinct classes of riboswitches that bind SAM or its derivative *S*-adenosylhomocysteine (SAH). One of them, the SAM-III riboswitch (Fuchs, Grundy, & Henkin, 2006) in the metK gene (encodes SAM synthetase) from *Lactobacillales* species, inhibits translation by sequestering the Shine–Dalgarno (SD) sequence (Fig. 5A) (Fuchs et al., 2006). When SAM is bound, the somewhat atypical SD sequence (GGGGG shown in the blue shaded area in Fig. 5A) is sequestered by base pairing with the anti-Shine–Dalgarno (ASD) sequence, thus hindering the binding of the 30S ribosomal subunits to mRNA. In the absence of SAM, the sequence outside the binding domain enables the riboswitch to adopt an alternative folding pattern, in which the SD sequence is exposed and free to engage the ribosomal subunit (Lu et al., 2011). Thus, SAM-III is an OFF switch for translation, in contrast to *add* A-riboswitch, which is an ON switch. Translation control is determined by competition between the SD–ASD pairing and loading of ribosomal subunit onto the SD sequence. In order to function as a switch, the SD sequence has to be exposed for ribosome recognition, which implies that at least part of the riboswitch structure accommodating SAM III has to unfold (Fig. 5). These considerations prompted us to quantitatively determine the folding landscape and the rates of conformational transitions between the ON and OFF states of the SAM-III riboswitch (Anthony, Perez, Garcia-Garcia, & Block, 2012).

In order to understand if SAM-III functions under thermodynamic control, we calculated the force-extension (f, z) or FEC as well as constant f-dependent free energy profiles as a function of z for SAM-III. The FEC (black curve in Fig. 5B) shows that there are two intermediate states, with extension $z \sim 14$ nm and $z \sim 9$ nm in the absence of SAM. Helices P1 and P4 rupture in a single step at ~ 9 pN and P2 unravels at $f \approx 10-11$ pN. Helix P3 unfolds fully only when $f > 12$pN. Interestingly, when SAM is bound, the riboswitch unfolds in an apparent all-or-none manner at $f \sim 15$ pN (Fig. 5B). The distribution z (blue curve in Fig. 5B), shows the presence of two intermediate states ahead of global unfolding. The

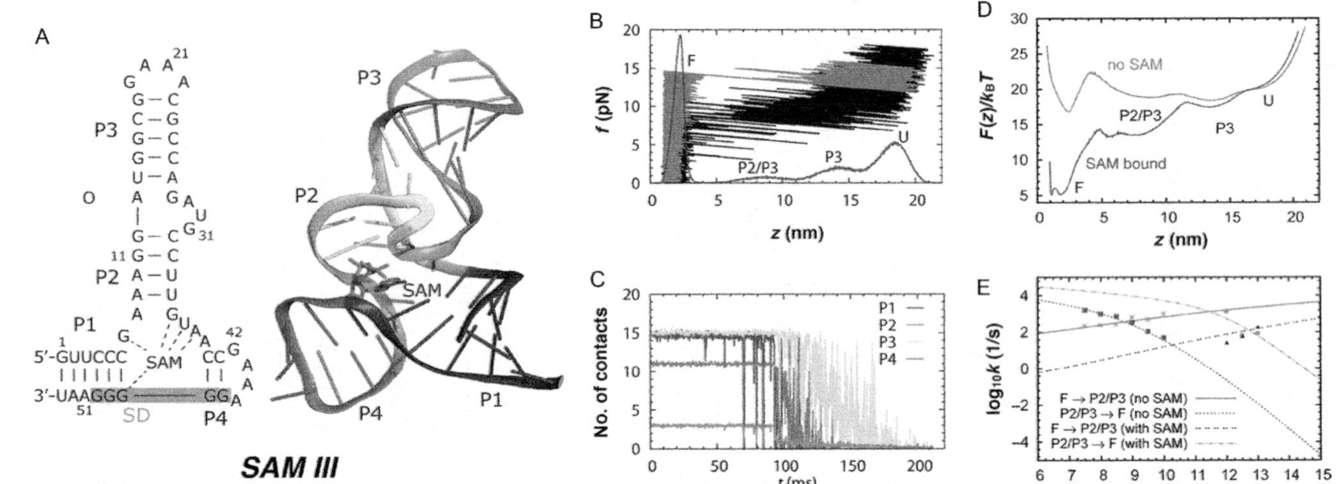

Figure 5 Dynamics of SAM-III riboswitch under force. (A) Structure of SAM-III RS. The blue shaded area on the left indicates the Shine–Dalgarno sequence recognized by the ribosome. (B) Simulated force-extension curve of SAM-III riboswitch in the absence of metabolite (black) produced at $r_f = 96$ pN/s. The distribution of molecular extension (z) during the pulling simulation is shown in blue at the bottom. FEC in red was produced in the presence of metabolite at the binding pocket. (C) Average number of contacts in each helix from P1 to P4. (D) Free energy profile at zero force calculated from streching simulation with and without metabolite (SAM) in the binding pocket. (E) Transition rates between F and P2/P3 states at varying forces. The data points are from explicit simulations. The lines were obtained from mean first passage time calculation on $F(z)$. *Figure adapted from J.-C. Lin and Thirumalai (2013).* (See the color plate.)

$z \sim 9$ nm peak corresponds to rupture of P1 and P4 helices. P3 unfolds in the later stages creating a peak at $z \sim 14$ nm. The hierarchical unfolding pathway of SAM-III riboswitch is $F \rightarrow \Delta P1 \Delta P4 \rightarrow \Delta P3 \rightarrow U$, where $\Delta P1 \Delta P4$ means helices P1 and P4 are ruptured, and $\Delta P2$ represents additional unfolding of P2. The observed order of unfolding is also reflected in the rupture of contacts, a more microscopic representation of unfolding dynamics (Fig. 5C). The intermediate states in Fig. 5B can be traced to the breaking of contacts within the helices. Just as in Fig. 5B the order of contact rupture corresponds to the order in which the helices unfold in the FEC in Fig. 5B.

Using simulations at constant force we also calculated the free energy profiles using $F(z,f) = -k_B T \log P(z)$ where $P(z)$ is the probability distribution of z. At $f < 9$ pN and in the absence of SAM the riboswitch is in the folded basin of attraction (Fig. 5D). The free energy profile in Fig. 5D also shows that binding of SAM consolidates the formation of helix P1 and P4, further stabilizing the folded state. At $f = 9$ pN, SAM binding stabilizes the folded state by $\sim 12 k_B T$, and increases the energy barrier for leaving the folded state by $\sim 3 k_B T$. The distance from the folded state to the first barrier in the absence of SAM is ~ 2 nm, which indicates unzipping of 2.5 base pairs, assuming a contour length increase of 0.4 nm/nt. In the presence of SAM, the position of the first barrier shifts to ~ 5 nm, implying that 4 base pairs of P1 next to the nucleotide G48 (Fig. 5A) that has direct contacts with SAM are ruptured at the transition state. Thus, disruption of contacts with SAM becomes the key barrier in the first unfolding step, and must be an important step in translational regulation.

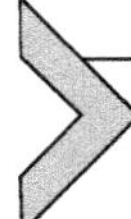

5. IS SAM RIBOSWITCH UNDER THERMODYNAMIC CONTROL?

As shown in Fig. 6A, the kinetic processes in riboswitches that control transcription are determined by a number of time scales. In the transcription process, the ability to function as an efficient switch depends on an interplay of the time scales: (i) metabolite binding rate (k_b), (ii) the folding times of the aptamer (k_f), (iii) the time scales to switch and adopt alternate conformations with the downstream expression platform (k_t), and (iv) the rate of transcription. In "OFF" riboswitches that shut down gene expression upon metabolite binding, a decision to terminate transcription has to be made before the terminator is synthesized, which puts bounds on the metabolite concentration, and the aptamer folding rate (k_f). For simplicity, $\gamma = k_t / k_f$ can serve as a

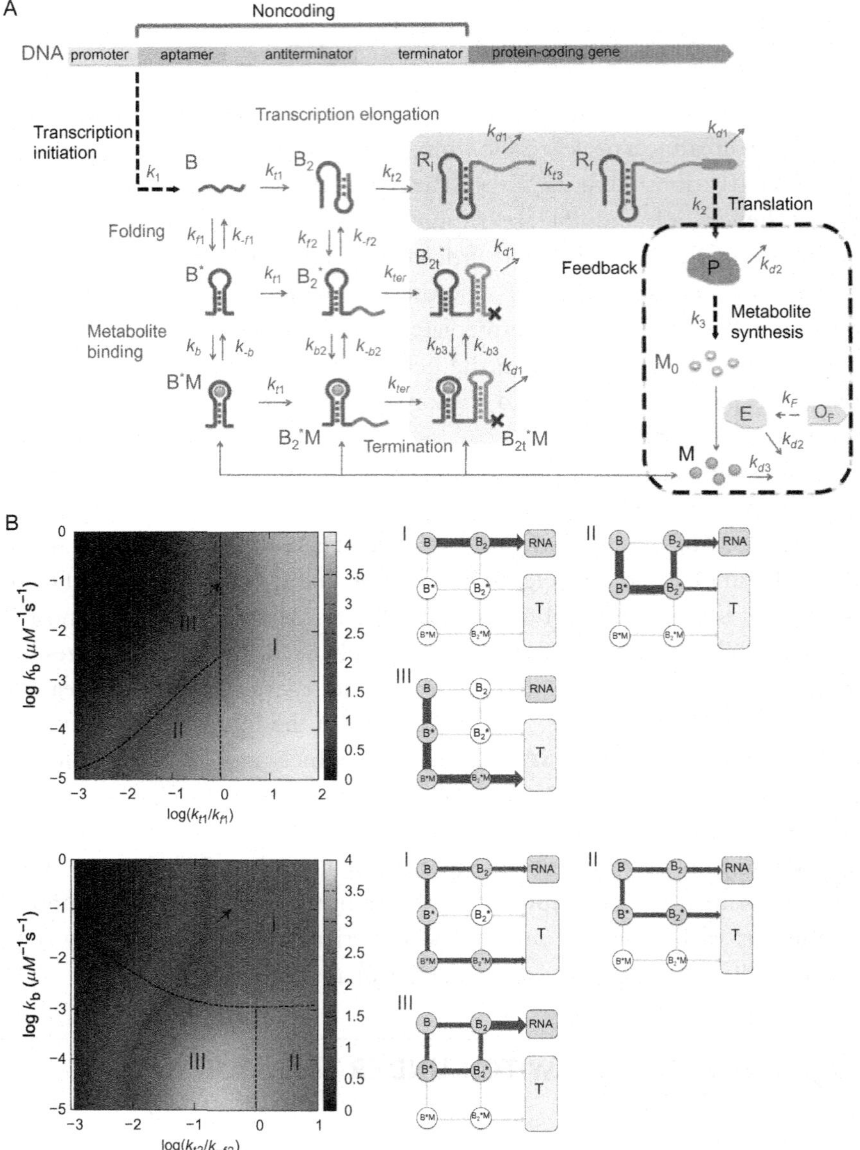

Figure 6 (A) Kinetic network model for transcription regulated by "OFF"-riboswitches. (B) Dependence of protein production on the network parameters with "negative feedback" ($k_{f1} = 0.1\ s^{-1}$, $k_{-f1} = 0.04\ s^{-1}$, $k_{f2} = 2.5 \times 10^{-3}\ s^{-1}$, $k_{-f2} = 0.04\ s^{-1}$, $k_{t1} = 0.1\ s^{-1}$, $k_{t2} = 0.016\ s^{-1}$, $k_b = 0.1\ \mu M^{-1}\ s^{-1}$, $k_{-b} = 10^{-3}\ s^{-1}$, $k_{t3} = 0.01\ s^{-1}$, $K_1 = 0.016$, $k_2 = 0.3\ s^{-1}$, $k_3 = 0.064\ s^{-1}$, $k_{d1} = 2.3 \times 10^{-3}\ s^{-1}$, $k_{d2} = 2.7 \times 10^{-4}\ s^{-1}$, $k_{d3} = 4.5 \times 10^{-3}\ s^{-1}$, and $\mu = 5 \times 10^{-4}\ s^{-1}$. (Top) Protein levels $[P]$ (color (different gray shades in the print version) coded) as functions of k_{t1}/k_{f1} and k_b). The dependence of $[P]$ on k_{t1} and k_b is categorized into three regimes. Points on the dashed line separating regime II and regime III satisfy $k_b[M] = k_{t1}$. The major pathway in the transcription process in each regime is shown on the right. The arrow indicates the data point from the value of $k_{t1} = 0.1\ s^{-1}$ and $k_b = 0.1\ s^{-1}$. (Bottom) $[P]$ as functions of k_{t2}/k_{f2} and k_b. Points on the dashed line separating regime I and II/III satisfy $k_{b1}[M] = k_{-f2}$. The data point corresponding to the arrow results from using the value of $k_{t2} = 0.016\ s^{-1}$ and $k_b = 0.1\ s^{-1}$. *Figure adapted from J.-C. Lin and Thirumalai (2012).*

simple criterion to determine whether the cotranscriptional folding of riboswitches is under thermodynamic or kinetic control. In the limit $\gamma \gg 1$ transcript synthesis is faster than the equilibration time of the riboswitch conformation. For typical values of these parameters in both FMN and pbuE A-riboswitch, efficient function mandates that the riboswitches be under kinetic control, which implies that the "OFF" and "ON" states of riboswitch are not in equilibrium.

In contrast, the function of SAM-III, which controls translation, is different. The major time scales that control the function of SAM-III RS, and those that regulate translation in general, are (i) bimolecular binding rate of SAM to RS (k_b), (ii) dissociation rate of SAM from the riboswitch complex (k_{-b}), (iii) the rate of mRNA degradation (k_{mRNA}). Thus, the only clear physical bound on the function of SAM-III is that binding of metabolite should occur multiple times before the mRNA degradation, which leads to $k_b[M] \gg k_{\mathrm{mRNA}}$, where $[M]$ is the concentration of SAM. Typical values of $k_b \sim 0.11 \mu M^{-1} s^{-1}$, $k_{mRNA} \approx 3\ min^{-1}$ and $k_{dis} \approx 0.089 \mathrm{s}^{-1}$ requires that $[M] \gtrsim 50$ nM. It is worth pointing out that our estimates of folding and unfolding times based on simulations at low forces, and other time scales are all much less than $k_{\mathrm{mRNA}}^{-1} \approx 20$ sec, which sets the longest time for translational control. Hence, the multiple transitions between the OFF and ON states can occur before mRNA is degraded, which gives additional credence to the argument that the function of SAM-III is under thermodynamic control. There is a caveat to this conclusion. It is known that in bacteria transcription and translation are coupled, which is likely to complicate our arguments. In order to provide a complicate description, we require a network model that includes transcription–translation coupling. Because k_{mRNA} is small it is still possible that the SAM riboswitch could be under thermodynamic control.

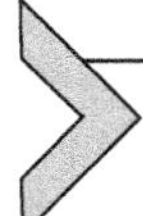

6. KINETIC NETWORK MODEL OF GENE REGULATION AND THE ROLE OF NEGATIVE FEEDBACK IN CONTROL OF TRANSCRIPTION

As stated earlier, gene expression is mediated by binding of metabolites to the conserved aptamer domain, which triggers an allosteric reaction in the downstream expression platform. However, the target metabolites are usually the products or their derivatives of the downstream gene that the riboswitches control. Hence, metabolite binding to riboswitches serves as a feedback signal to control RNA transcription or translation initiation.

The feedback through metabolite binding is naturally designed to be a fundamental network motif for riboswitches. In ON-riboswitches, metabolite binding thus stabilizes the aptamer structure during transcription and prevents the formation of the terminator stem before transcription is completed (pbuE A-riboswitch) or the formation of translation repressor stem before translation is initiated (add A-riboswitch). Whereas in OFF-riboswitch, metabolite binding shuts down the gene expression by promoting the formation of terminator stem (see Fig. 6). In order to understand the *in vivo* riboswitch, we developed a kinetic network model taking into account the interplay between the speed of RNA transcription, folding kinetics of the nascent RNA transcript, and the kinetics of metabolite binding to the nascent RNA transcript, and the role of feedback arising from interactions between synthesized metabolities and the transcript. The effects of speed of RNA transcription and metabolite binding kinetics have also been investigated experimentally *in vitro* in an insightful study involving the FMN riboswitches (Wickiser et al., 2005). They argued that FMN riboswitch is kinetically driven implying that the riboswitch does not reach thermodynamic equilibrium with FMN before a decision between continued transcription and transcription termination needs to be made. The mathematical solution of the kinetic network model, which uses as partial input the rates of switching obtained from the folding landscapes, show that in general riboswitches that control transcription are under kinetic control (J.-C. Lin & Thirumalai, 2012). A brief summary is presented here.

Efficient function of RS, implying a large dynamic range (quantified by response of the RS to varying metabolite concentration) without compromising the requirement to suppress transcription or translation, is determined by a balance between the transcription speed (k_{trxn}), the folding and unfolding rates of the apatmer (k_f and k_{-f}), and the binding and unbinding rates of the metabolite ($k_b[M]$ and k_{-b}, where $[M]$ is the metabolite concentration). In order to capture the physics behind the dynamics, it is necessary to consider kinetic network model describing the coupling between aptamer dynamics and transcription. In Fig. 6, demonstrating the kinetic network model for the transcription regulation by OFF-riboswitches, the upstream of the protein-coding gene consists of sequences involving the transcriptions of aptamer (B), antiterminator (B_2) and terminator (B_2^*) of the riboswitch. The transcription initiation is followed by elongation, folding of the RNA transcript, and metabolite binding. RNA polymerase first transcribes the aptamer (B), and moves on to the synthesis of the RNA transcript for anti terminator B_2 at a rate of k_{t1}, and terminator sequence at k_{t2}, resulting in the production of the regulatory region of RNA. R_i is the transcript with the sequence of the protein-coding region starting to be transcribed, and eventually grows to R_f, the full protein-coding region transcribed, with a rate of k_{t3}. During the process of transcription elongation, each of the transcript states, B and B_2, can form states with the aptamer domain folded (B^* and B_2^*) with a folding rate of k_{f1} and k_{f2}, respectively. The folded aptamers bound with metabolite (M) are B^*M and B_2^*M with binding rate constant k_b and k_{b2}, respectively. The transcripts in state B_2^* and B_2^*M can further elongate until the terminator sequence is transcribed with their expression platform forming a transcription terminator stem and dissociate from the DNA template with a rate of k_{ter}, forming B_{2t}^* and B_{2t}^*M. The fraction of transcription termination, f_{ter}, is determined from the amount of the terminated transcripts (in green block) relative to nonterminated transcripts (in blue block). The activated metabolte (M), produced from protein P and activated by the enzyme (E) encoded by the gene OF, can bind to the folded aptamer and can abort transcription, which imposes a negative feedback on the transcription process.

For a riboswitch to function with a large dynamic range, transcription levels should change significantly as the [M] increases from a low to high value. (i) In the high [M], RNA transcript in the aptamer folded state binds a metabolite with $k_b[M]$. In FNM-riboswitch, small k_{-b} value results in the formation of a terminator stem, which subsequently terminates transcription. (ii) In the low [M] limit, the aptamer folded state is mostly unbound and can remain folded until transcription termination or can fold to the antiterminator state, enabling the synthesis of full RNA transcript. The levels of transcription termination are thus controlled by the transition rates between the aptamer folded and unfolded states (k_{f1}, k_{-f1}, k_{f2}, k_{-f2}). Equilibrium between B_2 and B_2^* can be reached only if the rate of transcription is much slower than the rates of folding/unfolding and metabolite binding. By varying the rate of transcription, which can be experimentally realized by adding transcription factors such as NusA (Zhou, Ha, La Porta, Landick, & Block, 2011), it may be possible to drive the cotranscriptional folding of riboswitch from kinetic to thermodynamic control. However, for realistic values of the various rates in Fig. 6 we predict that transcription *in vivo* is under kinetic control.

In the presence of a negative feedback loop, the concentration of target metabolites is also regulated by gene expression. Under nominal operating conditions ($\gamma_2 = k_{t2}/k_{-f2} \sim 0.01 - 0.1$) binding of target metabolites, products of the downstream gene that riboswitches regulate, significantly suppresses the expression of proteins. Negative feedback suppresses the protein level by about half relative to the case without feedback. *In vivo*, the presence of RNA binding proteins, such as NusA (Zhou et al., 2011), may increase the pausing times, thus effectively reducing the transcription rates. Thus, the repression of the protein level by the riboswitch through metabolite binding may be up to 10-fold. Faster RNA folding and unfolding rates than those we obtained may also increase the suppression by negative feedback and broaden the range of transcription rates over which maximal suppression occurs. These predictions are amenable to experimental test.

In response to changes in the active operon level, the negative feedback speeds up the response time of expression and modestly reduces the percentage change in the protein level relative to change in the operon level. The steady-state level of expression for autoregulation varies as a square root of the DNA concentration. Adaptive biological systems may minimize the variation in gene expression to keep the systems functioning normally even when the environments change drastically. One may need to consider more complex networks than the single autoregulation in the transcription network to find near perfect adaptation to the environmental change (Ma, Trusina, El-Samad, Lim, & Tang, 2009).

The effect of negative feedback accounting for the binding of metabolites, which themselves are the product of genes that are being regulated. Our previous work showed that because of the interplay of a number of time scales determining the riboswitch function at the system there are many scenarios that can emerge, which can be encapsulated in terms of a dynamic phase diagram. An example dynamic phase diagram (for a full discussion see J.-C. Lin & Thirumalai, 2012) in terms of the transcription rates k_{trxn}, k_f, k_{-f}, $k_b[M]$ illustrates the complexity of the transcription process. The interplay between folding of RNA transcripts, transcription, and metabolite binding regulate the expression of P, which can be quantified using the production of the protein, $[P]$, on the transcription rates and the effective binding rate $k_b[M]$. The dynamic phase diagram in Fig. 4B, calculated by varying both k_{t1} (k_{t2}) and k_b with the equilibrium binding constant of the metabolite to the aptamer fixed to $K_D = 10$ nM, a value that is appropriate for FMN (Wickiser et al., 2005). We expect that after the aptamer sequence is transcribed, the formation of the aptamer structure is the key step in regulating transcription termination. Thus, regulation of $[P]$ should be controlled by the folding rate k_{f1}, the effective metabolite binding rate, and k_{t1} for regulation of $[P]$. Figure 4B shows three regimes for the dependence of $[P]$ on k_{t1} and $k_b[M]$. In regime I, $k_{t1} > k_{f1}$, the folding rate is slow relative to transcription to the next stage (Fig. 1), which implies that the aptamer structure does not form on the time scale set by transcription. The dominant flux is from B to B_2, which leads to high probability of fully transcribed RNA downstream because of the low transition rate from B_2 to B_2^*. The metabolite binding has little effect on protein expression in this regime, particularly for large k_{t1}/k_{f1}, and hence the protein is highly expressed. In regime II, $k_b[M] < k_{t1} < k_{f1}$, the aptamer has enough time to fold but metabolite binding is slow. The dominant flux is $B \rightarrow B^* \rightarrow B_2^*$, leading to formation of antiterminator stem ($B_2^* \rightarrow B_2$) or transcription termination ($B_2^* \rightarrow B_{2t}^*$). The expression level of protein is thus mainly determined by k_{-f2} and k_{t2}, and the protein production is partially suppressed in this regime. In regime III, $k_{t1} < k_{f1}$ and $k_{t1} < k_b[M]$, the aptamer has sufficient time to both fold and bind metabolite, the dominant pathway is $B \rightarrow B^* \rightarrow B^*M \rightarrow B_2^*M$, leading to transcription termination. The protein production is highly suppressed in this regime. The arrow shows that for parameters that are appropriate for FMN riboswitch (see tables 1 and 2 in J.-C. Lin & Thirumalai, 2012)k_{t1} fall on the interface of regime I and regime III. The metabolite binding fails to reach thermodynamic equilibrium due to low dissociation constant. However, the effective binding rate is high because the steady state concentration of metabolites ($\sim$ 25 μM) is in large excess over RNA transcripts. Thus, the riboswitch is kinetically driven under this condition even when feedback is included.

7. CONCLUDING REMARKS

Based on our previous works, we have provided broad overview, from atomic scale to systems level, of how the complex dynamics of riboswtiches emerge depending on many inter-related rates. At the atomic scale, dynamics of the surface water around riboswitches plays critical role in inducing the local fluctuation in the riboswtich. At the level of single riboswitch, we have shown that explicit simulations of riboswitches, in conjunction with single molecule experiment, is a powerful tool to understand the conformational dynamics of riboswtich both with and without metabolites. At the systems level, in which the minimal model of cellular environment is considered, the dynamics of riboswitches in isolation are modulated by a number of factors, which is made explicit by the kinetic network model of FMN riboswitch. Our collective works show that combination of theory, experiments, and simulations are needed to understand the function of riboswitches under cellular conditions.

Riboswitches also provide novel ways to engineer biological circuits to control gene expression by binding small molecules. As found in tandem riboswitches (Breaker, 2008; Sudarsan et al., 2006), multiple riboswitches can be engineered to control a single gene with greater regulatory complexity or increase in the dynamic range of gene control. Synthetic riboswitches have been successfully used to control the chemotaxis of bacteria (Topp & Gallivan, 2007). Our study provides a physical basis for not only analyzing future experiments but also in anticipating their outcomes.

ACKNOWLEDGMENTS

This work was supported by grants from the National Institutes of Health (GM 089685) and the National Science Foundation (CHE 13-61946).

REFERENCES

Allner, O., Nilsson, L., & Villa, A. (2013). Loop-loop interaction in an adenine-sensing riboswitch: A molecular dynamics study. *RNA*, *19*(7), 916–926.

Anthony, P. C., Perez, C. F., Garcia-Garcia, C., & Block, S. M. (2012). Folding energy landscape of the thiamine pyrophosphate riboswitch aptamer. *Proceedings of the National Academy of Sciences of the United States of America*, *109*(5), 1485–1489.

Breaker, R. R. (2008). Complex riboswitches. *Science*, *319*, 1795–1797.

Cho, S., Pincus, D., & Thirumalai, D. (2009). Assembly mechanisms of RNA pseudoknots are determined by the stabilities of constituent secondary structures. *Proceedings of the National Academy of Sciences of the United States of America*, *106*(41), 17349.

Dethoff, E. A., Chugh, J., Mustoe, A. M., & Al-Hashimi, H. M. (2012). Functional complexity and regulation through RNA dynamics. *Nature*, *482*(7385), 322–330.

Feng, J., Walter, N. G., & Brooks, C. L., III. (2011). Cooperative and Directional Folding of the preQ(1) Riboswitch Aptamer Domain. *Journal of the American Chemical Society*, *133*(12), 4196–4199.

Frieda, K. L., & Block, S. M. (2012). Direct observation of cotranscriptional folding in an adenine riboswitch. *Science*, *338*(6105), 397–400.

Fuchs, R. T., Grundy, F. J., & Henkin, T. M. (2006). The SMK box is a new SAM-binding RNA for translational regulation of SAM synthetase. *Nature Structural & Molecular Biology*, *13*(3), 226–233.

Greenleaf, W. J., Frieda, K. L., Foster, D. A. N., Woodside, M. T., & Block, S. M. (2008). Direct observation of hierarchical folding in single riboswitch aptamers. *Science*, *319*, 630–633.

Hofacker, I. (2003). Vienna RNA secondary structure server. *Nucleic Acids Research*, *31*(13), 3429.

Hyeon, C., Dima, R. I., & Thirumalai, D. (2006). Pathways and kinetic barriers in mechanical unfolding and refolding of RNA and proteins. *Structure*, *14*, 1633–1645.

Hyeon, C., Morrison, G., & Thirumalai, D. (2008). Force dependent hopping rates of RNA hairpins can be estimated from accurate measurement of the folding landscapes. *Proceedings of the National Academy of Sciences of the United States of America*, *105*, 9604–9606.

Hyeon, C., & Thirumalai, D. (2005). Mechanical unfolding of RNA hairpins. *Proceedings of the National Academy of Sciences of the United States of America*, *102*, 6789–6794.

Hyeon, C., & Thirumalai, D. (2007). Mechanical unfolding of RNA : From hairpins to structures with internal multiloops. *Biophysical Journal*, *92*, 731–743.

Jacob, W. F., Santer, M., & Dahlberg, A. E. (1987). A single base change in the Shine-Dalgarno region of 16S rRNA of Escherichia coli affects translation of many proteins.

Proceedings of the National Academy of Sciences of the United States of America, 84(14), 4757–4761.
Kim, J. N., & Breaker, R. R. (2008). Purine sensing by riboswitches. *Biology of the Cell, 100*(1), 1–11.
Lin, J., & Thirumalai, D. (2008). Relative stability of helices determines the folding landscape of adenine riboswitch aptamers. *Journal of the American Chemical Society, 130*, 14080–14081.
Lin, J.-C., Hyeon, C., & Thirumalai, D. (2014). Sequence-dependent folding landscapes of adenine riboswitch aptamers. *Physical Chemistry Chemical Physics, 16*, 6376.
Lin, J.-C., & Thirumalai, D. (2012). Gene regulation by riboswitches with and without negative feedback loop. *Biophysical Journal, 103*(11), 2320–2330.
Lin, J.-C., & Thirumalai, D. (2013). Kinetics of allosteric transitions in S-adenosylmethionine riboswitch are accurately predicted from the folding landscape. *Journal of the American Chemical Society, 135*(44), 16641–16650.
Lu, C., Smith, A. M., Ding, F., Chowdhury, A., Henkin, T. M., & Ke, A. (2011). Variable sequences outside the SAM-binding core critically influence the conformational dynamics of the SAM-III/SMK box riboswitch. *Journal of Molecular Biology, 409*(5), 786–799.
Luzar, A., & Chandler, D. (1996). Effect of environment on hydrogen bond dynamics in liquid water. *Physical Review Letters, 76*, 928–931.
Ma, W., Trusina, A., El-Samad, H., Lim, W. A., & Tang, C. (2009). Defining network topologies that can achieve biochemical adaptation. *Cell, 138*(4), 760–773.
Mandal, M., Boese, B., Barrick, J. E., Winkler, W. C., & Breaker, R. R. (2003). Riboswitches control fundamental biochemical pathways in bacillus subtilis and other bacteria. *Cell, 113*, 577–586.
Montange, R. K., & Batey, R. (2008). Riboswitches: Emerging themes in RNA structure and function. *Annual Review of Biophysics, 37*, 117–133.
Neupane, K., Daniel, H. Y., Foster, A. N., Wang, F., & Woodside, M. T. (2011). Single-molecule force spectroscopy of the add adenine riboswitch relates folding to regulatory mechanism. *Nucleic Acids Research, 39*, 7677–7687.
Nikolova, E. N., & Al-Hashimi, H. M. (2010). Thermodynamics of RNA melting, one base pair at a time. *RNA, 16*(9), 1687–1691.
Quarta, G., Sin, K., & Schlick, T. (2012). Dynamic energy landscapes of riboswitches help interpret conformational rearrangements and function. *PLoS Computational Biology, 8*(2), e1002368.
Rinnenthal, J., Klinkert, B., Narberhaus, F., & Schwalbe, H. (2010). Direct observation of the temperature-induced melting process of the Salmonella fourU RNA thermometer at base-pair resolution. *Nucleic Acids Research, 38*(11), 3834–3847.
Serganov, A., & Nudler, E. (2013). A decade of riboswitches. *Cell, 152*(1), 17–24.
Song, J., Franck, J., Pincus, P., Kim, M. W., & Han, S. (2014). Specific ions modulate diffusion dynamics of hydration water on lipid membrane surfaces. *Journal of the American Chemical Society, 136*(6), 2642–2649.
Sudarsan, N., Hammond, M. C., Block, K. F., Welz, R., Barrick, J. E., Roth, A., et al. (2006). Tandem riboswitch architectures exhibit complex gene control functions. *Science, 314*(5797), 300–304.
Thirumalai, D., Mountain, R. D., & Kirkpatrick, T. R. (1989). Ergodic behavior in supercooled liquids and in glasses. *Physical Review A, 39*, 3563–3574.
Topp, S., & Gallivan, J. P. (2007). Guiding bacteria with small molecules and RNA. *Journal of the American Chemical Society, 129*(21), 6807–6811.
Whitford, P. C., Schug, A., Saunders, J., Hennelly, S. P., Onuchic, J. N., & Sanbonmatsu, K. Y. (2009). Nonlocal helix formation is key to understanding S-adenosylmethionine-1 riboswitch function. *Biophysical Journal, 96*(2), L7–L9.

Wickiser, J. K., Winkler, W. C., Breaker, R. R., & Crothers, D. M. (2005). The speed of RNA transcription and metabolite binding kinetics operate an FMN riboswitch. *Molecular cell, 18*(1), 49–60.

Yoon, J., Lin, J.-C., Hyeon, C., & Thirumalai, D. (2014). Dynamical transition and heterogeneous hydration dynamics in RNA. *The Journal of Physical Chemistry B, 118*, 7910–7919.

Yoon, J., Thirumalai, D., & Hyeon, C. (2013). Urea-induced denaturation of preQ1-riboswitch. *Journal of the American Chemical Society, 135*, 12112–12121.

Zhang, Q., Sun, X., Watt, E. D., & Al-Hashimi, H. M. (2006). Resolving the motional modes that code for RNA adaptation. *Science, 311*(5761), 653–656.

Zhou, J., Ha, K. S., La Porta, A., Landick, R., & Block, S. M. (2011). Applied force provides insight into transcriptional pausing and its modulation by transcription factor NusA. *Molecular cell, 44*(4), 635–646.

SECTION III

Ions, Ligands, and RNA Interactions

CHAPTER ELEVEN

Computational Methods for Prediction of RNA Interactions with Metal Ions and Small Organic Ligands

Anna Philips*^,1, Grzegorz Łach†, Janusz M. Bujnicki†,‡

*European Center for Bioinformatics and Genomics, Institute of Bioorganic Chemistry, Polish Academy of Science, Poznan, Poland
†International Institute of Molecular and Cell Biology, Warsaw, Poland
‡Faculty of Biology, Institute of Molecular Biology and Biotechnology, Adam Mickiewicz University, Poznan, Poland
[1]Corresponding author: e-mail address: aphilips@ibch.poznan.pl

Contents

Abstract

In the recent years, it has become clear that a wide range of regulatory functions in bacteria are performed by riboswitches—regions of mRNA that change their structure upon external stimuli. Riboswitches are therefore attractive targets for drug design, molecular engineering, and fundamental research on regulatory circuitry of living cells. Several mechanisms are known for riboswitches controlling gene expression, but most of them perform their roles by ligand binding. As with other macromolecules, knowledge of the 3D structure of riboswitches is crucial for the understanding of their function. The development of experimental methods allowed for investigation of RNA structure and its complexes with ligands (which are either riboswitches' substrates or inhibitors) and metal cations (which stabilize the structure and are also known to be

Methods in Enzymology, Volume 553
ISSN 0076-6879
http://dx.doi.org/10.1016/bs.mie.2014.10.057

riboswitches' inhibitors). The experimental probing of different states of riboswitches is however time consuming, costly, and difficult to resolve without theoretical support. The natural consequence is the use of computational methods at least for initial research, such as the prediction of putative binding sites of ligands or metal ions. Here, we present a review on such methods, with a special focus on knowledge-based methods developed in our laboratory: LigandRNA—a scoring function for the prediction of RNA–small molecule interactions and MetalionRNA—a predictor of metal ions-binding sites in RNA structures. Both programs are available free of charge as a Web servers, LigandRNA at http://ligandrna.genesilico.pl and MetalionRNA at http://metalionrna.genesilico.pl/.

1. INTRODUCTION

Functions of RNA molecules depend on their interactions with other molecules in the cell (Dieterich & Stadler, 2013; Fulle & Gohlke, 2010; Rivas & Eddy, 2001; Thomas & Hergenrother, 2008). In particular, many RNAs whose function extends beyond coding of protein sequence exert their role by interacting with metal ions and/or small organic ligands. The RNA backbone is negatively charged and the neutralization of the electrostatic repulsion by the binding of cations is essential for the formation of compact tertiary structures that are functionally important [reviews: Draper, 2004, 2008; Serra et al., 2002]. Moreover, metal ions also often serve as essential cofactors in the active sites of ribozymes; i.e., RNAs that function as enzymes like the hammerhead ribozyme, self-splicing introns, and ribonuclease P (RNaseP) (Schnabl & Sigel, 2010; Sigurdsson & Eckstein, 1995). Binding of ligands is also an important function of some riboswitches; e.g., mRNA-embedded noncoding elements that regulate the translation of the coding part of RNA by undergoing conformational changes (Montange & Batey, 2008). Most riboswitches comprise two domains: an aptamer and an expression platform. The aptamer directly binds the ligand, and in response to that binding, the expression platform undergoes structural changes. Ligands that bind to riboswitches range from single atoms such as metal ions (Baker et al., 2012), to amino acids (Mandal, Boese, Barrick, Winkler, & Breaker, 2003; Mandal et al., 2004; Rodionov, Vitreschak, Mironov, & Gelfand, 2003), to very complex organic metabolites like pyrophosphate (TPP) (Mironov et al., 2002), vitamin B_{12} (Warner, Savvi, Mizrahi, & Dawes, 2007), thiamine flavine mononucleotide (FMN) (Winkler, Cohen-Chalamish, & Breaker, 2002), and many others (Garst, Edwards, & Batey, 2011). It is worth to mention that, in the recent years riboswitches have been identified as a potential target for the development of antibacterial drugs due to the fact that they are common in bacteria and rare in eukaryotic cells (Mulhbacher, Brouillette, et al., 2010; Mulhbacher, St-Pierre, & Lafontaine, 2010).

The analyses of the atomic details of RNA–ligand interactions are greatly facilitated by the availability of high-resolution structures of RNA–ligand complexes. However, the experimental structure determination for RNA and its complexes is challenging, and currently cannot be accomplished in a high-throughput manner. As an alternative, computational predictive methods can be used. It must be emphasized here that the experimental determination of structures for RNAs and their complexes is technically more difficult than protein structure determination. For this reason, RNA structure determination has lagged behind analogous works on protein structures, and in macromolecular structure databases there are considerably more structures of proteins (also with ligands) than such structures for RNA. Likewise, the development of computational methods for protein 3D structure prediction preceded the development of such methods for RNA. Recently, however, a number of computational methods were developed for RNA structure analyses, often inspired by the previous works on protein structure, which enabled predictions of RNA complexes with ligands. In this chapter, we discuss the most typical approaches and present in more detail methods developed in our laboratory.

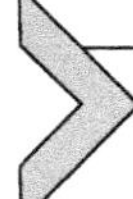

2. COMPUTATIONAL MODELING OF RNA–LIGAND COMPLEX STRUCTURES

In principle, the interactions of RNA molecules with both metal ions and small organic molecules can be modeled using methods of quantum mechanics or molecular mechanics. Interaction energies are, however, nontrivial to compute as they involve large cancelations between the energies of the complex and of the constituents. The interaction free energies are also highly dependent on the entropic effects and interactions with solvent and, as such, are unlikely to be computed using single-point calculations with and all-atom force field, and the computations require sampling of a large number of configurations even for fixed positions of both RNA and ligand. Such calculations, while possible, are time consuming, even for a single RNA and single ligand and are difficult to set up in high-throughput applications, such as virtual screening of possible RNA inhibitors.

Intermolecular interaction potential between RNA and other molecules can also be approximated using statistical potentials. They are a class of empirical potentials functions derived from the statistics gathered from a large number of experimentally solved structures.

The magnitude of interaction between molecules can be quantified using the binding constant K_b, which is defined as:

$$K_b = \frac{[\text{RNA} + \text{ligand}]}{[\text{RNA}][\text{ligand}]},$$

where [RNA + ligand] is the molar concentration of the RNA–ligand complex, while [RNA] and [ligand] denote the molar concentrations of the constituents of this complex, all in thermodynamics equilibrium. The binding constant is related to the change of the Gibbs free energy ΔG^0 (Sippl, 1993, 1995):

$$\Delta G^0 = -RT \ln K_b,$$

where R is the universal gas constant and T is the temperature of the system.

The statistical potentials are meant to approximate the changes in the free energy of the system and their derivation is based on several assumptions: that the change of the free energy can be decomposed into pairwise interactions between atoms or groups of atoms, and that these terms can themselves be approximated based on statistics of atom–atom contacts observed in known structures. The decomposition of the potential into pairwise interactions can be done in several ways. If $W^{(n)}(\vec{r_1}, \vec{r_2}, \ldots, \vec{r_n})$ is the total potential for the system, then the most commonly used decomposition is:

$$W^{(n)}(\vec{r_1}, \vec{r_2}, \ldots, \vec{r_n}) = \sum_{a<b=\{1,\ldots,n\}} W^{(2)}_{T(a)T(b)}(r_{ab}), \quad r_{ab} = \vec{r_a} - \vec{r_b},$$

where $\vec{r_a}$ is the position of the atom a, r_{ab} is the distance between atoms a and b, and $T(a)$ and $T(b)$ refer to the atom type. The sum is only done over pairs of atoms, which do not belong to the same molecule. In this case, the potential is a sum of pairwise interactions between atoms dependent only on the distance between them and on their types. The problem of finding a single function of 3N variables is therefore reduced to finding a number of 1D functions (radial potentials)—one for each of the pair of atom types considered in the model. The radial statistical potential is computed using the so-called inverse Boltzmann Ansatz, as a logarithm of the ratio:

$$W^{(2)}_{T(a)T(b)}(r) = -RT \ln \frac{g^{\text{obs}}_{T(a)T(b)}(r,\ r+\Delta r)}{g^{\text{est}}_{T(a)T(b)}(r,\ r+\Delta r)},$$

between $g^{\text{obs}}_{T(a)T(b)}(r, r+\Delta r)$—the observed number of pairs of atoms of types $T(a)$ and $T(b)$, in distance range between r and $r+\Delta r$ and g^{es}_{ab}—the estimated number of such pairs in the absence of interactions:

$$g^{\text{est}}_{T(a)T(b)}(r,\ r+\Delta r) = \chi_{T(a)} \cdot \chi_{T(b)} \cdot V(r,\ r+\Delta r) \sum_{ab} \sum_{r=n\cdot\Delta r} g^{\text{obs}}_{ab}(r,\ r+\Delta r),$$

where $\chi_{T(a)}$ and $\chi_{T(b)}$ are the fractions of atoms of the types $T(a)$ and $T(b)$, and $V(r, r+\Delta r)$ is the volume of a single bin (spherical shell):

$$V(r,\ r+\Delta r) = \frac{4\pi}{3} |(r+\Delta r)^3 - r^3|.$$

2.1. Computational methods for prediction of metal ion-binding sites in RNA structures

In the X-ray crystallography of macromolecules, it is difficult to identify the position of ions, apart from those that are bound very stably to their protein or nucleic acid partners. For this reason, most of models of macromolecular structures determined experimentally contain a much smaller number of defined ions than expected from the actual concentration of ions in the solution (Zheng, Chruszcz, Lasota, Lebioda, & Minor, 2008). Furthermore, determination of identity of ions is not straightforward. For example, Mg^{2+}, Na^{+}, and H_2O all have 10 electrons each and can be distinguished from each other only in high-resolution crystal structures. Hence, many bound cations can be easily mistaken for water molecules or may be missing from crystal structures. The positions of metal ions are also difficult to determine by NMR.

The correct placement of ions in models of macromolecular structure is essential not only for the functional interpretation of such interactions but also for studies on stability and conformational changes of biological molecules, such as in computational simulations of molecular dynamics and thermodynamics. Therefore, a number of predictive computational methods have been developed for adding explicit ions to the structural model of a macromolecule (a protein or a nucleic acid) as to reflect the actual experimental conditions.

One group of methods have been developed to model the physico-chemical properties of the system under investigation. In the simplest approach, methods such as CHARMM (Brooks et al., 2009) can be used to compute an electrostatic field around a macromolecular structure in solution and metal ions can be placed in locations that neutralize the local concentration of opposite charge. This approach proved useful for proteins because their surfaces are usually quite rugged and exhibit diverse physic-chemical characteristics, including variation of electrostatic charge that can manifest itself in very specific "attractors" for ions. However, in nucleic acids, the net charge is negative and the overall distribution of charge is more uniform, hence the placement of ions in specific positions based on electrostatic complementarity is more problematic (Baker, Anisimov, & MacKerell, 2011). To address this problem, (Hermann & Westhof, 1998) applied Brownian-dynamics simulations of cations diffusing under the influence of random Brownian motion within the electrostatic field. (Misra & Draper, 2000) presented an analytical model based on the nonlinear Poisson–Boltzmann equation that describes the energetic and stoichiometric linkage between the Mg^{2+} binding and RNA folding. (Tan & Chen, 2005, 2010) developed a statistical mechanical model based on ensemble of discrete ion distributions; it models electrostatics and steric interactions for tightly bound ions and uses the mean-field fluid model to describe the diffuse ions. The advantage of these methods is that they model not only the structure alone but also the physico-chemistry of the system and therefore can be used to infer the dynamics and thermodynamic parameters of the systems under study. An important feature of these and similar methods is the examination of the system under physical conditions defined by the user, such as temperature, concentration of different ions, possible presence of other molecules, etc. These methods are, however, computationally very costly, and require specialized expertise to set up and run the simulations, and to interpret their results. They are difficult to automate and to make predictions for a large number of structures.

Another group of methods for predicting metal ion-binding sites rely on information extracted from experimentally determined structures of macromolecules. In particular, these methods infer the statistical characteristics of typical ion-binding sites from a large collection of structures deposited in the Protein Data Bank. Often a machine learning method is used to generate a predictive model that can be used in a fully automated way to identify potential ion-binding sites in a query structures. Such methods are usually computationally much less demanding than those based on the physical properties of the system such as structure and dynamics. Typical disadvantages of the knowledge-based methods include a strong bias toward detection of the most typical binding sites and inability to take the conformational dynamics into account. One example of such an approach is FEATURE (Banatao, Altman, & Klein, 2003), a knowledge-based predictor that can predict the most typical metal ion-binding sites in RNA structures. FEATURE employs supervised learning on a training set consisting of positive and negative examples of Mg^{2+} ion-binding sites to create a statistical model that describes the micro-environments surrounding site-bound and diffusely bound cations. To create a statistical model, 126 physico-chemical and structural properties that influence or take part in RNA–Mg^{2+} ion interactions were used. This method is also available as a Web server (webFEATURE) (Liang, Banatao, Klein, Brutlag, & Altman, 2003).

2.2. Computational methods for prediction of RNA–small molecule complex structures

The structures of organic molecules complexed with RNA are easier to determine experimentally than positions of metal ions, because the former are larger and typically easier to distinguish from other components of the complex. However, successful crystallization of RNA–ligand complexes is far from trivial. Furthermore, even for RNAs and RNA–ligand complexes with known 3D structures, it is often desired to elucidate the potential binding pose and interaction energy for a number of different small organic molecules; e.g., potential inhibitors of the given RNA activity. One example is the structure determination for complexes of the ribosome (which is a large RNA–protein complex with ribozyme activity) with various antibiotics that bind to one of its functionally important sites. Such task cannot be performed experimentally in a high-throughput fashion, which has prompted the development and application of computational methods for RNA–small molecule docking.

In general, methods for prediction of interactions between RNA molecules and small organic compounds utilize similar approaches to those described above for prediction of RNA–metal ion complexes, namely, the consideration of physico-chemical properties of the molecules or the application of a statistical method to develop a knowledge-based predictor. One difference between the predictions of RNA–ion complexes and RNA–small organic molecule complexes is the fact that the positions of the ions can be approximated by single points or by isotropic spheres, while organic molecules are bigger and more complex and have anisotropic properties and therefore their internal structures must be taken into account in addition to their position and orientation with respect to the macromolecular receptor structure. As a result, prediction of organic compound binding is more complex and usually involves at least two steps, namely, generation of ligand poses and pose scoring and ranking. While the treatment of the small-molecule compound that binds to RNA is essentially the same as in the case of protein–ligand binding, the interactions between the RNA receptor and the small organic ligand required the development of new scoring functions.

Morley and Afshar were among the first to create a scoring function specific for RNA–small molecule ligand complexes. Initially, they expanded their proprietary high-throughput protein–ligand docking program to deal with RNA–ligand complexes (Morley & Afshar, 2004). Their scoring function accounts for a wide range of molecular interactions, including hydrogen bonds, attractive lipophilic interactions, repulsive steric interactions, positively charged carbon–acceptor interactions, aromatic stacking interactions, donor–donor repulsion, acceptor–acceptor repulsion, and an estimate of the entropic cost of binding. The most recent version of their software, rDock, is an open-source program that has evolved from molecular docking program RiboDock developed at the Vernalis Company for high-throughput virtual screening applications. rDock can be used against proteins and nucleic acids, and allows the user to incorporate additional constraints and information as a bias to guide docking (Ruiz-Carmona et al., 2014).

Moitessier, Westhof, and Hanessian (2006) developed another scoring function dedicated exclusively to the prediction of interactions between RNA and aminoglycoside antibiotics. That function was implemented in the AutoDock program (Morris et al., 2009). It is based on the calculation of the intermolecular interaction energy that accounts for the presence of dynamically bound water molecules to the RNA. An important feature of this method is that it allows for both ligand and RNA flexibility.

DrugScoreRNA is a general-purpose knowledge-based function for scoring complexes of RNA–small organic molecule complexes; it was developed by the Gohlke group (Pfeffer & Gohlke, 2007). It employs a distance-dependent potential developed on the basis of contacts between the ligand and receptor atoms, as in the DrugScore method developed earlier by the same group for scoring protein–ligand complexes (Gohlke, Hendlich, & Klebe, 2000). This approach presumes that the relative strength of interactions between an atom of type x in the ligand and an atom of type y in the nucleic acid separated by the distance *r* can be predicted from the normalized radial pair-distribution function. The distribution function was derived from known complexes in the form of contact statistics. The original DrugScoreRNA potential was derived from 670 crystallographically determined nucleic acid–ligand and nucleic acid–protein complexes. Ligand and nucleic acid atom types were patterned following the Tripos atom types notation (SYBYL Molecular Modeling Software, 7.3; Tripos Inc.: St. Louis, MO, 2006).

Dock6 is a docking suite of programs originally developed for docking small organic ligands to protein structures, whose functionalities were later extended to include RNA–ligand docking (Lang et al., 2009). Four important steps of the Dock6 procedure are as follows: (1) Generate spheres representing the receptor surface. (2) Select spheres that define the binding site(s). It is possible to choose the largest sphere cluster or spheres within some radius of a desired location defined by the user. Manual sphere selection is possible as well. The last two steps are as follows: (3) carry out the grid calculation and (4) determine the actual docking of ligand conformers. Dock6 is a highly configurable program with many options, so expert knowledge is required to run calculations. There are several approaches to the sampling of the poses (e.g., using chemical matching that associates the chemical properties to the spheres) and there are nine built-in scoring functions, differing in speed and theoretical foundations. The default scoring function is a grid-based score, based on the nonbonded terms of the molecular mechanics force field (Kuntz, Blaney, Oatley, Langridge, & Ferrin, 1982). The force field type is defined by the user, as both the receptor and the ligand require an initial preparation with external tools (e.g., Chimera; Pettersen et al., 2004).

Guilbert and James have also addressed the RNA–ligand docking problem by applying a classical molecular mechanics force field to the receptor and the ligand in their docking procedure MORDOR (Guilbert & James, 2008); similar to the methodology used by Dock6. Their method requires receptor and ligand preparation and allows for both ligand and receptor flexibility. The predictive power of both Dock6 and MORDOR was reported to be comparable, but Dock6 is 3–10 times faster (Lang et al., 2009).

3. MetalionRNA AND LigandRNA

3.1. General principles of MetalionRNA and LigandRNA predictors

MetalionRNA and LigandRNA serve different purposes: MetalionRNA predicts metal ions in any RNA 3D structure; LigandRNA ranks and scores ligand poses generated by any third-party docking program. Nonetheless, both programs are based on a very similar approach.

In our work, we applied a statistical approach to create an anisotropic statistical potential dependent on both the distance and the angle between interacting pairs of atoms of the RNA and the metal ions (MetalionRNA) or individual atoms of the small-molecule ligand (LigandRNA). In both programs, the potential is decomposed into a sum of three-body terms:

$$W^{(n)}\left(\vec{r_1}, \vec{r_2}, \ldots, \vec{r_n}\right) = \sum_{abc} W^{(3)}_{T(a)T(b)T(c)}\left(r_{bc}, \vartheta_{abc}\right),$$

$$\cos \vartheta_{abc} = \frac{\left(\vec{r_a} - \vec{r_b}\right)\cdot\left(\vec{r_b} - \vec{r_c}\right)}{\left|\left|\vec{r_a} - \vec{r_b}\right|\right|\cdot\left|\left|\vec{r_b} - \vec{r_c}\right|\right|}.$$

The sum is taken over covalently bonded a, b atoms in RNA and c of the ligand. We also limit our predictor of RNA–metal ion interactions to contacts formed by the metal cations with oxygen and nitrogen atoms that are most electronegative and are known to make the strongest contribution to RNA–metal binding. In the case of MetalionRNA potential—first, we defined a list of covalently bonded atom pairs [a, b] in nucleotide residues of which b is an electronegative O or N atom that may directly interact with a metal ion. Under the above assumption, the derivation of the components of the statistical potential follows analogously to the isotropic potential:

$$W^{(3)}_{T(a)T(b)T(c)}(r, \vartheta) = -RT \ln \frac{g^{\mathrm{obs}}_{T(a)T(b)T(c)}(r, r+\Delta r; \vartheta, \vartheta+\Delta\vartheta)}{g^{\mathrm{est}}_{T(a)T(b)T(c)}(r, r+\Delta r; \vartheta, \vartheta+\Delta\vartheta)}.$$

The estimated number of counts is computed as:

$$g^{\mathrm{est}}_{T(a)T(b)T(c)}(r, r+\Delta r; \vartheta, \vartheta+\Delta\vartheta) = \chi_{[T(a),T(b)]}\cdot\chi_{T(c)}\cdot V(r, r+\Delta r; \vartheta, \vartheta+\Delta\vartheta)\cdot g^{\mathrm{obs}},$$

where $\chi_{[T(a),T(b)]}$ is the molar fraction of covalently bonded atom pairs of types $T(a)$ and $T(b)$ as fraction of all covalently bonded pairs of atoms, and g^{obs} is the total observed number of counts for all types of RNA and ligand atoms and all geometries. The volume of the individual counting bins is now that of the conical sections of spherical shells (see Fig. 1B):

$$V(r, r+\Delta r; \vartheta, \vartheta+\Delta\vartheta) = \frac{2\pi}{3}\left|(r+\Delta r)^3 - r^3\right|\cdot\left|\cos(\vartheta+\Delta\vartheta) - \cos\vartheta\right|.$$

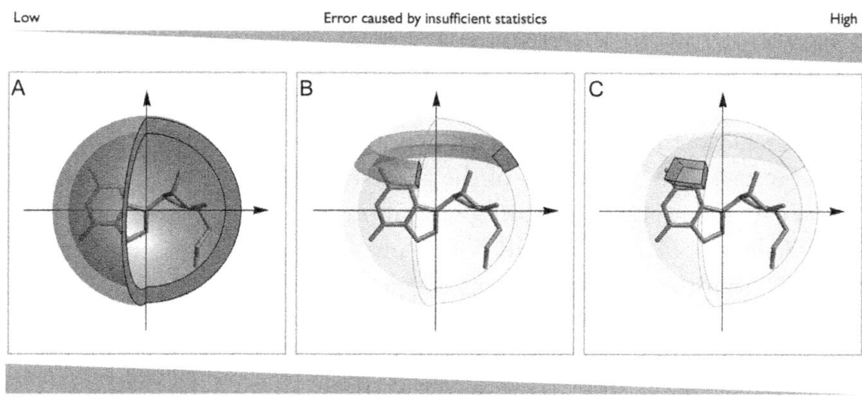

Figure 1 Comparison of three variants of the statistical potential: spherically symmetric (A), axially symmetric (B), and fully anisotropic (C). They model atom–molecule interactions with a gradually increasing number of details, but the statistical errors of the derived potential (resulting from a limited number of know RNA structures) increases as well. The spherically symmetric statistical potentials (A) have been used in DrugScore (for proteins), RiboDock, and DrugScoreRNA. The axially symmetric statistical potential (B) between pair of atoms of the RNA and the atom of the ligand, which is a trade-off between systematic and statistical error, has been used in MetalionRNA and LigandRNA.

It is noteworthy that our choice of the sampling grid is a trade-off between having a detailed enough model of the intermolecular interactions and having large enough counting bins to avoid accumulating statistical error due to an insufficient number of experimentally determined structures of RNAs and RNA–ligand complexes. More detailed representations of the interaction potential are possible, for example, instead of atom–atom or bond–atom interaction, one can decompose the potential into bond–bond interactions:

$$W^{(n)}\left(\vec{r_1}, \vec{r_2}, \ldots, \vec{r_n}\right) = \sum_{abcd} W^{(4)}_{T(a)T(b)T(c)T(d)}\left(r_{ac}, \vartheta_{acb}, \vartheta_{acd}\right),$$

or interactions between atom-triples (of the RNA) and atoms (ions or ligand atoms):

$$W^{(n)}\left(\vec{r_1}, \vec{r_2}, \ldots, \vec{r_n}\right) = \sum_{abcd} W^{(4)}_{T(a)T(b)T(c)T(d)}\left(r_{ad}, \vartheta_{abd}, \varphi_{abcd}\right).$$

Both of these choices would be advantageous over our choices in terms of how well the potential is represented and may be used once a sufficient number of experimentally determined RNA structures will become available.

3.2. Running MetalionRNA and LigandRNA

Both programs have their own freely accessible Web servers at http://metalionrna.genesilico.pl/ and http://ligandrna.genesilico.pl/. The Web interface allows the user to run the computations online without the necessity of installing local versions of the programs.

Step 1: Input RNA structure preparation

MetalionRNA and LigandRNA accept as an input an RNA 3D structure in PDB format. Both programs accept and work with modified residues and can process RNA–protein complexes (however, the protein component is ignored in the calculations). It is also possible to provide a PDB code (instead of a pdb file) of the RNA molecule in the submission forms.

Step 2: Metal ion prediction settings or ligands preparations

On MetalionRNA Web server, the user can specify the metal ion type to be predicted. At present, predictions for magnesium, sodium, and potassium are possible. The ionic radius of the cation can be specified by the user or the default value can be used. The submission form also allows the user to define how many ions MetalionRNA will predict, or one may use the default value, which is calculated on the basis of the number of residues in the target RNA structure. The minimal distance between predicted cations is set to the default value, which is the minimal distance value observed in known structures, but it is also possible to provide this value arbitrarily.

LigandRNA submission form requires providing ligand(s) poses. The poses can be generated by any docking program or their position can be modeled manually. The only condition is that the poses must be in *.mol2 format, which is a typical docking tool input/output. Some programs have their own output formats, in such a case it must be transformed into the *.mol2 format. Poses generated for different ligands (e.g., VS output) are acceptable in the same *.mol2 file.

Step 3: RNA's statistical potential computation

MetalionRNA and LigandRNA use a grid-based function to calculate the potential around RNA molecule submitted by the user. The most important advantage of using a grid is that the discretization of space obviates the need to solve the potential function analytically and allows mapping of the statistical data into well-defined portions of space.

The LigandRNA and MetalionRNA Websites allow the user to select whether to perform predictions with the potentials described in the original articles or with the ones that are being continually updated. The potential update is automatically performed once per week. The servers download structures of RNA–cation and RNA–ligand complexes newly released in the PDB, which fulfill the conditions, described in the section derivation of the statistical potential—practical information. These structures are added to the original training sets and the statistical potentials are recalculated. This way, the potentials are continually improved and their statistical power continually increases.

To compute the RNA molecule potential, our programs divide the search space around RNA into a cubic grid with a grid width of 0.5 Å. This value is small enough to cover the space between the RNA atom and the chelated cation/ligand atom with at least a few grid cells. For a larger grid width of the cubic grid, the statistics would be biased by the cell boundaries, negatively affecting the prediction quality. For smaller cells, the computational power and time increase significantly, without noticeable influence on the accuracy of the results (for more details, see MetalionRNA original article (Philips et al., 2012)). However, MetalionRNA server offers a grid width of 0.25 Å to choose for analysis of a special purpose, but we do not recommend this for common practice. Subsequently, for each RNA atom pair, the program computes the potential value in all cells of the cubic grid within the radius of 9 Å around the atom (MetalionRNA), or in case of LigandRNA within the radius of 6 Å. Next, the potential values are computed for a selected ion type (MetalionRNA), or for all atom types present in the submitted ligands (LigandRNA). Whenever the grid cell coincidences with any RNA atom (is within its van der Waals radius), the grid cell is discarded as cations/ligands cannot overlap with the RNA molecule. This prevents MetalionRNA from placing cations too close to the RNA atoms and LigandRNA from favoring incorrect ligand poses with steric clashes with RNA.

Step 4: Generation of the output

MetalionRNA ranks grid cells and, for the top-scored, all cells within a radius corresponding to half of the minimal distance between two cations of the same type are examined. The radius of the new candidate cation cannot overlap with the radius of a previously proposed cation with a better score. If this condition is fulfilled, MetalionRNA places a cation in the center of the top-scored cell and removes the cells covered by the new cation from further consideration. This procedure is repeated until a default or user-defined number of predicted cation positions is reached.

LigandRNA does not place ligand posed by itself, but scores ligand poses submitted by the user. The pose score is the sum of its atoms' potentials (all cells within the van der Waals radius of a certain ligand atom type are examined) divided by the number of its heavy atoms. The scoring scheme allows comparison between scores obtained for different ligands. For poses generated by the Dock6 program, there is an additional option to compute a LigandRNA–Dock6 combined score, which significantly increases the accuracy of both methods (Philips, Milanowska, Lach, & Bujnicki, 2013). The added value of the combination of LigandRNA and Dock6 results most likely from the very different, and hence complementary, character of both scoring functions. LigandRNA is a knowledge-based statistical potential, while Dock6 scores ligand poses on the basis of a physics-based force field.

The time required for MetalionRNA and LigandRNA to return predictions depends mainly on the size of the RNA molecule. Currently, our Web service uses a simple queuing system that allows running one prediction at a time. For the majority of riboswitches with known structures, where a typical length is around ~60 to ~120 nt, it takes about 5–10 min to obtain the results. The result files are available as separate Web pages. The pages with the output files are kept on the server for 1 week. MetalionRNA output consists of (1) a predicted cation positions in text and (2) a PDB formatted file and a script to display the predicted cations in the PyMOL viewer. LigandRNA output consists of (1) a file with ranked ligand poses in text format and (2) the file that contains ligand poses' ids and their LigandRNA score. For the combined LigandRNA–Dock6 score, there is an additional text file with the ligand poses ranked according to the consensus score. On both servers, a Jmol applet is available for displaying the best results for immediate verification by the user. Moreover, MetalionRNA and LigandRNA return PDB files containing the RNA receptor structure with the statistical potential mapped on the individual atoms of the receptor (averaged for all grid cells within 2 Å from a given atom). MetalionRNA returns one PDB file with the statistical potential for the RNA structure of the previously selected ion type. LigandRNA returns four PDB files with the four variants of the LigandRNA potential for O, C, and N atoms of the ligand separately, and for all atoms combined. With such modified PDB files, the distribution of potential values on surface atoms can be easily displayed in all commonly used structure visualization systems by coloring atoms according to the B-factor field. The Jmol applet also allows for potential visualization.

3.3. Use of MetalionRNA and LigandRNA for riboswitches

3.3.1 Prediction of metal ion-binding sites in riboswitches with MetalionRNA

Originally, the predictive power of MetalionRNA was proved by fivefold cross-validation test using a set of 290 RNA–metal ion complexes and illustrated in the form of receiver-operating characteristic (ROC) plots. This benchmark showed the high accuracy of MetalionRNA, as it reached for RNA–Mg^{2+} ~96% AUC (AUC—area under ROC curve) for the cut-off distance (the maximum distance between a predicted and a real metal ion in which the prediction is marked as correct) up to 3 Å; for RNA–K^+ ~97% AUC for cut-off distance of 4 Å (this is because the ionic radius of K^+ (1.38 Å) is much larger than the Mg^{2+} ionic radius, 0.72 Å); for RNA–Na^+ ~91% for a maximum distance of 3 Å between the predicted and the real Na^+ ion in which the prediction is marked as correct. (The Na^+ ionic radius is 1 Å.) Additionally, our test proved that MetalionRNA can be used also to predict ion-binding sites in DNA, as it reached ~93% AUC for cut-off distance of 3 Å for DNA–Mg^{2+} complexes. For details of this benchmark, see the original paper on MetalionRNA (Philips et al., 2012).

To show the MetalionRNA tool's ability of identifying metal ion-binding site in riboswitches, we chose two complexes deposited after the MetalionRNA release, which were not included in the original MetalionRNA test and training sets.

A crystal structure of cobalamin riboswitch (PDB ID: 4FRG) was solved at the resolution of 2.9 Å. The structure contains seven Mg^{2+} ions, of which one (no. 114) is not directly bound to the structure. MetalionRNA calculated that for a structure of this size (84 nt) eight ions are expected to be observed in a crystal structure solved under "average" conditions, hence the eight top-scored predictions are considered in our test. MetalionRNA correctly predicted three out of six magnesium-binding sites using the basic version of the potential (see Table 1); Mg^{2+} no. 111 (RMSD to the real ion 2 Å), no. 113 (RMSD 1 Å), and no. 116 (RMSD 3.1 Å). Using the updated variant of the potential (one that took new RNA structures into account) it managed to find two binding sites with very good accuracy of 1 and 3.1 Å (Mg^{2+} 113 and 116, respectively) and one with the accuracy of 4.2 Å (111). Using the potential derived from a manually curated dataset of Mg^{2+}-binding sites, MetalionRNA found four ions with very high accuracy of 0.5, 1.8, 2.1, and 1 Å, ions number: 111, 115, 116, and 117. The "curated" potential identified two Mg^{2+}-binding sites missed by two other variants of the potential: Mg^{2+} no. 115 and 116 (RMSD of 1.8 and 1.0 Å, respectively). However, it missed ion no. 113, correctly identified with high accuracy by basic and updated potentials. None of the potentials identified ion-binding sites 112 and 114 (see Fig. 2).

The SAM-I riboswitch structure (PDB ID: 4B5R) contains seven K^+ ions. Its crystal structure was solved at the resolution of 2.95 Å. It is 94 nt

Table 1 A list of Mg^{2+} ions in the cobalamin riboswitch structure (PDB ID: 4FRG) for which predictions using MetalionRNA were done

Mg^{2+} (atom no.)	Results for the original potential (Philips et al., 2012)	Results for potential based on the updated dataset	Results for potential derived from curated binding sites (based on the Minor group dataset)
111	2.0 Å (rank 8)	4.2 Å (rank 7)	0.5 Å (rank 2) 2.3 Å (rank 1)
112	X	X	X
113	1.0 Å (rank 4)	1.0 Å (rank 5)	X
114	5.7 Å (rank 4)	5.7 Å (rank 5)	X
115	X	X	1.8 Å (rank 4)
116	3.1 Å (rank 2)	3.1 Å (rank 3)	2.1 Å (rank 7)
117	X	X	1.0 Å (rank 3)

The first column from the left lists real Mg^{2+} ions' identifiers as labeled in the PDB file 4FRG. Column 2 describes the predictions made by MetalionRNA using the original Mg^{2+} potential. Column 3 describes the MetalionRNA predictions made using the updated potential, column 4 describes the MetalionRNA predictions made using the potential derived from binding sites, curated by the Minor. In all columns, the first value is a prediction distance to the respective Mg^{2+} ion, the value in the brackets is the prediction rank by score with respect to the total number of all generated binding sites (eight).

long, so MetalionRNA computed that an "average" structure of that size may have nine bound K^+ ions. The results for this riboswitch are presented in Table 2 and Fig. 3. Again, we run MetalionRNA prediction for this structure using the original and updated potentials. The usage of "curated" potential was not possible, as it is relevant only for Mg^{2+} ions-binding sites. Four K^+ cations (no. 1100, 1102, 1103, and 1113) were identified by the updated potential with an accuracy of 1.8, 2.0, 3.3, and 3.4 Å, respectively. The original potential failed to find K^+ cation 1103. Two K^+-binding sites were identified neither by basic nor updated potential (K^+ no. 1097 and 1098). These two cations are on the RNA surface, so they may be diffusely bound (if at all).

3.3.2 *Prediction of small organic ligand-binding sites in riboswitches with LigandRNA*

To test the ability of LigandRNA to discriminate between native-like and non-native-like poses of small-molecule ligands with respect to their riboswitch receptors, and to compare its performance to other methods, we

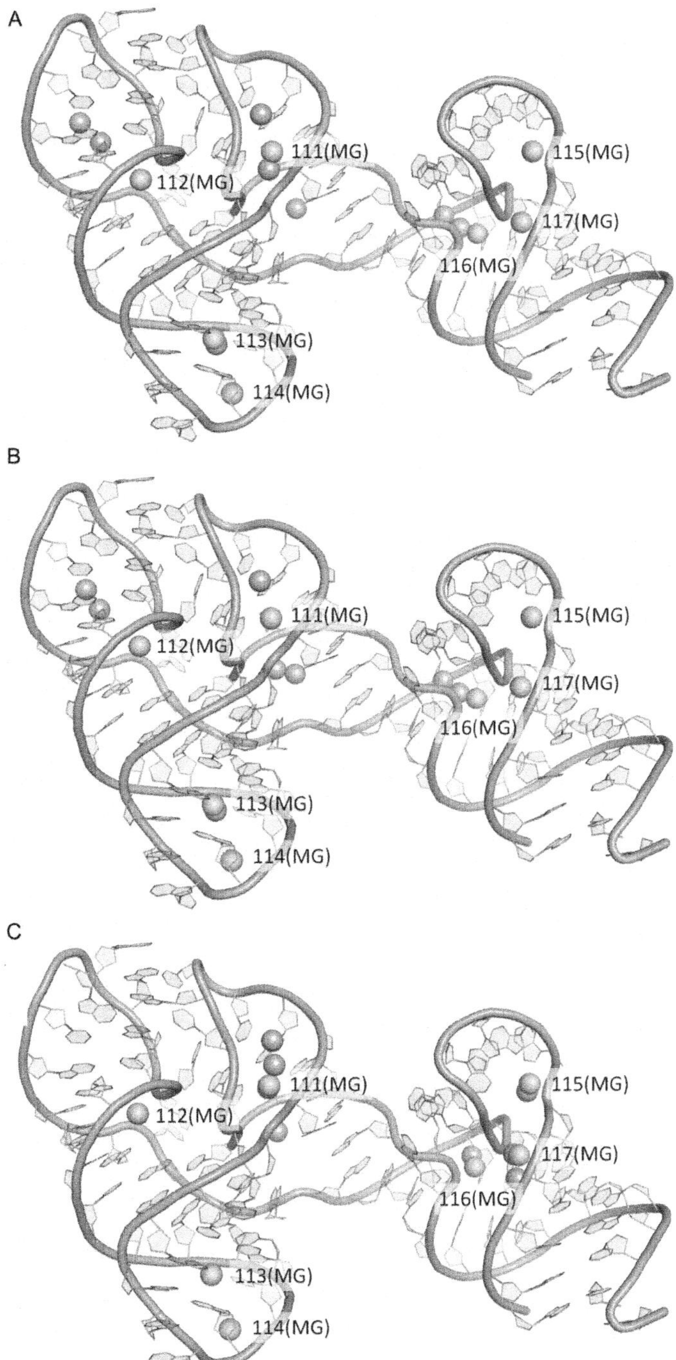

Figure 2 Cobalamin riboswitch structure (PDB ID: 4FRG) with the experimentally determined positions of Mg^{2+} cations indicated by green (dark gray in the print version) labeled balls. Top-scoring Mg^{2+} cations predicted by MetalionRNA are shown as gray balls; (A) using the original potential, (B) using the updated potential, and (C) using the curated potential; for more details, see Table 1.

Table 2 A list of K^+ ions in the SAM-I riboswitch structure (PDB ID: 4B5R) for which predictions using MetalionRNA were done

K^+ (atom no.)	Results for the original potential (Philips et al., 2012)	Results for potential based on the updated dataset
1097	X	X
1098	X	X
1100	1.8 (rank 2)	1.8 (rank 2)
1102	2.1 (rank 5)	2.0 (rank 3)
1103	X	3.3 (rank 5)
1113	2.6 (rank 8)	3.4 (rank 8)
1114	5.1 (rank 4)	6.4 (rank 6)

The first column from the left lists real K^+ ions' identifiers as labeled in the PDB file 4B5R. Column 2 describes the predictions made by MetalionRNA using the original K^+ potential. Column 3 describes the MetalionRNA predictions made using the updated potential. In both the columns, the first value is a prediction distance to the respective K^+ ion, the value in the brackets is the prediction rank by score with respect to the total number of all generated binding sites (nine).

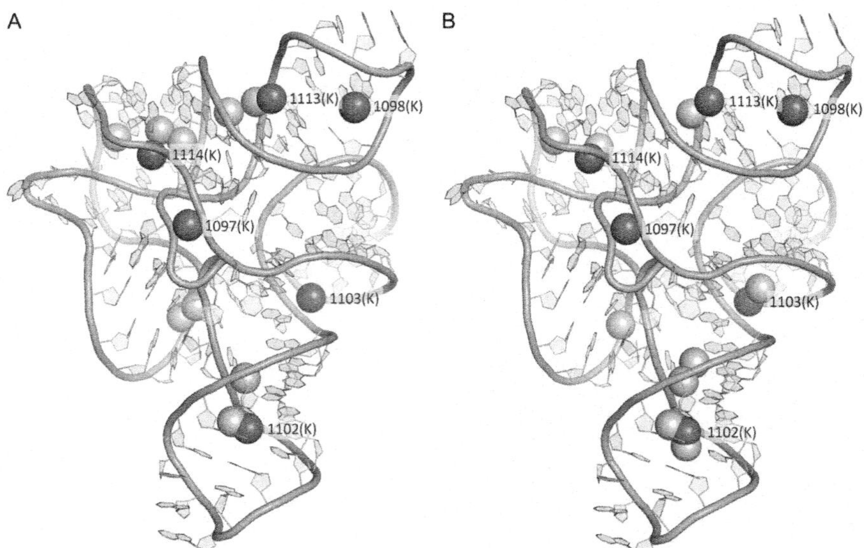

Figure 3 SAM-I riboswitch structure (PDB ID: 4B5R) with the experimentally determined positions of K^+ cations indicated by purple labeled balls. The top-scoring K^+ cations predicted by MetalionRNA are shown as gray balls (A) using the original potential and (B) using the updated potential; for more details, see Table 2. (See the color plate.)

ran a benchmark test with separate training and test datasets (Philips et al., 2013). For each structure from the test set (15 complexes), a few hundred ligand poses were generated with Dock6 (with default parameters that include the following settings: a grid-score function, a flexible ligand docking, all atom model, automated matching, and an internal energy calculation. The internal energy settings included the following special options: bump filter, chemical matching, secondary scoring, etc., disabled. The additional criterion was that at least one pose had to have a root mean square deviation of all heavy atoms (RMSD) ≤2 Å to the reference experimental structure. This restriction resulted from the fact that LigandRNA is a scoring function for evaluation of ligand poses and would never find a near-native in the dataset, where such poses are not present. The poses generated by docking were then scored with LigandRNA, DrugScoreRNA, and combinations of the aforementioned potentials (including the Dock6 scoring function).

LigandRNA found the best solution with RMSD ≤ 3 Å to the native structure in 13 cases, DrugScoreRNA in 9 cases, and Dock6 in 11 cases. The combined potential of LigandRNA and Dock6 gave the best result among all methods, as it found solutions with RMSD ≤ 3 Å to the reference structure in 15 out of 15 cases (see Table 3). Our benchmarks demonstrated that the LigandRNA potential is effective in identifying poses that are close to the experimentally determined structures. Among 15 structures of riboswitch–ligand complexes, for which the docking procedure was able to generate at least one pose with RMSD ≤ 2 Å to the native structure, top-scored solutions proposed by LigandRNA, DrugScoreRNA, and Dock6, had RMSD ≤ 3 Å in 87%, 60%, and 73% of the cases, respectively. Moreover, we tested various combinations of the individual scoring functions (data not shown) and we found that a "meta-predictor" comprising Dock6 and LigandRNA improves the accuracy of individual predictions. The combination of the Dock6 and LigandRNA scoring achieves 100% correctly identified ligand poses (with RMSD ≤ 3 Å to the native structure).

Our riboswitch test set comprises mostly complexes included in test sets of other methods described in Section 1 (RiboDock, scoring function for aminoglycosides, the original DrugScoreRNA and MORDOR), so we were able to compare the results obtained in our study for a subset of these complexes to the results reported in the original publications of the aforementioned methods. RiboDock found solution with RMSD ≤ 3 Å in 5 cases out of 7, the original DrugScoreRNA potential in 7 out of 9 cases, and MORDOR generated near-native poses in 11 out of 12 cases. Moitessier's potential for aminoglycosides was not tested on riboswitch structures.

Table 3 Results of test for 15 riboswitch–ligand complexes

PDB ID	Ligand type	LigandRNA	DrugScore RNA	Dock6	LigandRNA+Dock6	RiboDock according to Morley and Afshar (2004)	Original DrugScore RNA according to Pfeffer and Gohlke (2007)	MORDOR (Guilbert & James, 2008)
1Q8N	Malachite green	1.3	1.1	7.7	2.0	–	3.7	1.9
1F1T	Malachite green	1.1	1.2	1.2	1.2	–	–	0.3
1NBK	Argininamide	3.0	1.5	2.6	2.1	–	8.2	3.6
1KOC	Arginine	2.3	2.1	2.6	2.2	2.7	1.6	1.6
1Y26	Adenine	0.5	30.5	0.5	0.5	–	–	0.5
1FMN	Flavin mononucleotide	2.9	2.2	2.9	2.9	0.8	1.6	1.4
1TOB	Aminoglycoside	2.6	3.5	4.8	2.3	9.8	1.5	2.6
2TOB	Aminoglycoside	1.0	1.0	1.5	1.5	1.5	1.5	0.9
1NEM	Neomycin	9.3	1.1	0.7	1.2	8.7	0.7	1.0
1AM0	Adenosine monophosphate	4.6	5.7	1.4	1.7	1.8	2.9	0.9
3GX2	s-Adenosylmethionine	0.6	5.3	4.7	0.7	–	–	–
3D2X	Thiamine pyrophosphate	1.6	6.9	1.9	1.9	–	–	
1KOD	Citrulline	1.9	5.1	5.3	2.0	3.2	1.9	2.0
2GDI	Thiamine pyrophosphate	1.8	2.1	2.1	2.1	–	–	2.1
3SUX	5-Hydroxymethylene-6-hydrofolic acid	0.4	0.4	0.5	0.4	–	–	–

The first column presents the PDB codes of the reference RNA–ligand structures determined experimentally, the second ligand type. Columns three to six show the RMSD values (in Angstroms) for top-scored poses returned by the methods that we tested: LigandRNA, DrugScoreRNA, Dock6, and the combined potential of LigandRNA and Dock6. Columns seven to nine show the RMSD values for top-scored poses reported by the authors of the other methods for prediction of RNA–ligand interactions (RiboDock, the original DrugScoreRNA, and MORDOR). We present results only for complexes which were included in both test sets (ours and the other method's).

Table 4 Results of the tests on newly deposited 15 riboswitch–ligand complexes

PDB ID	Ligand type	LigandRNA	Dock6	LigandRNA + Dock6
4FEP	9H-Purine-2,6-diamine	0.7	0.1	0.1
4LVZ	9H-Purine-2,6-diamine	3.4	3.5	3.5
4FEO	9H-Purine-2,6-diamine	0.4	0.3	0.3
4LVW	7-Deazaguanine	2.7	0.4	2.7
4LW0	Adenine	0.2	0.4	0.3
4LVX	Tetrahydrobiopterin	3.8	3.8	3.8
4FEL	Hypoxanthine	0.6	0.2	0.2
4FEJ	Hypoxanthine	2.4	0.1	0.1
4FEN	Hypoxanthine	2.5	0.1	0.1
4LVY	Pemetrexed	4.4	4.4	4.4
2MIY	**7-Deaza-7-aminomethyl-guanine**	**0.7**	**3.7**	**0.7**
4JF2	7-Deaza-7-aminomethyl-guanine	0.5	0.3	0.4
4B5R	**s-Adenosylmethionine**	**1.5**	**7.2**	**5.6**
4KQY	**s-Adenosylmethionine**	**1.0**	**8.8**	**5.0**

The first column presents the PDB codes of the reference RNA–ligand structures determined experimentally, the second ligand type. Columns three to five show RMSD values (in Angstroms) for top-scored poses returned by the methods that we tested: LigandRNA, Dock6, and the combined potential of LigandRNA, and Dock6. Hits with RMSD ≤ 3 Å to the reference structure, identified only by LigandRNA are indicated in bold.

For an additional test, we decided to run LigandRNA for riboswitch–ligand complexes deposited in PDB after the original publication of LigandRNA. We collected a set of 17 structures, for which we generated a few hundred ligand poses with Dock6 (with defined ligand-binding site and default parameters). For 15 complexes, we obtained at least one pose with RMSD ≤ 2 Å. Next, we scored all poses using LigandRNA and the LigandRNA–Dock6 combined score. Unfortunately, we do not consider the results obtained by Dock6 to be completely unbiased, as the docking procedure was conducted in such a way as to force the program to generate at least a few near-native ligand poses (a ligand-binding site was defined in a place where the native ligand was). Table 4 summarizes the results. LigandRNA found solutions with RMSD ≤ 3 Å in 11 cases, Dock6 in 8 cases, the combination of LigandRNA–Dock6 score in 9 cases. In three

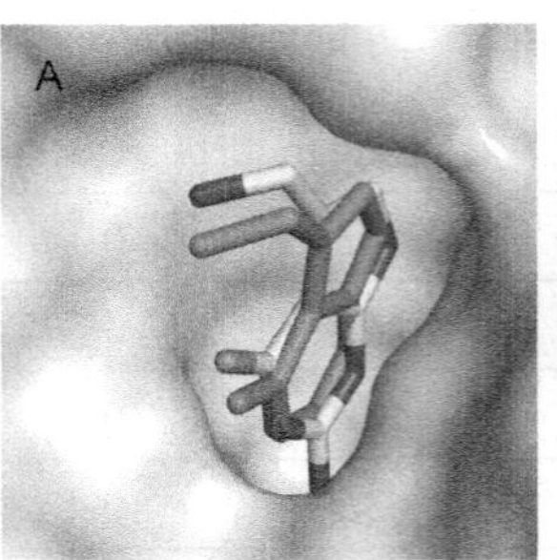

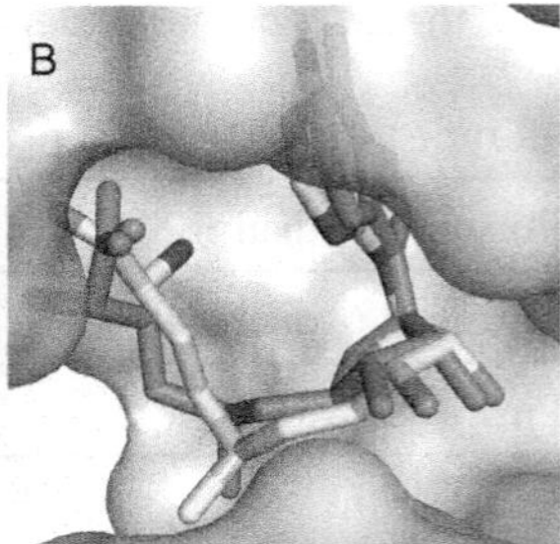

Figure 4 (A) Class II preQ1 riboswitch (PDB ID: 2MIY) in complex with 7-deaza-7-aminomethyl-guanine (PRF) and (B) SAM-I riboswitches (PDB ID: 4KQY) in complex with s-adenosylmethionine (SAM). The experimentally identified ligand poses are shown in light yellow, and the poses from computational docking identified as best-scored by LigandRNA are depicted in red. (See the color plate.)

cases, LigandRNA was the only scoring function that found a near-native ligand pose. This was for class II preQ1 riboswitch (PDB ID: 2MIY) in complex with 7-deaza-7-aminomethyl-guanine (PRF) and for two SAM-I riboswitches in complex with s-adenosylmethionine (SAM), but from different bacteria (*Thermoanaerobacter tengcongensis* (PDB ID: 4B5R) and *Bacillus subtilis* (PDB ID: 4KQY)). The results are additionally presented in the Fig. 4.

We also tested the updated potential, for which 17 new RNA–small molecule complexes were added to the original dataset. The results were essentially identical, in agreement with the limited influence of a small number of structures on the performance of a statistical potential.

ACKNOWLEDGMENTS

We thank the current and former members of the Bujnicki group for discussions and constructive comments; in particular, we thank those who contributed to the development of MetalionRNA and LigandRNA: Kaja Milanowska, Kristian Rother, and Michał Boniecki. We also thank Wayne Dawson for extensive discussions on physics- and statistics-based predictive methods. We thank Wladek Minor for collaboration on the identification and characterization of Mg^{2+}-binding sites. We also thank Wayne Dawson, Łukasz Malczak, and Kaja Milanowska for proofreading the chapter. This work was funded primarily by Foundation for Polish Science (grant TEAM/2009-4/2 and "Ideas for Poland" fellowship to J. M. B.); the development of LigandRNA was additionally funded by the Polish National Science Centre (grant NCN 2011/03/N/NZ2/01428 to A. P.); the development and maintenance of Web servers in the Bujnicki laboratory is supported by the Polish Ministry of Science and Higher Education (grant POIG.02.03.00-00-003/09 to J. M. B.).

REFERENCES

Baker, C. M., Anisimov, V. M., & MacKerell, A. D., Jr. (2011). Development of CHARMM polarizable force field for nucleic acid bases based on the classical Drude oscillator model. *The Journal of Physical Chemistry B*, *115*(3), 580–596. http://dx.doi.org/10.1021/jp1092338.

Baker, J. L., Sudarsan, N., Weinberg, Z., Roth, A., Stockbridge, R. B., & Breaker, R. R. (2012). Widespread genetic switches and toxicity resistance proteins for fluoride. *Science*, *335*(6065), 233–235. http://dx.doi.org/10.1126/science.1215063.

Banatao, D. R., Altman, R. B., & Klein, T. E. (2003). Microenvironment analysis and identification of magnesium binding sites in RNA. *Nucleic Acids Research*, *31*(15), 4450–4460.

Brooks, B. R., Brooks, C. L., 3rd., Mackerell, A. D., Jr., Nilsson, L., Petrella, R. J., Roux, B., et al. (2009). CHARMM: The biomolecular simulation program. *Journal of Computational Chemistry*, *30*(10), 1545–1614.

Dieterich, C., & Stadler, P. F. (2013). Computational biology of RNA interactions. *Wiley Interdisciplinary Reviews: RNA*, *4*(1), 107–120. http://dx.doi.org/10.1002/wrna.1147.

Draper, D. E. (2004). A guide to ions and RNA structure. *RNA*, *10*(3), 335–343.

Draper, D. E. (2008). RNA folding: Thermodynamic and molecular descriptions of the roles of ions. *Biophysical Journal*, *95*(12), 5489–5495.

Fulle, S., & Gohlke, H. (2010). Molecular recognition of RNA: Challenges for modelling interactions and plasticity. *Journal of Molecular Recognition*, *23*(2), 220–231. http://dx.doi.org/10.1002/jmr.1000.

Garst, A. D., Edwards, A. L., & Batey, R. T. (2011). Riboswitches: Structures and mechanisms. *Cold Spring Harbor Perspectives in Biology*, *3*(6). http://dx.doi.org/10.1101/cshperspect.a003533, a003533.003510.001101/cshperspect.a003533 [pii].

Gohlke, H., Hendlich, M., & Klebe, G. (2000). Knowledge-based scoring function to predict protein-ligand interactions. *Journal of Molecular Biology*, *295*(2), 337–356.

Guilbert, C., & James, T. L. (2008). Docking to RNA via root-mean-square-deviation-driven energy minimization with flexible ligands and flexible targets. *Journal of Chemical Information and Modeling*, *48*(6), 1257–1268. http://dx.doi.org/10.1021/ci8000327.

Hermann, T., & Westhof, E. (1998). Exploration of metal ion binding sites in RNA folds by Brownian-dynamics simulations. *Structure*, *6*(10), 1303–1314.

Kuntz, I. D., Blaney, J. M., Oatley, S. J., Langridge, R., & Ferrin, T. E. (1982). A geometric approach to macromolecule-ligand interactions. *Journal of Molecular Biology*, *161*(2), 269–288, 0022-2836(82)90153-X [pii].

Lang, P. T., Brozell, S. R., Mukherjee, S., Pettersen, E. F., Meng, E. C., Thomas, V., et al. (2009). DOCK 6: Combining techniques to model RNA-small molecule complexes. *RNA*, *15*(6), 1219–1230. http://dx.doi.org/10.1261/rna.1563609.

Liang, M. P., Banatao, D. R., Klein, T. E., Brutlag, D. L., & Altman, R. B. (2003). WebFEATURE: An interactive web tool for identifying and visualizing functional sites on macromolecular structures. *Nucleic Acids Research*, *31*(13), 3324–3327.

Mandal, M., Boese, B., Barrick, J. E., Winkler, W. C., & Breaker, R. R. (2003). Riboswitches control fundamental biochemical pathways in Bacillus subtilis and other bacteria. *Cell*, *113*(5), 577–586.

Mandal, M., Lee, M., Barrick, J. E., Weinberg, Z., Emilsson, G. M., Ruzzo, W. L., et al. (2004). A glycine-dependent riboswitch that uses cooperative binding to control gene expression. *Science*, *306*(5694), 275–279. http://dx.doi.org/10.1126/science.1100829.

Mironov, A. S., Gusarov, I., Rafikov, R., Lopez, L. E., Shatalin, K., Kreneva, R. A., et al. (2002). Sensing small molecules by nascent RNA: A mechanism to control transcription in bacteria. *Cell*, *111*(5), 747–756.

Misra, V. K., & Draper, D. E. (2000). Mg(2+) binding to tRNA revisited: The nonlinear Poisson-Boltzmann model. *Journal of Molecular Biology*, *299*(3), 813–825. http://dx.doi.org/10.1006/jmbi.2000.3769.

Moitessier, N., Westhof, E., & Hanessian, S. (2006). Docking of aminoglycosides to hydrated and flexible RNA. *Journal of Medicinal Chemistry, 49*(3), 1023–1033.

Montange, R. K., & Batey, R. T. (2008). Riboswitches: Emerging themes in RNA structure and function. *Annual Review of Biophysics, 37*, 117–133. http://dx.doi.org/10.1146/annurev.biophys.37.032807.130000.

Morley, S. D., & Afshar, M. (2004). Validation of an empirical RNA-ligand scoring function for fast flexible docking using Ribodock. *Journal of Computer-Aided Molecular Design, 18*(3), 189–208.

Morris, G. M., Huey, R., Lindstrom, W., Sanner, M. F., Belew, R. K., Goodsell, D. S., et al. (2009). AutoDock4 and AutoDockTools4: Automated docking with selective receptor flexibility. *Journal of Computational Chemistry, 30*(16), 2785–2791. http://dx.doi.org/10.1002/jcc.21256.

Mulhbacher, J., Brouillette, E., Allard, M., Fortier, L. C., Malouin, F., & Lafontaine, D. A. (2010). Novel riboswitch ligand analogs as selective inhibitors of guanine-related metabolic pathways. *PLoS Pathogens, 6*(4), e1000865. http://dx.doi.org/10.1371/journal.ppat.1000865.

Mulhbacher, J., St-Pierre, P., & Lafontaine, D. A. (2010). Therapeutic applications of ribozymes and riboswitches. *Current Opinion in Pharmacology, 10*(5), 551–556. http://dx.doi.org/10.1016/j.coph.2010.07.002.

Pettersen, E. F., Goddard, T. D., Huang, C. C., Couch, G. S., Greenblatt, D. M., Meng, E. C., et al. (2004). UCSF Chimera—A visualization system for exploratory research and analysis. *Journal of Computational Chemistry, 25*(13), 1605–1612.

Pfeffer, P., & Gohlke, H. (2007). DrugScoreRNA—Knowledge-based scoring function to predict RNA-ligand interactions. *Journal of Chemical Information and Modeling, 47*(5), 1868–1876. http://dx.doi.org/10.1021/ci700134p.

Philips, A., Milanowska, K., Lach, G., Boniecki, M., Rother, K., & Bujnicki, J. M. (2012). MetalionRNA: Computational predictor of metal-binding sites in RNA structures. *Bioinformatics, 28*(2), 198–205. http://dx.doi.org/10.1093/bioinformatics/btr636.

Philips, A., Milanowska, K., Lach, G., & Bujnicki, J. M. (2013). LigandRNA: Computational predictor of RNA-ligand interactions. *RNA, 19*(12), 1605–1616. http://dx.doi.org/10.1261/rna.039834.113.

Rivas, E., & Eddy, S. R. (2001). Noncoding RNA gene detection using comparative sequence analysis. *BMC Bioinformatics, 2*, 8.

Rodionov, D. A., Vitreschak, A. G., Mironov, A. A., & Gelfand, M. S. (2003). Regulation of lysine biosynthesis and transport genes in bacteria: Yet another RNA riboswitch? *Nucleic Acids Research, 31*(23), 6748–6757.

Ruiz-Carmona, S., Alvarez-Garcia, D., Foloppe, N., Garmendia-Doval, A. B., Juhos, S., Schmidtke, P., et al. (2014). rDock: A fast, versatile and open source program for docking ligands to proteins and nucleic acids. *PLoS Computational Biology, 10*(4), e1003571. http://dx.doi.org/10.1371/journal.pcbi.1003571.

Schnabl, J., & Sigel, R. K. (2010). Controlling ribozyme activity by metal ions. *Current Opinion in Chemical Biology, 14*(2), 269–275, S1367-5931(09)00195-1 [pii].

Serra, M. J., Baird, J. D., Dale, T., Fey, B. L., Retatagos, K., & Westhof, E. (2002). Effects of magnesium ions on the stabilization of RNA oligomers of defined structures. *RNA, 8*(3), 307–323.

Sigurdsson, S. T., & Eckstein, F. (1995). Structure-function relationships of hammerhead ribozymes: From understanding to applications. *Trends in Biotechnology, 13*(8), 286–289, S0167-7799(00)88966-0 [pii].

Sippl, M. (1993). Boltzmann's principle, knowledge-based mean fields and protein folding. An approach to the computational determination of protein structures. *Journal of Computer-Aided Molecular Design*, 7, 473–501.

Sippl, M. J. (1995). Knowledge-based potentials for proteins. *Current Opinion in Structural Biology, 5*(2), 229–235.

Tan, Z. J., & Chen, S. J. (2005). Electrostatic correlations and fluctuations for ion binding to a finite length polyelectrolyte. *The Journal of Chemical Physics, 122*, 044903.

Tan, Z. J., & Chen, S. J. (2010). Predicting ion binding properties for RNA tertiary structures. *Biophysical Journal, 99*(5), 1565–1576, S0006-3495(10)00772-1 [pii].

Thomas, J. R., & Hergenrother, P. J. (2008). Targeting RNA with small molecules. *Chemical Reviews, 108*(4), 1171–1224. http://dx.doi.org/10.1021/cr0681546.

Warner, D. F., Savvi, S., Mizrahi, V., & Dawes, S. S. (2007). A riboswitch regulates expression of the coenzyme B12-independent methionine synthase in Mycobacterium tuberculosis: Implications for differential methionine synthase function in strains H37Rv and CDC1551. *Journal of Bacteriology, 189*(9), 3655–3659.

Winkler, W. C., Cohen-Chalamish, S., & Breaker, R. R. (2002). An mRNA structure that controls gene expression by binding FMN. *Proceedings of the National Academy of Sciences of the United States of America, 99*(25), 15908–15913. http://dx.doi.org/10.1073/pnas.212628899.

Zheng, H., Chruszcz, M., Lasota, P., Lebioda, L., & Minor, W. (2008). Data mining of metal ion environments present in protein structures. *Journal of Inorganic Biochemistry, 102*(9), 1765–1776. http://dx.doi.org/10.1016/j.jinorgbio.2008.05.006.

CHAPTER TWELVE

Computational Prediction of Riboswitches

P. Clote[1]
Biology Department, Boston College, Boston, Massachusetts, USA
[1]Corresponding author: e-mail address: clote@bc.edu

Contents

Abstract

Riboswitches present a ubiquitous genetic regulatory mechanism for prokaryotes and have been found in HIV1, fungi, plants, and even *H. sapiens*. We present an overview of approaches to predict riboswitch aptamers and, more generally, RNA conformational switches.

1. INTRODUCTION

The detection and functional annotation of noncoding RNA (ncRNA) genes remains a task of great biological importance, since it is now understood that the human genome is "pervasively transcribed," where most transcripts have no known function. Indeed, analysis of the ENCODE Consortium data (Birney et al., 2007) indicates that 93% of the human genome may be transcribed in multiple RNAs (Birney et al., 2007), and that "given sufficient sequencing depth the whole genome may appear as transcripts" (Van Bakel, Nislow, Blencowe, & Hughes, 2010). ncRNA is

Methods in Enzymology, Volume 553
ISSN 0076-6879
http://dx.doi.org/10.1016/bs.mie.2014.10.063

now known to be involved in a wide range of previously unsuspected roles in many biological processes, including *retranslation* of the genetic code (selenocysteine insertion (Böck, Forschhammer, Heider, & Baron, 1991), ribosomal frameshift (Bekaert et al., 2003)), transcriptional and translational gene regulation (*riboswitches* (Lim, Glasner, Yekta, Burge, & Bartel, 2003; Mandal, Boese, Barrick, Winkler, & Breaker, 2003)), temperature-sensitive conformational switches (repression of heat-shock gene expression (ROSE) elements, fourU thermometers (Chowdhury, Ragaz, Kreuger, & Narberhaus, 2003; Tucker & Breaker, 2005)), chemical modification of specific nucleotides in the ribosome (Omer et al., 2000), regulation of alternative splicing (Cheah, Wachter, Sudarsan, & Breaker, 2007), long ncRNA (Kung, Colognori, & Lee, 2013), etc.

Since nucleotide statistics, such as codon usage frequencies, distinguish protein-coding genes from background genomic regions, hidden Markov models (HMMs) and generalized HMMs have proven to be efficient, accurate gene finders, e.g., `Glimmer` (Delcher, Harmon, Kasif, White, & Salzberg, 1999) for prokaryotic genomes and `GenScan` (C. Burge & Karlin, 1997) for eukaryotic genomes. The situation is quite different with ncRNA, i.e. transcribed RNA that does not code for a protein. Examples of ncRNA are transfer RNA, ribosomal RNA (rRNA), microRNA (miRNA), piRNA, and *cis*-regulatory elements, which are portions of the $5'$ or $3'$ untranslated region (UTR) of messenger RNA that control the transcription or translation of the mRNA. In such ncRNAs, an extension of HMMs known as *stochastic context-free grammars* (SCFGs) has been introduced in Eddy and Durbin (1994) to recognize RNAs, whose alignment with a training set of RNAs, indicates a covariation compatible with the common consensus secondary structure. The transfer RNA gene finder `tRNAscan-SE` (Lowe & Eddy, 1997) is an example application of SCFG.

Other specialized gene finders exist, trained using support vector machines (SVM) (Vapnik, 1998) to recognize precursor miRNAs (Hertel & Stadler, 2006; Ng & Mishra, 2007; Xue et al., 2005) and small nucleolar RNAs (snoRNAs) (Hertel, Hofacker, & Stadler, 2008). Here, miRNAs are processed from $\sim$ 70 nt stem–loop precursors by the RNase III nuclease Dicer. miRNAs hybridize to a portion of target messenger RNA and cause the degradation of target mRNA (plants) or prevent the translation of target mRNA (plants and animals) (Lim et al., 2003; Tuschl, 2003). snoRNAs guide the methylation or pseudouridylation of rRNAs and splicesomal RNAs. Some general ncRNA and *cis*-regulatory module gene finders have been developed, using either SCFGs or thermodynamics-based methods. In particular, `Infernal` (Nawrocki & Eddy, 2013b; Nawrocki, Kolbe, & Eddy, 2009) is a SCFG that can be trained on any structural alignment of RNAs—indeed, the Rfam database (S. W. Burge et al., 2013) is maintained and extended by `Infernal` software. Using thermodynamics-based methods, Washietl et al. (Gruber, Neubock, Hofacker, & Washietl, 2007; Washietl, Hofacker, & Stadler, 2005) developed the algorithm `RNAz`, which works as follows. Given a multiple sequence alignment of related RNAs and a novel RNA to classify, `RNAz` compares the average minimum free energy of the RNAs, as computed by `RNAfold` from the Vienna RNA Package (Hofacker, 2003), with the minimum free energy of the entire alignment, as computed by an extension of `RNAfold` that includes an energy bonus for compensatory mutations (covariation of base-paired positions). A support vector machine (C.-C. Chang & Lin, 2001; Vapnik, 1998) is then trained to recognize this ratio, along with GC content and several other parameters.

2. RIBOSWITCHES

A bacterial riboswitch is a portion of the 5′ UTR of messenger RNA that can undergo a conformational change, ultimately regulating protein production (Serganov et al., 2004). Riboswitches are often found upstream of the first gene of an operon— for example, purine riboswitches (Serganov et al., 2004) are found in the 5′ UTR of bacterial operons responsible for purine synthesis and metabolism. Depending on riboswitch type, the conformational change is caused when a specific ligand (guanine, thiamine pyrophosphate, lysine, etc.) binds to the *aptamer*, a ligand-binding region of the riboswitch. Upon binding, a region downstream of the aptamer, called *expression platform*, undergoes secondary and tertiary structure modification.[1] This conformational change may either turn on or off the corresponding gene by either transcriptional or translational regulation of the messenger RNA (Tucker & Breaker, 2005), depending on the particular riboswitch. The common feature shared by all riboswitches is that a gene is regulated by conformational change upon ligand binding. Bacterial riboswitches are often found upstream of operons, regulating groups of genes, as in purine *de novo* synthesis and salvage (Mandal et al., 2003). More recently, a eukaryotic riboswitch (the thiamine pyrophosphate, TPP,

[1] RNA thermometers, such as the repression of ROSE element, function analogously to riboswitches, except that the conformational change is triggered by temperature (Chowdhury et al., 2003).

riboswitch—the most common bacterial riboswitch) has been found that resides in an intronic region and controls alternative messenger RNA splicing by conformational change (Cheah et al., 2007). Riboswitches that control alternative splicing have also been found in the 3′ UTR, as in the case of the thiamine biosynthetic gene THIC of all plant species examined by Wachter et al. (2007).

A riboswitch consists of two equally important parts: an upstream *aptamer*, capable of highly discriminative binding to a particular ligand, and a downstream *expression platform*, capable of undergoing a radical conformational change upon binding of a ligand with the discriminating aptamer. Since aptamers have been under strong evolutionary pressure to bind with high affinity (K_D around 5 nM (Wickiser, Cheah, Breaker, & Crothers, 2005)), there is very strong sequence conservation found in the aptameric region of orthologous riboswitches. It is presumably due to this fact that *all* current Rfam riboswitch families are in fact only families of riboswitch *aptamers*, without the expression platform.

Riboswitch-regulated bacterial genes may be controlled by either *transcriptional regulation*, as in the case of purine riboswitches (guanine and adenine, also called G-box and A-box riboswitches), or by *translational regulation*, as in the case of the TPP riboswitch, the most commonly found bacterial riboswitch as determined by the genomic survey of Sudarsan, Barrick, and Breaker (2003). Indeed, TPP riboswitches have been found in bacteria, such as *Escherichia coli* and *Bacillus subtilis* (Rodionov, Vitreschak, Mironov, & Gelfand, 2002; Winkler, Cohen-Chalamish, & Breaker, 2002), in the plants *Arabidopsis thaliana*, *Oryza sativa*, and *Poa secunda* (Kubodera et al., 2003; Miranda-Ríos, Navarro, & Soberón, 2001; Sudarsan et al., 2003), and in the fungus, *Neurospora crassa*, where instead of controlling gene expression, the TPP riboswitch regulates alternative splicing (Cheah et al., 2007). Figure 1A depicts the predicted gene ON and gene OFF structure of the guanine riboswitch yxjA (renamed nupG) (Johansen, Nygaard, Lassen, Agerso, & Saxild, 2003) from *B. subtilis*. Though these structures are computed, they are highly consistent with the experimentally determined gene ON and gene OFF structures of the xanthine phosphoribosyltransferase (XPT) guanine riboswitch from *B. subtilis* (Serganov et al., 2004). Default status of the guanine riboswitch is ON; binding of the guanine ligand stabilizes the aptamer and forms the *terminator stem–loop*, which prematurely terminates transcription of the mRNA. In contrast, Fig. 1B depicts the mechanism determined in Winkler et al. (2002) for the gene ON and gene OFF structure of the

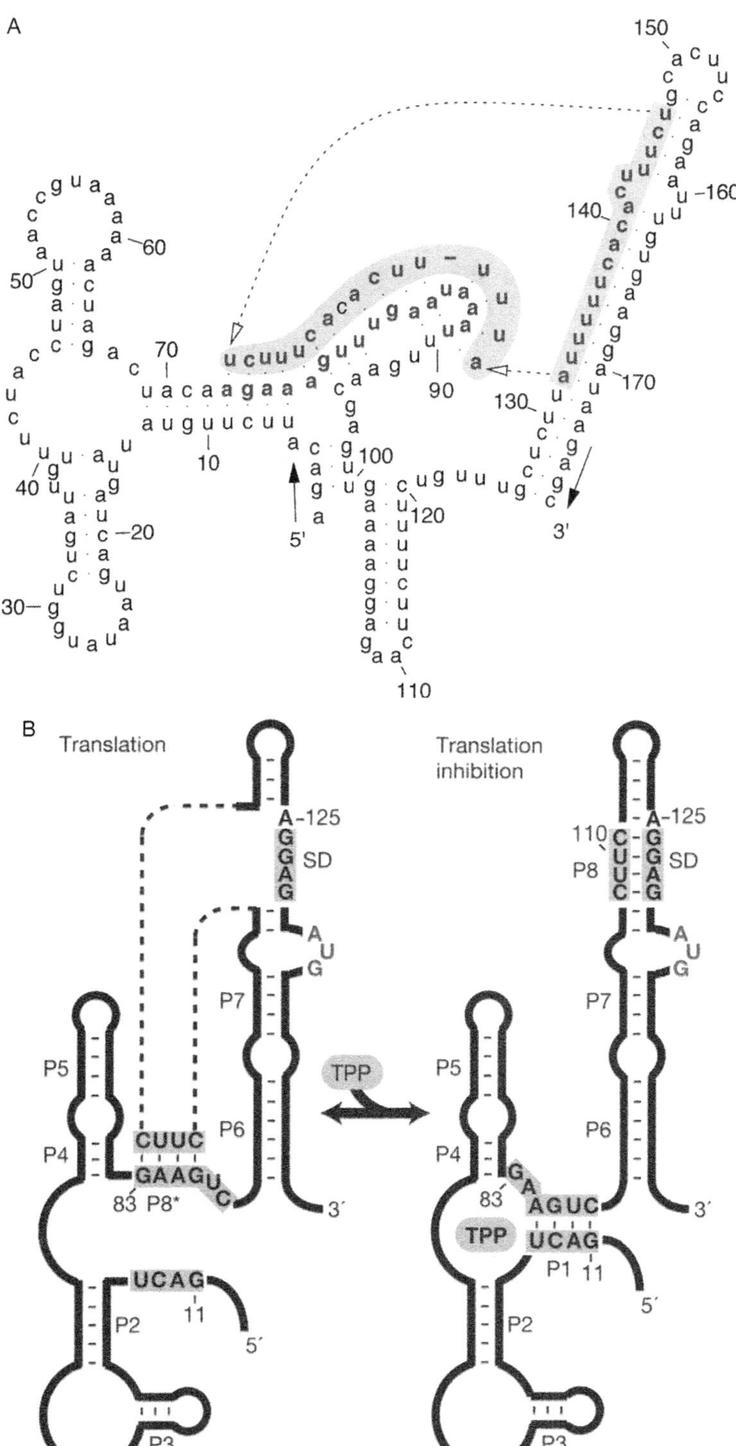

Figure 1 See legend on next page.

TPP riboswitch from *E. coli*. The transcribed thiM mRNA cannot be translated when the Shine–Dalgarno ribosome-binding sequence is sequestered in a terminal stem–loop shown in the figure.

To illustrate the fact that there is strong sequence conservation in the aptamer, in contrast to very weak sequence conservation in the expression platform, Fig. 2 depicts a *structural alignment* between two xanthine phosphoribosyltransferase (XPT) riboswitches, both G-box purine (guanine) riboswitches from *B. subtilis* and *Bacillus halodurans*. Using aptamers given in Mandal et al. (2003), we extracted flanking nucleotides from the EMBL genomic data, up to the coding sequence (CDS); in this manner, we were sure to contain the expression platform. Subsequently, we ran `Dynalign` (Mathews & Turner, 2002) on the extracted sequences, where `Dynalign` is a dynamic programming thermodynamics-based algorithm that determines the common secondary structure of two RNA sequences having minimum free energy. The aptamer is the three-way junction in Fig. 2A, while the expression platform is part of the long (putative) *terminator loop* portion in Fig. 2B, which corresponds to the gene OFF signal in this case involving transcriptional regulation. In contrast, Fig. 3 shows the structural alignment obtained from the `Foldalign` Web server (Havgaard, Torarinsson, & Gorodkin, 2007), using default settings (global alignment, Delta 25, Gap elongation cost −55, Gap opening cost −110 Lambda 200). The `Dynalign` aptamer structures are consistent with the experimentally determined structure of Serganov et al. (2004), while the aptamer structures obtained by `Foldalign` are not. However, it should be mentioned that we ran the `Dynalign` software for several hours before the program converged.

Figure 1 (A) G-box riboswitch yxjA (renamed nupG) from *B. subtilis*, whose leader was determined in Johansen et al. (2003). In absence of guanine, the riboswitch has a (putative) antiterminator, indicated by the arrow and green (gray in the print version) color for hybridization of positions $5'-73\cdots89-3'$ with $3'-147\cdots132-5'$. Upon binding of guanine, the aptamer (nt 1–76) is stabilized, and the (putative) terminator loop (nt 126–177) is formed with the hybridization of $5'-126\cdots148-3'$ with $3'-177\cdots155-5'$. The secondary structure and alternative hybridizations were obtained by an unpublished method of our lab (manuscript in preparation); however, the computed structure is in agreement with experimentally determined structures of other purine riboswitches (Mandal & Breaker, 2004; Serganov et al., 2004)—compare with Fig. 4A. (B) Gene ON and gene OFF structure for the TPP riboswitch from *E. coli*, as determined by Winkler et al. (2002) and Serganov, Polonskaia, Phan, Breaker, and Patel (2006). Note that the Shine–Dalgarno ribosome-binding sequence is sequestered in the gene OFF structure. Panel (B) is a reproduction of Figure 5 of Winkler et al. (2002), reprinted by permission from Macmillan Publishers Ltd.: Nature **419**, 31 October 2002. (See the color plate.)

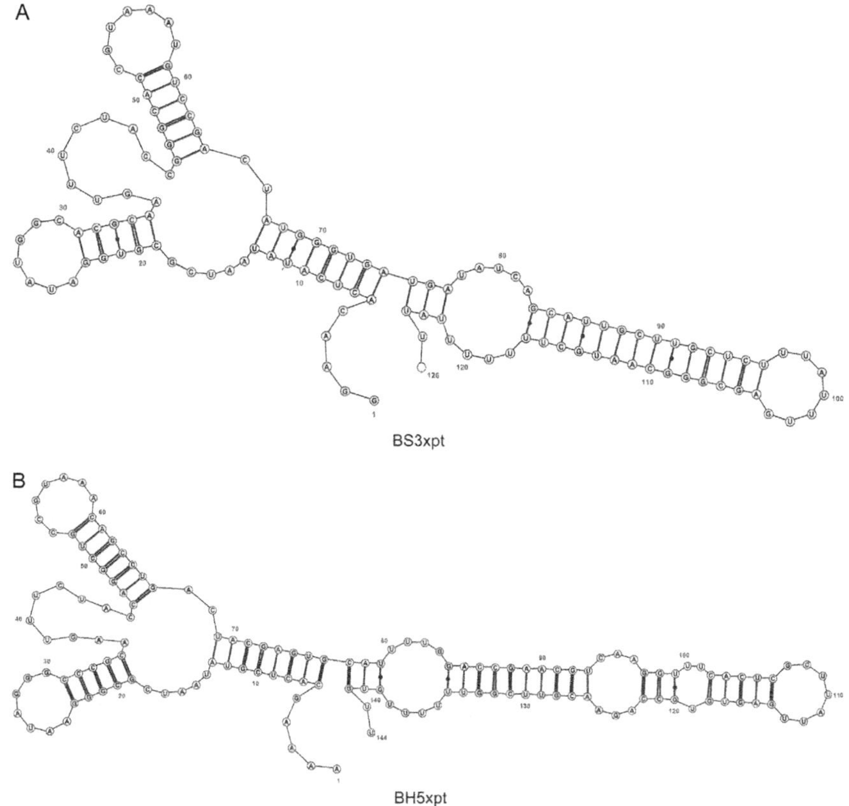

Figure 2 Structural alignment produced by the program `Dynalign` (Mathews & Turner, 2002) for two G-box purine riboswitches: ((A) BS3xpt) XPT from *Bacillus subtilis* with GenBank accession number X83878.1 and ((B) BH5xpt) XPT from *Bacillus halodurans* with GenBank accession number BA000004.3:1593062-1592870. RNA sequences were obtained by BLASTing the aptamers given in Mandal et al. (2003) and subsequently extracting sequences beginning with aptamer and ending immediately before coding region (CDS), thus ensuring inclusion of expression platform.

Analyzing the structural alignment of Fig. 3, we see that in contrast to the strong evolutionary pressure for *sequence conservation* of the aptamer, there is only *secondary structure conservation* in the expression platform. Indeed, in the most conserved portion of the expression platform, the terminator loop, only 12 out of 68 (17.6%) of the nucleotides are conserved.

3. RIBOSWITCH GENE FINDERS

As we have seen, riboswitches constitute a ubiquitous genetic regulation system for bacteria and are known to regulate alternative splicing in

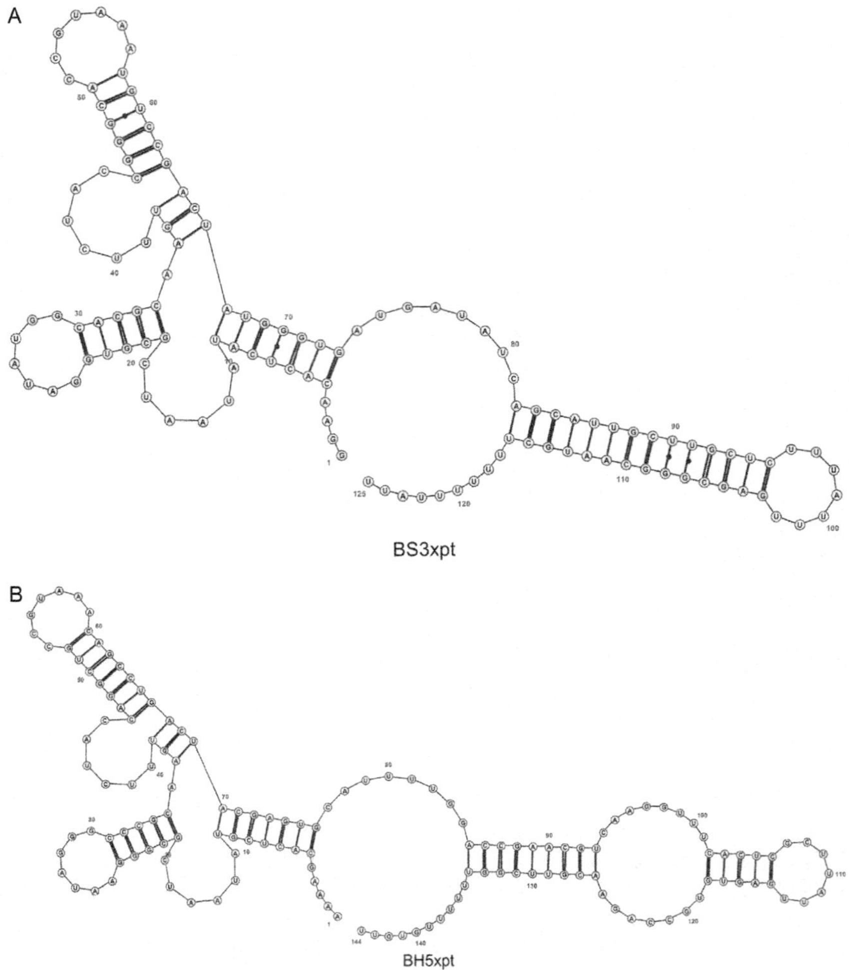

Figure 3 Structural alignment produced by the program `Foldalign` (Havgaard, Lyngso, & Gorodkin, 2005) for XPT riboswitch sequences from *B. subtilis* and *B. halodurans*—the same sequence as described in the caption to Fig. 2.

certain eukaryotes—for more background on riboswitches, see the review articles Wakeman, Winkler, and Dann (2007) and Edwards, Klein, and Ferre-D'Amare (2007). Understanding and manipulating riboswitches could impact public health, since Ooms, Huthoff, Russell, Liang, and Berkhout (2004) have discovered a riboswitch that regulates RNA dimerization and packaging in human immunodeficiency virus type 1 virions. In this context, it is noteworthy that a first step toward such deliberate manipulation of riboswitches was taken by Kim, Gusti, Pillai, and Gaur (2005), who described an artificial riboswitch to control pre-mRNA splicing. More recently, a stress-related riboswitch was found in *H. sapiens* (Ray et al., 2009), thus reinforcing the importance of this ubiquitous molecule and its various regulatory roles.

We now describe some computational approaches for riboswitch detection. All current riboswitch finders employ machine learning methods, often using riboswitch aptamers from seed alignments in the Rfam database (S. W. Burge et al., 2013; Gardner et al., 2011) as training set. While early versions of the Rfam database (Griffiths-Jones, Bateman, Marshall, Khanna, & Eddy, 2003) included 12 riboswitch families (Purine, Lysine, FMN, TPP, Glycine, SAM, Cobalamin, yybP-ykoY, ykkC-yxkD, SAM, PreQ1, glmS), version 12.0 of the Rfam database (July 2014, 2450 families) (S. W. Burge et al., 2013) contains 34 riboswitch families. This count includes eight "leader" sequences, including known riboswitches such as ykoK leader, which is an Mg^{++}-sensing riboswitch, and others believed to be putative regulatory RNAs, such as ylbH leader found in *B. subtilis* (Barrick et al., 2004a).

In our opinion, the most important tool for riboswitch prediction is the SCFG `Infernal` (Nawrocki & Eddy, 2013b; Nawrocki et al., 2009), in large part due to the fact that the important Rfam database depends on `Infernal` for maintenance and extension. The recently released version 1.1 of `Infernal` (Nawrocki & Eddy, 2013b) is reported to be 100 times faster than earlier versions and has been used for the identification of functional RNA homologues in metagenomic data (Nawrocki & Eddy, 2013a). See also El Korbi, Ouellet, Naghdi, and Perreault (2014) for additional examples of the application of `Infernal` in homology searches for riboswitches and ribozymes.

Nevertheless, due to the importance of riboswitches, numerous other methods have been developed, some of which are mentioned below. In Sudarsan et al. (2003), Barrick et al. (2004b), and Barrick (2009), a simple search program, `Sequence Sniffer`, is used to detect riboswitch aptamers, subsequently validated experimentally. Weinberg and Ruzzo (2004) describe riboswitch aptamer `Covariance Models` (CMs), leading to the program `CMfinder` (Yao, Weinberg, & Ruzzo, 2006), a CM-based RNA motif algorithm, used in a computational pipeline for the discovery of novel *cis*-regulatory ncRNA in prokaryotes (Weinberg et al., 2007, 2010; Yao et al., 2007). `CMfinder` extends the SCFG approach of Eddy and Durbin (1994) in order to determine conserved RNA secondary structure in genomes of related species. By exploiting comparative genomics of aligned bacterial genomes, Torarinsson et al. (2008) used the `CMfinder` pipeline to investigate the 1% of human genome regions experimentally probed by the encode Consortium (Birney et al., 2007). They found that 84% of the `CMfinder` candidates were not covered by the algorithm `RNAz` of Washietl et al. Moreover, in a group of 11 ncRNA candidates tested by RT-PCR, 10 were confirmed to be transcribed in *H. sapiens*. Weinberg et al. (2007) applied `CMfinder` to bacterial genomes and identified 22 novel candidate RNA motifs, of which 6 were predicted to be riboswitches, 2 of which were found to be novel riboswitches. Essentially, the pipeline of Weinberg et al. proceeds as follows: (i) apply `CMfinder` (Yao et al., 2007) to the 5′ UTR upstream of known operons, as annotated by the Gene Ontology (Ashburner et al., 2000), and report conserved sequence/secondary structural elements as potential riboswitches, (ii) validate putative riboswitches by using in-line probing and other biochemical methods (Barrick et al., 2004b; Cheah et al., 2007; Mandal et al., 2004; Winkler et al., 2002).

Using metagenomics techniques, Kazanov, Vitreschak, and Gelfand (2007) provide an overview of the abundance and functional diversity of riboswitches in microbial communities. Zhang, Borovok, Aharonowitz, Sharan, and Bafna (2006) describe a sequence-based filtering method for ncRNA identification and apply their method in searching for riboswitch aptamers. Bengert and Dandekar (2004) describe Riboswitch Finder, a Web server and downloadable software to detect riboswitch aptamers. Their method uses PERL string matching to detect specific (consensus) sequence motifs known to be important in a small collection of experimentally validated *purine riboswitches* available to the authors at that time. Subsequently, subsequences that match the consensus pattern are folded, and minimum free energy values are returned with predicted secondary structure. Bengert and Dandekar benchmark their algorithm against a set of known purine riboswitches, but claim that their method can in principle be extended to other riboswitch families.

Abreu-Goodger and Merino (2005) describe the Web server RibEx for locating *Riboswitch Like Elements* and other conserved bacterial regulatory elements. Based on sequence conservation in the regulatory regions of orthologous groups of genes, RibEx searches for the aptamer using sequence-based weight matrices as implemented in the program Motif-Alignment Search Tool (MAST) (Balley & Gribskov, 1998). Though the authors argue that the highly conserved aptamer should suffice to locate riboswitches correctly, nevertheless, the authors apply a method of Merino and Yanofsky (2005) to detect the *attenuator* loop.

T.-H. Chang et al. (2009) describe the RiboSW Web server, capable of detecting riboswitch aptamers in 12 riboswitch families in the Rfam database (Gardner et al., 2011). The algorithm underlying RiboSW is based on two characteristics: (1) RNA secondary structure and (2) sequence conservation within the functional region. The authors describe four types of structural components that suffice to describe secondary structure motifs (e.g., hairpin, internal stem) that are representative of a particular riboswitch aptamer family (e.g., purine riboswitch aptamers are described as a three-way junction). In the first step of the algorithm, a search for structural components in the input sequence is performed. In the second step, the functional region is detected using the HMM software `HMMER` (Eddy, 1998). The authors benchmark their software against the two previously described methods, Riboswitch Finder (Bengert & Dandekar, 2004) and RibEx (Abreu-Goodger & Merino, 2005); nevertheless, the authors point out that TPP and Cobalamin riboswitch aptamers are more difficult to detect with their method.

Singh, Bandyopadhyay, Bhattacharya, Krishnamachari, and Sengupta (2009) describe a profile hidden Markov model (pHMM) to detect riboswitch aptamers. Using the UCSC software SAM (Karplus, Barrett, & Hughey, 1998), the authors build family-specific pHMMs from the sequences in the Rfam seed alignment of riboswitch (aptamer) families, then use `HMMER` (Eddy, 1998) to perform a genomic search for riboswitch aptamers, using the model produced by SAM. The authors claim comparable performance with `Infernal` 1.0 (Nawrocki et al., 2009), yet with much less computation time required. However, note that `Infernal` 1.1 runs 100 times faster than `Infernal` 1.0 and hence should run substantially faster than the method of Singh et al. (2009).

In Havill, Bhatiya, Johnson, Sheets, and Thompson (2014), very recently another genomic riboswitch prediction method has been developed, relying on a dynamic programming algorithm to determine the heaviest path in a multipartite graph. The authors report 88–99% sensitivity and greater than 99.99% specificity on 13 riboswitch families.

3.1. Using `Infernal`

In order to search for purine riboswitch aptamers in the *B. subtilis* genome using `Infernal`, type the command *cmsearch –notextw –acc ../covariance:models/RF00167_seed.cm NC_000964.faa*

where `cmsearch` denotes a search for hits in the FASTA format GenBank file NC_000964 for the entire genome of *B. subtilis* (subsp. subtilis str. 168), using the trained *covariance model* RF00167_seed.cm. This CM was derived by previously training `Infernal` on the sequences from the seed alignment of Rfam family RF00167 of purine riboswitch aptamers. The flags `--notextw` and `--acc`, respectively, indicate not to wrap output text and to use accession codes in reporting hits found—this facilitates parsing the output.

There are six purine riboswitches in *B. subtilis*, all correctly identified in the following output:

```
rank     E-value  score  bias  sequence                            start     end   mdl trunc   gc  description
----   --------- ------ -----  ---------------------------- ------- -------   --- ----- ----  -----------
 (1) !     1.2e-16   87.2   0.0  gi|255767013|ref|NC_000964.3|  698369  698470 +  cm    no 0.42  Bacillus subtilis subsp.  subtilis
str.  168 chromosome, complete genome
 (2) !     3.1e-16   85.6   0.0  gi|255767013|ref|NC_000964.3|  694425  694526 +  cm    no 0.38  Bacillus subtilis subsp.  subtilis
str.  168 chromosome, complete genome
 (3) !     3.8e-16   85.2   0.5  gi|255767013|ref|NC_000964.3| 4005523 4005624 +  cm    no 0.28  Bacillus subtilis subsp.  subtilis
str.  168 chromosome, complete genome
 (4) !     2.8e-15   81.7   0.0  gi|255767013|ref|NC_000964.3| 2320213 2320114 -  cm    no 0.46  Bacillus subtilis subsp.  subtilis
str.  168 chromosome, complete genome
 (5) !     2.7e-13   73.6   0.1  gi|255767013|ref|NC_000964.3|  626446  626347 -  cm    no 0.30  Bacillus subtilis subsp.  subtilis
str.  168 chromosome, complete genome
 ------ inclusion threshold ------
 (6) ?            0.93   22.8   0.0  gi|255767013|ref|NC_000964.3|  732805  732712 -  cm    no 0.43  Bacillus subtilis subsp.  subtilis
str.  168 chromosome, complete genome
```

By default, hits are reported if the *E*-value is less than 10; significant hits, with *E*-value less than 0.01, are denoted by an exclamation point, while less significant hits are denoted by a question mark. The secondary structure predicted by `Infernal` can be parsed from the remaining output:

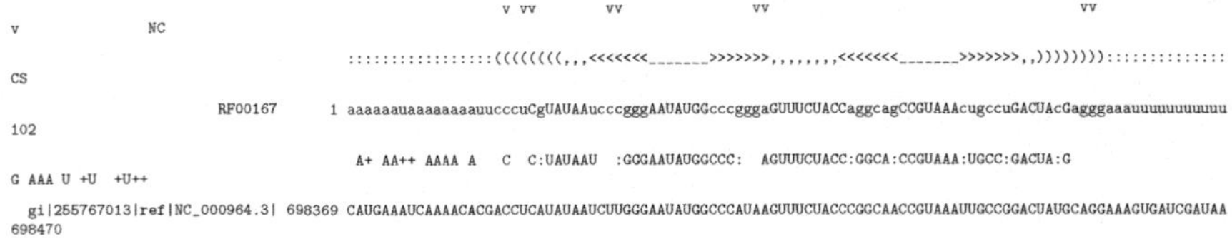

```
                                              v vv       vv          vv                          vv
v             NC
                                   ::::::::::::::::(((((((,,,<<<<<<<_______>>>>>>>,,,,,,,,<<<<<<<_______>>>>>>>,,))))))):::::::::::::
CS
                       RF00167      1 aaaaaauaaaaaaaauucccuCgUAUAAucccgggAAUAUGGcccgggaGUUUCUACCaggcagCCGUAAAcugccuGACUAcGagggaaauuuuuuuuuu
102
                                    A+ AA++ AAAA A   C  C:UAUAAU  :GGGAAUAUGGCCC:  AGUUUCUACC:GGCA:CCGUAAA:UGCC:GACUA:G
G AAA U +U  +U++
  gi|255767013|ref|NC_000964.3| 698369 CAUGAAAUCAAAACACGACCUCAUAUAAUCUUGGGAAUAUGGCCCAUAAGUUUCUACCCGGCAACCGUAAAUUGCCGGACUAUGCAGGAAAGUGAUCGAUAA
698470
```

In line 2, base pairs are indicated by parentheses and angle brackets, and in more complicated structures, by square brackets and curly brackets—different types of parentheses/brackets indicate different levels of nesting in the case of multiloops. However, the usual dot–bracket notation for secondary structure is obtained by replacing all types of brackets by matching parentheses, removing dots (which indicate gaps), and replacing all remaining symbols by dots. Noncanonical base pairs are flagged by matching occurrences of "v" in line 1. The consensus sequence for the CM is given in line 3, and the sequence of the hit is given in line 5. Remaining details are explained in the `Infernal` documentation. Finally, it should be noted that the developers of `Infernal` make available a collection of all 2208 trained CMs for RFam 11.0—see http://infernal.janelia.org/.

3.2. Using HMMER

The conserved sequence information from aptamer regions is so strong that it is possible to train the hidden Markov model HMMER 1.8.5 (Eddy, Mitchison, & Durbin, 1995) on Rfam seed alignments of purine riboswitches (after removing those purine riboswitches present in *B. subtilis*). We obtained a HMMER model for the training set by the command

hmmb RF00167.hmm RF00167seed.stockholm

where RF00167.hmm denotes the trained HMMER model (a binary file), and RF00167seed.stockholm is the training set in Stockholm format (whereby instances from *B. subtilis* were previously removed). We then performed a whole genome scan of all 5′ UTRs of *B. subtilis* (400 nt upstream of CDS) using the trained HMM, by the command

hmmls RF00167.hmm NC_000964.faa

All six known purine riboswitches of *B. subtilis* were immediately found with HMMER scores of 49.12, 49.12, 62.34, 63.09, 65.81, and 70.07, while the next highest score of a nonriboswitch was 15.71 and average HMMER score for all 5′ UTRs of *B. subtilis* was − 155.18. The significantly faster speed of HMMER over that of Infernal 1.0 (prior to Infernal 1.1) is the reason some of the previously reviewed riboswitch aptamer prediction methods had used HMMs.

4. CONFORMATIONAL SWITCHES

In this chapter, we have so far discussed computational methods for detecting riboswitch aptamers. Riboswitches are one instance (and probably the most important instance) of *conformational switches*. Other examples of biologically important conformational switches are, for instance, the *hok/sok* (host-killing/suppression of killing) system in *E. coli* to check for sufficient plasmid copy number and the conformational switch in spliced leader RNA from *Leptomonas collosoma*, in which a portion of the 5′ exon is donated to another mRNA by *trans*-splicing using a kinetically determined conformational switch.

We now discuss some programs capable of predicting conformational switches—specifically, we mention paRNAss (Voss, Meyer, & Giegerich, 2004), RNAshapes (Voss, Giegerich, & Rehmsmeier, 2006), RNAbor (Freyhult, Moulton, & Clote, 2007), FFTbor (Senter, Sheik, Dotu, Ponty, & Clote, 2012), and the recent, entropy-based method of

Manzourolajdad (2014). In contrast to the machine-learning riboswitch finders discussed in the previous section, it seems fair to say that the methods discussed in this section are capable of *suggesting* the presence of a conformational switch, but as yet cannot compete with machine-learning methods that rely on sequence/structure signals of curated Rfam seed alignments of riboswitch aptamers. Another issue is that some conformational switches may in fact be switches between kinetically trapped structures and hence not detectable by thermodynamics-based (or related SCFG entropy) methods. The research in this field would benefit greatly from the existence of a database containing more than the handful of currently known non-riboswitch conformational switches—see the Web site http://bibiserv.techfak.uni-bielefeld.de/parnass/examples.html for a list of conformational switches.

In `paRNAss` (Voss et al., 2004), an RNA switch is predicted by means of studying properties of the energy landscape of the RNA. Secondary structures are sampled from the structure space using `RNAbor` (Wuchty, Fontana, Hofacker, & Schuster, 1999). Pairwise distances are calculated between the sampled structures using two different distance measures (e.g., base pair distance, energy barrier distance, tree alignment distance). Points in the corresponding scatterplot are clustered; if the RNA is a conformational switch, then it should have two stable structures so that `paRNAss` output should display two clouds of points—one near the x-axis and one near the y-axis. Note that since `paRNAss` calls `RNAbor` from the Vienna RNA Package program (Hofacker, 2003), it requires a user-defined energy bound, E, in order to generate all secondary structures within E (kcal/mol) of the minimum free energy. Although the `paRNAss` engine is no longer in service, since the authors consider it to be obsolete with the introduction of `RNAshapes`, historically `paRNAss` was the first computational method for conformational switch prediction; hence, it is of interest to have sketched the method.

`RNAshapes` (Voss et al., 2006) partitions the collection of secondary structures into disjoint equivalence classes, or *shapes*, depending on bifurcation topology; for instance, the *shape* of the familiar cloverleaf secondary structure of transfer RNA is `[[][][]]`. For each shape considered, the *Boltzmann probability* of that shape is computed, along with the minimum free energy structure, called *shrep*, of that shape. In the case of a conformational switch, one would expect to see two or more shapes having large Boltzmann probabilities.

Given a starting *reference* structure, `RNAbor` (Freyhult et al., 2007) partitions the collection of secondary structures into collections C_k of structures having base pair distance k to the reference structure. The Boltzmann probability and the minimum free energy of C_k are computed simultaneously for all values of k. In the case of a conformational switch, if the reference structure is taken to be the minimum free energy structure (or a known metastable structure), then one would expect to see a *peak* at $k = 0$ and a second *peak* at a value $k \gg 0$. For input sequence of length n, `RNAbor` runs in time $O(n^5)$ and space $O(n^3)$. Since these time and space requirements may be prohibitive, the newer algorithm `FFTbor` (Senter et al., 2012) employs polynomial interpolation using the fast Fourier transform, in order to compute the Boltzmann probabilities in time $O(n^4)$ and space $O(n^2)$.

Finally, Manzourolajdad's recent PhD dissertation (Manzourolajdad, 2014) describes how to compute the *structural entropy* $H = -\sum_i p_i \ln p_i$ (Manzourolajdad, Wang, Shaw, & Malmberg, 2013), where the sum is taken over all secondary structures of an input sequence, and probabilities of secondary structure derivations are computed by the *inside* and *outside* algorithms for SCFGs (Durbin, Eddy, Krogh, & Mitchison, 1998). Manzourolajdad's thesis additionally describes logistic regression classifiers for riboswitches that depend on various features, including structural entropy, sequence length, GC content, and minimum free energy. In one of the many computational benchmarkings, Table 3.5 of Manzourolajdad (2014) indicates a true positive rate TP% [resp. false positive rate FT%] of 91.1 [resp. 54.5] for detection of riboswitches in *B. subtilis* using a logistic classifier with the previously mentioned features.

We now mention a couple of sample applications. In mitosis, the *hok/sok* system ensures that daughter cells contain R1 plasmids in sufficient copy number as follows. The *hok* gene of *E. coli* codes a small (52-amino acid) toxin causing irreversible damage to the cell membrane. If plasmids are sufficiently present, then a portion of the 64 nt *sok*-RNA, which is complementary to *hok*-mRNA leader region, binds to the active conformation of *hok*-mRNA, thus causing degradation of the complex by RNase III. If plasmids are not present in sufficient copy number, then the cell is killed by *hok* toxin. For details on the hok/sok system, see Franch, Gultyaev, and Gerdes (1997) and Shapiro, Bengali, Kasprzak, and Wu (2001). Figure 4B shows how `RNAbor` (Freyhult et al., 2007) and `RNAshapes` (Voss et al., 2006) both detect a conformational switch in *E. colihok*-RNA.

For the 76 nt conformational switch which controls hepatitis delta virus ribozyme catalysis with PDB code `1SJ3:R`, `RNAbor` indicates a sharp second peak, suggesting a conformational change (data not shown); in contrast, `RNAshapes` (Giegerich, Voss, & Rehmsmeier, 2004) computes a

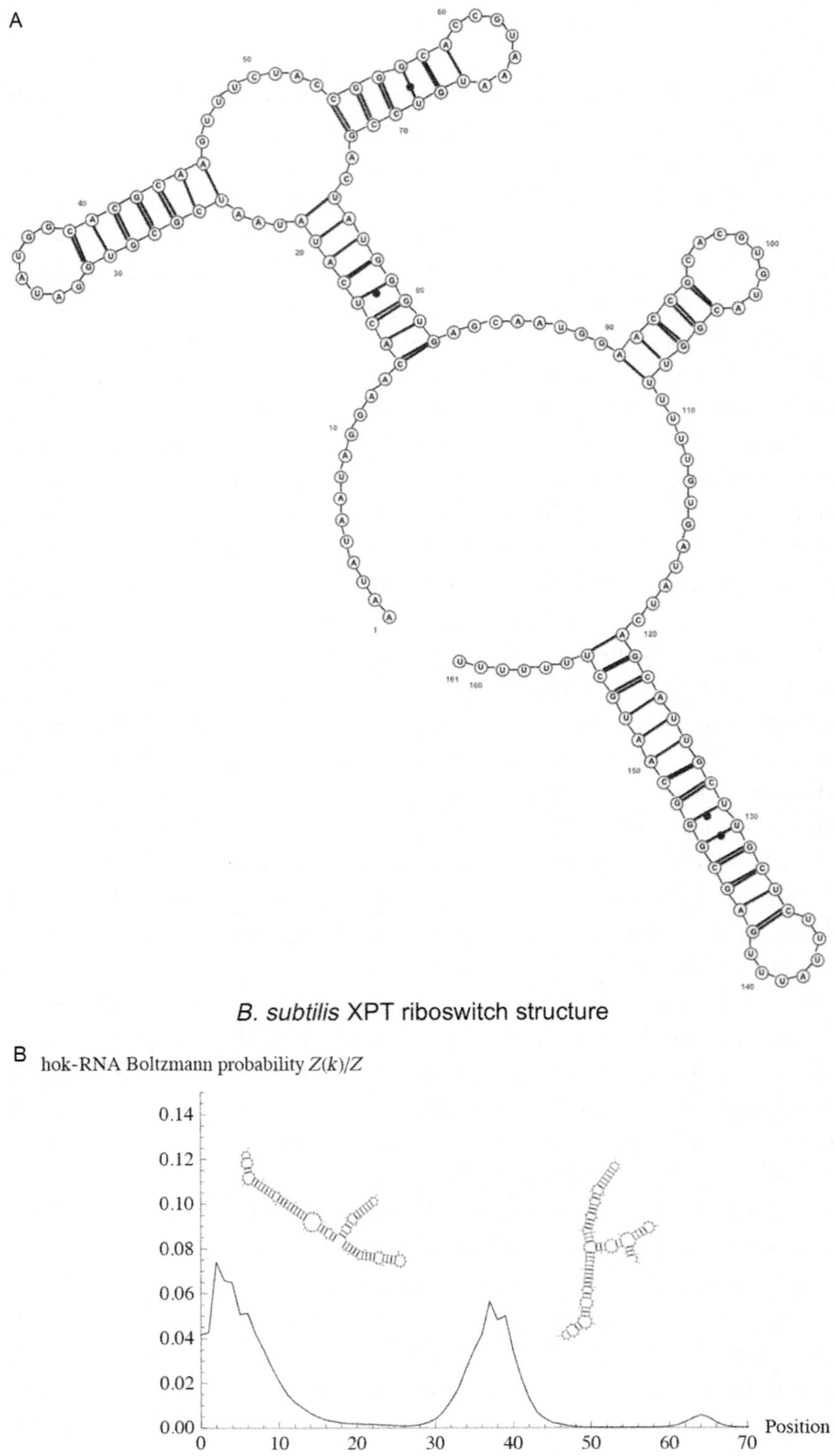

A

B. subtilis XPT riboswitch structure

B

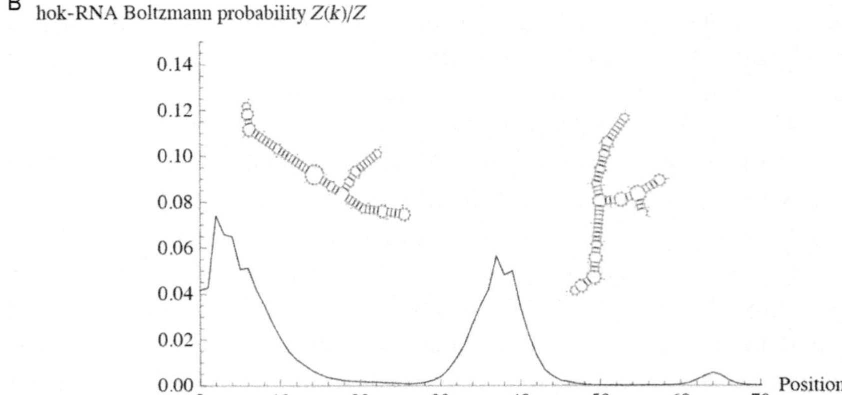

Figure 4 See legend on next page.

probability of 0.9978298 for the shape [ ] [ ] (two external stem–loops), thus suggesting that there is no conformational change. Finally, it should be noted that many conformational switches have a pseudoknotted structure, while paRNAss, RNAshapes, RNAbor, FFTbor, and the structural entropy-based method of Manzourolajdad (2014) all consider *non*-pseudoknotted secondary structures. This suggests that it could be of benefit to extend conformational switch prediction software to include at least certain types of commonly occurring pseudoknots.

4.1. Computational experiments

Four XPT genes are annotated in the complete genome of the *Staphylococcus aureus* subsp. aureus strain MRSA252 (Holden et al., 2004) (GenBank accession code BX571856.1). The first XPT gene is annotated at position 441383..441961, as predicted by amino acid sequence similarity of the translated coding region (E-value 1.5×10^{-33}, 53.403% id in 191 aa with *B. halodurans* XPT). The upstream region at position 441050..441153 is annotated as a putative XPT riboswitch, as predicted by Infernal (Rfam) with score 73.10. For the genomic region from 441033..441383,

```
>gi—49240382:441033-441383 Staphylococcus aureus subsp. aureus strain MRSA252, complete genome
GAAAATTTGCTATTATCGTTAAATAATTTACATAAACTCATATAATCTAAAGAATATGGCTTTAGAAGTT
TCTACCATGTTGCCTTGAACGACATGACTATGAGTAACAACACAATACTAGGAGTAGCTTCAGCCATTAA
ATTGTAACCATGATGGGTGATTTATATCATTTTATATGATGGTCACAGTTTATTTGATGAAACTTCTTTT
ACATTGATTGCATGACCAATACGTGATGCATGTTCGTTCACTCATAAACCCTGAAACTATTATTTAGTTT
GGGGATTTTTTTGTATCTAGCACCAATTTAAGAGCAAAATGTTTCACACAAATCTGAGGAGGTTTTAAGA
G
```

Figure 4 (A) *B. subtilis* secondary structure, determined by in-line probing—image produced with VARNA (Darty, Denise, & Ponty, 2009), using base pairs of gene OFF XPT riboswitch structure from Serganov et al. (2004). Note the similarity with the predicted structure from Fig. 1A. (B) *hok* (*ho*st *k*illing) mRNA folds into two different conformations (Franch et al., 1997). Using 142 nt hok-RNA sequence GGCGCUUGAG GCUUUCUGCC UCAUGACGUG AAGGUGGUUU GUUGCCGUGU UGUGUGGCAG AAAGAAGAUA GCCCCGUAGU AAGUUAAUUU UCAUUAACCA CCACGAGGCA UCCCUAUGUC UAGUCCACAU CAGGAUAGCC UC from http://bibiserv.techfak.uni-bielefeld.de/parnass/examples.html, RNAbor produced the displayed density plot, with peaks at 2 and 37 bp from the MFE structure. Also shown are superimposed structures for the MFE(2) [resp. MFE(37)] structure taken over all structures having base pair distance 2 [resp. 37] from the minimum free energy structure. In contrast, RNAshapes (Voss et al., 2004) has two large probability shapes: [ ] [ ] [ ] with probability 0.6067672 (three stem–loops) and [[ ] [ ]] with probability 0.2516662 (three-way junction multiloop).

starting 350 nt upstream of the start codon, we performed three computational experiments, explained in the following.

From the secondary structure of the XPT riboswitch of *B. subtilis*, as displayed in Fig. 4A, we *trimmed* off 12 of the 13 leading [resp. 5 of the 6 trailing] *unpaired* nucleotides, to obtain our *target* 144 nt XPT gene OFF secondary structure. In computational experiment 1, we ran `FFTbor` on all 144 nt windows of the genomic region 441033–441383 of *S. aureus*. For each window, `FFTbor` computed, simultaneously for each k, the probability $p(k) = \frac{Z_k}{Z}$ of secondary structures having base pair distance k from the target trimmed XPT riboswitch structure. Using the values obtained, Fig. 5A depicts the *expected base pair distance* (Garcia-Martin, Clote, & Dotu, 2013; Senter et al., 2012) $\sum_k p(k)\cdot k$ as a function of starting position for each 144 nt window of the relevant genomic region. For these values, we have mean 80.64, standard deviation 3.92, maximum 87.50, minimum 68.78, and the Z-score of the minimum is $\frac{68.78-80.64}{3.92} = -3.03$. The minimum expected base pair distance of 68.78 occurs at position 34 (corresponding to genomic position 441066). Note that the `Infernal` prediction of the *untrimmed* XPT riboswitch occurs at position 441050, and adding 12 nt that we had trimmed off means that `Infernal` predicts the start of the trimmed XPT riboswitch to occur at position 441062—virtually the same location as our prediction using `FFTbor`. `Infernal` 1.1 is faster than `FFTbor`; however, note that while `Infernal` heavily exploits sequence homology, in addition to structural homology, `FFTbor` does not.

In computational experiment 2, we used `FFTbor` to compute the probability $p(k)$, for each $1 \leq k \leq 144$ on each 144 nt window of the genomic region, of secondary structures to have base pair distance k with the *minimum free energy structure* of the current 144 nt window. Define a *peak*, to be the ordered pair $(x, p(x))$, such that $p(x)$ is greater than $p(x')$ for any $x' \in [x-10, x+10]$; i.e., a peak is the position x and its `FFTbor` probability $p(x)$, for which $p(x)$ is greater than any neighbor in the flanking $\pm$ 10 nt. Figure 5B depicts the *ratio* $\frac{p_2}{p_1}$ of the second largest probability peak p_2 divided by the largest probability peak p_1. We reasoned that if the current 144 nt window were to contain a *conformational switch*, then $p_1 \approx p_2$, since there should be two distinct large peaks in the Boltzmann density plot (analogous to Fig. 4B). For the ratios obtained, we found the mean 0.416730, standard deviation 0.369394, maximum 0.977564, and minimum 0. The five largest ratios and their locations are given by (0.977564102564, 139),

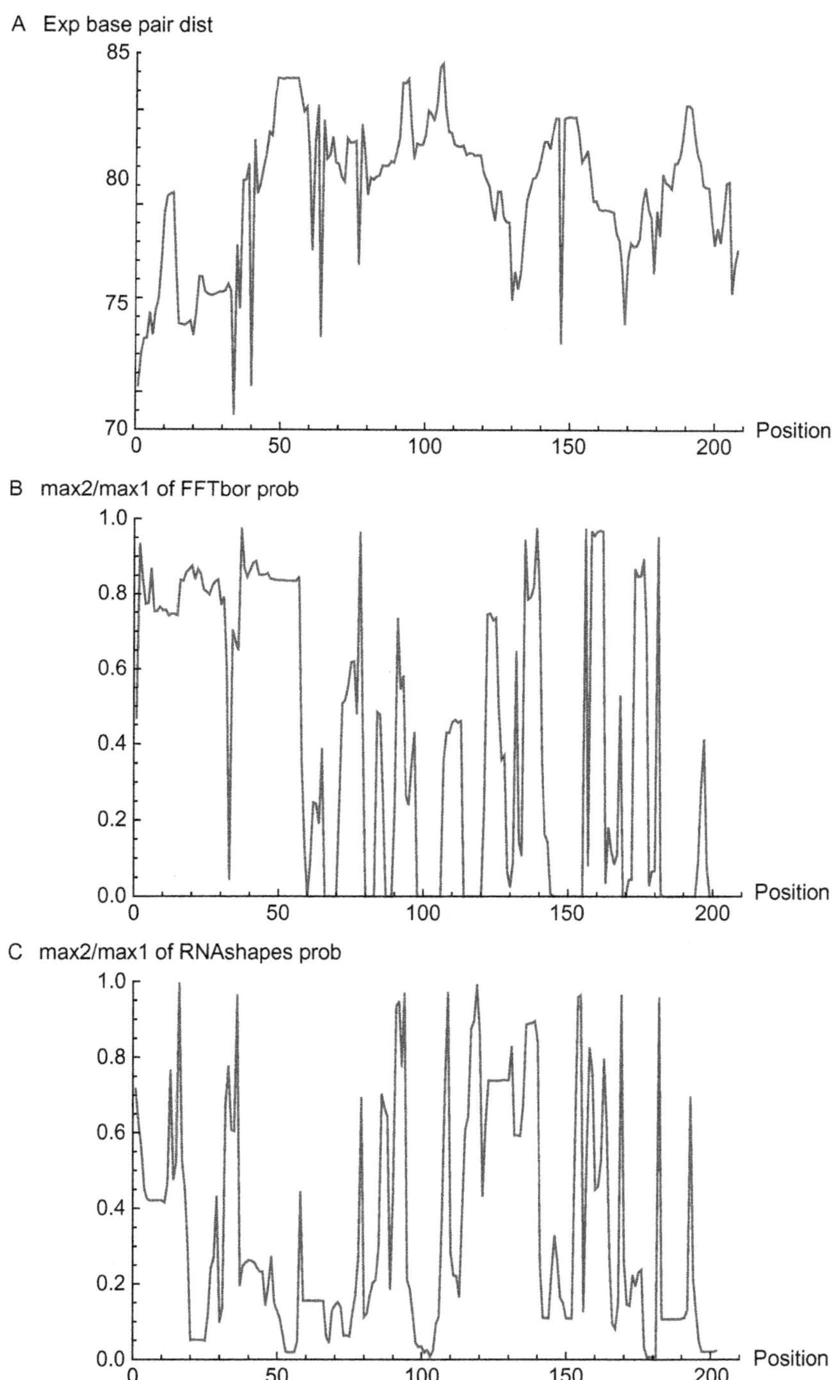

Figure 5 Computational experiments with the 5′ UTR, 350 nt upstream of the homologue of the XPT gene of *Staphylococcus aureus* subsp. aureus strain MRSA252 *(Continued)*

(0.977186311787, 37), (0.977011494253, 156), (0.971264367816, 161), and (0.969696969697, 158). Note that the ratio $\frac{p_2}{p_1} = 0.977186311787$ occurs at nt 37 of the genomic region, not far from the position predicted by `Infernal`, and that the *Z*-score for 0.977186311787 is 1.51723.

In computational experiment 3, we ran `RNAshapes` (Voss et al., 2004) on each 144 nt window and determined the ratio of the largest and second largest probabilities p_1 [resp. p_2] of a shape. Figure 5C depicts the *ratio* $\frac{p_2}{p_1}$ of the second largest probability p_2 divided by the largest probability p_1 as a function of starting position in the extracted genomic region. We reasoned that if the current 144 nt window were to contain a *conformational switch*, then $p_1 \approx p_2$; hence $\frac{p_2}{p_1} \approx 1$. For the ratios obtained, we have mean 0.343431, standard deviation 0.296639, maximum 0.997328, and minimum 0.006866. The maximum 0.997328, corresponding to a *Z*-score of 2.204, is obtained at position 16, corresponding to genomic position 441048—again, very close to the `Infernal` prediction of 441050. Note that the five highest probability ratios $\frac{p_2}{p_1}$ and their position in the genomic region are given as follows: (0.997328256409, 16), (0.992625075672, 119), (0.972082783167, 109), (0.971491714367, 94), and (0.966006926247, 169).

Figure 5—Cont'd (Holden et al., 2004) (GenBank accession code BX571856.1). Note that in all three panels, the position of the predicted XPT riboswitch (A) and the position of the predicted *conformational switch* (B and C) are very close to the `Infernal` prediction; however, unlike `Infernal` that exploits sequence homology as well as secondary structures, `FFTbor` and `RNAshapes` do not. (A) Using `FFTbor`, for each 144 nt window of the 5′ UTR, the *expected base pair distance* was computed to the target trimmed XPT secondary structure (see text). The `FFTbor` prediction is within 2 nt of the `Infernal` prediction. (B) For each 144 nt window, and each $1 \leq k \leq 144$, the probability $p(k)$ of secondary structures of the window contents to have base pair distance k to the MFE structure of the window contents is computed. The ratio $\frac{p_2}{p_1}$ of the second largest *peak* probability divided by the largest peak probability is graphed as a function of position in the genomic region. Reasoning that if the window contents were to constitute a *conformational switch*, then the ratio $\frac{p_2}{p_1} \approx 1$, and we determine that the second highest ratio lies within the `Infernal` predicted XPT riboswitch. (C) For each 144 nt window, the ratio $\frac{p_2}{p_1}$ of the second largest *shape* probability divided by the largest shape probability is computed, as determined by `RNAshapes`. Reasoning that if the window contents were to constitute a *conformational switch*, then the ratio $\frac{p_2}{p_1} \approx 1$, and we determine that the highest ratio occurs within 2 nt of the beginning of the `Infernal` predicted XPT riboswitch.

5. CONCLUSION

In this chapter, we have reviewed software for the prediction of riboswitches and, more generally, of conformational switches. Due to the immense utility of the Rfam database, and the fact that `Infernal` is used to maintain and extend Rfam, we believe that `Infernal` 1.1 is a tool of choice in the prediction of ncRNA, and more specifically of riboswitch *aptamers*. We have discussed several approaches for the prediction of conformational switches, the most recent of which is the logistic classifier of Manzourolajdad (2014). The computational experiments reported in Fig. 5 suggest that more powerful machine learning software, such as neural networks and support vector machines, could be developed that include additional features such as peak or shape probability ratios p_2/p_1. Nevertheless, this would multiply the number of parameters necessary to be learned, making it perhaps premature before a larger database of bonafide conformational switches becomes available.

6. ACKNOWLEDGMENTS

We would like to thank Evan Senter for discussions about usage of `Infernal`, and both Evan Senter and Juan Antonio Garcia-Martin for feedback from a preliminary draft of this chapter. Thanks to Yann Ponty for contributing Fig. 1. This research was funded in part by National Science Foundation grant DBI-1262439. Any opinions, findings, and conclusions or recommendations expressed in this chapter are those of the authors and do not necessarily reflect the views of the National Science Foundation.

REFERENCES

Abreu-Goodger, C., & Merino, E. (2005). RibEx: A web server for locating riboswitches and other conserved bacterial regulatory elements. *Nucleic Acids Research*, *33*, W690–W692. Retrieved from, http://dx.doi.org/10.1093/nar/gki445.

Ashburner, M., Ball, C. A., Blake, J. A., Botstein, D., Butler, H., Cherry, J. M., et al. (2000). Gene Ontology: Tool for the unification of biology. The Gene Ontology Consortium. *Nature Genetics*, *25*(1), 25–29.

Balley, T. L., & Gribskov, M. (1998). Combining evidence using p-values: Application to sequence homology searches. *Bioinformatics*, *14*, 48–54.

Barrick, J. E. (2009). Predicting riboswitch regulation on a genomic scale. *Methods in Molecular Biology*, *540*, 1–13.

Barrick, J. E., Corbino, K. A., Winkler, W. C., Nahvi, A., Mandal, M., Collins, J., et al. (2004). New RNA motifs suggest an expanded scope for riboswitches in bacterial genetic control. *Proceedings of the National Academy of Sciences of the United States of America, 101*(17), 6421–6426.

Barrick, J. E., Corbino, K. A., Winkler, W. C., Nahvi, A., Mandal, M., Collins, J., et al. (2004). New RNA motifs suggest an expanded scope for riboswitches in bacterial genetic control. *Proceedings of the National Academy of Sciences of the United States of America, 101*(17), 6421–6426.

Bekaert, M., Bidou, L., Denise, A., Duchateau-Nguyen, G., Forest, J. P., Froidevaux, C., et al. (2003). Towards a computational model for − 1 eukaryotic frameshifting sites. *Bioinformatics, 19*, 327–335.

Bengert, P:, & Dandekar, T. (2004). Riboswitch finder—A tool for identification of riboswitch RNAs. *Nucleic Acids Research, 32*(Suppl. 2), W154–W159. Retrieved from, http://dx.doi.org/10.1093/nar/gkh352.

Birney, E., Stamatoyannopoulos, J. A., Dutta, A., Guigo, R., Gingeras, T. R., Margulies, E. H., et al. (2007). Identification and analysis of functional elements in 1% of the human genome by the ENCODE pilot project. *Nature, 447*(7146), 799–816.

Böck, A., Forschhammer, K., Heider, J., & Baron, C. (1991). Selenoprotein synthesis: An expansion of the genetic code. *Trends in Biochemical Sciences, 16*, 463–467.

Burge, C., & Karlin, S. (1997). Prediction of complete gene structures in human genomic DNA. *Journal of Molecular Biology, 268*, 78–94.

Burge, S. W., Daub, J., Eberhardt, R., Tate, J., Barquist, L., Nawrocki, E. P., et al. (2013). Rfam 11.0: 10 years of RNA families. *Nucleic Acids Research, 41*(Database), D226–D232.

Chang, C.-C., & Lin, C.-J. (2001). *LIBSVM: A library for support vector machines [Computer software manual]*. Software available at, http://www.csie.ntu.edu.tw/~cjlin/libsvm.

Chang, T.-H., Huang, H.-D., Wu, L.-C., Yeh, C.-T., Liu, B.-J., & Horng, J. T. (2009). Computational identification of riboswitches based on RNA conserved functional sequences and conformations. *RNA, 15*(7), 1426–1430. Retrieved from, http://dx.doi.org/10.1261/rna.1623809.

Cheah, M. T., Wachter, A., Sudarsan, N., & Breaker, R. R. (2007). Control of alternative RNA splicing and gene expression by eukaryotic riboswitches. *Nature, 447*(7143), 497–500.

Chowdhury, S., Ragaz, C., Kreuger, E., & Narberhaus, F. (2003). Temperature-controlled structural alterations of an RNA thermometer. *The Journal of Biological Chemistry, 278*(48), 47915–47921.

Darty, K., Denise, A., & Ponty, Y. (2009). VARNA: Interactive drawing and editing of the RNA secondary structure. *Bioinformatics, 25*(15), 1974–1975.

Delcher, A. L., Harmon, D., Kasif, S., White, O., & Salzberg, S. L. (1999). Improved microbial gene identification with GLIMMER. *Nucleic Acids Research, 27*(23), 4636–4641.

Durbin, R., Eddy, S., Krogh, A., & Mitchison, G. (1998). *Biological sequence analysis: Probabilistic models of proteins and nucleic acids.* Cambridge, UK: Cambridge University Press.

Eddy, S. R. (1998). Profile hidden Markov models. *Bioinformatics, 14*, 755–763.

Eddy, S. R., & Durbin, R. (1994). RNA sequence analysis using covariance models. *Nucleic Acids Research, 22*, 2079–2088.

Eddy, S. R., Mitchison, G., & Durbin, R. (1995). Maximum discrimination hidden Markov models of sequence consensus. *Journal of Computational Biology, 2*(1), 9–24.

Edwards, T. E., Klein, D. J., & Ferre-D'Amare, A. R. (2007). Riboswitches: Small-molecule recognition by gene regulatory RNAs. *Current Opinion in Structural Biology, 17*(3), 273–279.

El Korbi, A., Ouellet, J., Naghdi, M. R., & Perreault, J. (2014). Finding instances of riboswitches and ribozymes by homology search of structured RNA with Infernal. *Methods in Molecular Biology, 1103*, 113–126.

Franch, T., Gultyaev, A. P., & Gerdes, K. (1997). Programmed cell death by hok/sok of plasmid R1: Processing at the hok mRNA 3H-end triggers structural rearrangements that allow translation and antisense RNA binding. *Journal of Molecular Biology*, *273*, 38–51.

Freyhult, E., Moulton, V., & Clote, P. (2007). Boltzmann probability of RNA structural neighbors and riboswitch detection. *Bioinformatics*, *23*(16), 2054–2062.

Garcia-Martin, J. A., Clote, P., & Dotu, I. (2013). RNAiFold: A constraint programming algorithm for RNA inverse folding and molecular design. *Journal of Bioinformatics and Computational Biology*, *11*(2), 1350001. http://dx.doi.org/10.1142/S0219720013500017.

Gardner, P. P., Daub, J., Tate, J., Moore, B. L., Osuch, I. H., Griffiths-Jones, S., et al. (2011). Rfam: Wikipedia, clans and the "decimal" release. *Nucleic Acids Research*, *39*(Database), D141–D145.

Giegerich, R., Voss, B., & Rehmsmeier, M. (2004). Abstract shapes of RNA. *Nucleic Acids Research*, *32*(16), 4843–4851.

Griffiths-Jones, S., Bateman, A., Marshall, M., Khanna, A., & Eddy, S. R. (2003). Rfam: An RNA family database. *Nucleic Acids Research*, *31*(1), 439–441.

Gruber, A. R., Neubock, R., Hofacker, I. L., & Washietl, S. (2007). The RNAz web server: Prediction of thermodynamically stable and evolutionarily conserved RNA structures. *Nucleic Acids Research*, *35*(Web), W335–W338.

Havgaard, J. H., Lyngso, R. B., & Gorodkin, J. (2005). The FOLDALIGN web server for pairwise structural RNA alignment and mutual motif search. *Nucleic Acids Research*, *33*(Suppl. 2), W650–W653. Retrieved from, http://dx.doi.org/10.1093/nar/gki473.

Havgaard, J. H., Torarinsson, E., & Gorodkin, J. (2007). Fast pairwise structural RNA alignments by pruning of the dynamical programming matrix. *PLoS Computational Biology*, *3*(10), 1896–1908.

Havill, J. T., Bhatiya, C., Johnson, S. M., Sheets, J. D., & Thompson, J. S. (2014). A new approach for detecting riboswitches in DNA sequences. *Bioinformatics*, *30*(21), 3012–3019.

Hertel, J., Hofacker, I. L., & Stadler, P. F. (2008). Snoreport: Computational identification of snoRNAs with unknown targets. *Bioinformatics*, *24*(2), 158–164.

Hertel, J., & Stadler, P. F. (2006). Hairpins in a Haystack: Recognizing microRNA precursors in comparative genomics data. *Bioinformatics*, *22*(14), e197–e202.

Hofacker, I. L. (2003). Vienna RNA secondary structure server. *Nucleic Acids Research*, *31*, 3429–3431.

Holden, M. T., Feil, E. J., Lindsay, J. A., Peacock, S. J., Day, N. P., Enright, M. C., et al. (2004). Complete genomes of two clinical Staphylococcus aureus strains: Evidence for the rapid evolution of virulence and drug resistance. *Proceedings of the National Academy of Sciences of the United States of America*, *101*(26), 9786–9791.

Johansen, L. E., Nygaard, P., Lassen, C., Agerso, Y., & Saxild, H. H. (2003). Definition of a second Bacillus subtilis pur regulon comprising the pur and xpt-pbuX operons plus pbuG, nupG (yxjA), and pbuE (ydhL). *Journal of Bacteriology*, *185*(17), 5200–5209.

Karplus, K., Barrett, C., & Hughey, R. (1998). Hidden Markov models for detecting remote protein homologies. *Bioinformatics*, *14*, 846–856.

Kazanov, M. D., Vitreschak, A. G., & Gelfand, M. S. (2007). Abundance and functional diversity of riboswitches in microbial communities. *BMC Genomics*, *8*, 347.

Kim, D. S., Gusti, V., Pillai, S. G., & Gaur, R. K. (2005). An artificial riboswitch for controlling pre-mRNA splicing. *RNA*, *11*(11), 1667–1677.

Kubodera, T., Watanabe, M., Yoshiuchi, K., Yamashita, N., Nishimura, A., Nakai, S., et al. (2003). Thiamine-regulated gene expression of Aspergillus oryzae thiA requires splicing of the intron containing a riboswitch-like domain in the 5′-UTR. *FEBS Letters*, *555*(3), 516–520. Retrieved from, http://www.sciencedirect.com/science/article/B6T36-4B29TG2-2/2/88075d8fb77e49f4bb4d846fafd1664c.

Kung, J. T., Colognori, D., & Lee, J. T. (2013). Long noncoding RNAs: Past, present, and future. *Genetics*, *193*(3), 651–669.

Lim, L. P., Glasner, M. E., Yekta, S., Burge, C. B., & Bartel, D. P. (2003). Vertebrate microRNA genes. *Science*, *299*(5612), 1540.

Lowe, T., & Eddy, S. (1997). tRNAscan-SE: A program for improved detection of transfer RNA genes in genomic sequence. *Nucleic Acids Research*, *25*(5), 955–964.

Mandal, M., Boese, B., Barrick, J. E., Winkler, W. C., & Breaker, R. R. (2003). Riboswitches control fundamental biochemical pathways in Bacillus subtilis and other bacteria. *Cell*, *113*(5), 577–586.

Mandal, M., & Breaker, R. R. (2004). Adenine riboswitches and gene activation by disruption of a transcription terminator. *Nature Structural & Molecular Biology*, *11*(1), 29–35.

Mandal, M., Lee, M., Barrick, J. E., Weinberg, Z., Emilsson, G. M., Ruzzo, W. L., et al. (2004). A glycine-dependent riboswitch that uses cooperative binding to control gene expression. *Science*, *306*(5694), 275–279.

Manzourolajdad, A. (2014). *Ab initio identi_cation of regulatory RNAs using information-theoretic uncertainty*. University of Georgia.

Manzourolajdad, A., Wang, Y., Shaw, T. I., & Malmberg, R. L. (2013). Information-theoretic uncertainty of SCFG-modeled folding space of the non-coding RNA. *Journal of Theoretical Biology*, *318*, 140–163.

Mathews, D. H., & Turner, D. H. (2002). Dynalign: An algorithm for finding the secondary structure common to two RNA sequences. *Journal of Molecular Biology*, *317*, 191–203.

Merino, E., & Yanofsky, C. (2005). Transcription attenuation: A highly conserved regulatory strategy used by bacteria. *Trends in Genetics*, *21*, 249–305.

Miranda-Ríos, J., Navarro, M., & Soberón, M. (2001). A conserved RNA structure (thi box) is involved in regulation of thiamin biosynthetic gene expression in bacteria. *Proceedings of the National Academy of Sciences of the United States of America*, *98*(17), 9736–9741. Retrieved from, http://dx.doi.org/10.1073/pnas.161168098.

Nawrocki, E. P., & Eddy, S. R. (2013). Computational identification of functional RNA homologs in metagenomic data. *RNA Biology*, *10*(7), 1170–1179.

Nawrocki, E. P., & Eddy, S. R. (2013). Infernal 1.1: 100-fold faster RNA homology searches. *Bioinformatics*, *29*(22), 2933–2935.

Nawrocki, E. P., Kolbe, D. L., & Eddy, S. R. (2009). Infernal 1.0: Inference of RNA alignments. *Bioinformatics*, *25*(10), 1335–1337.

Ng, K. L., & Mishra, S. K. (2007). De novo SVM classification of precursor microRNAs from genomic pseudo hairpins using global and intrinsic folding measures. *Bioinformatics*, *23*(11), 1321–1330.

Omer, A. D., Lowe, T. M., Russell, A. G., Ebhardt, H., Eddy, S. R., & Dennis, P. P. (2000). Homologues of small nucleolar RNAs in Archaea. *Science*, *288*, 517–522.

Ooms, M., Huthoff, H., Russell, R., Liang, C., & Berkhout, B. (2004). A riboswitch regulates RNA dimerization and packaging in human immunodeficiency virus type 1 virions. *Journal of Virology*, *78*(19), 10814–10819.

Ray, P. S., Jia, J., Yao, P., Majumder, M., Hatzoglou, M., & Fox, P. L. (2009). A stress-responsive RNA switch regulates VEGFA expression. *Nature*, *457*(7231), 915–919.

Rodionov, D. A., Vitreschak, A. G., Mironov, A. A., & Gelfand, M. S. (2002). Comparative genomics of thiamin biosynthesis in procaryotes. *Journal of Biological Chemistry*, *277*(50), 48949–48959. Retrieved from, http://www.jbc.org/content/277/50/48949.abstract.

Senter, E., Sheik, S., Dotu, I., Ponty, Y., & Clote, P. (2012). Using the Fast Fourier Transform to accelerate the computational search for RNA conformational switches. *PLoS One*, 7(12), e50506.

Serganov, A., Polonskaia, A., Phan, A. T., Breaker, R. R., & Patel, D. J. (2006). Structural basis for gene regulation by a thiamine pyrophosphate-sensing riboswitch. *Nature*, *441*(7097), 1167–1171.

Serganov, A., Yuan, Y. R., Pikovskaya, O., Polonskaia, A., Malinina, L., Phan, A. T., et al. (2004). Structural basis for discriminative regulation of gene expression by adenine- and guanine-sensing mRNAs. *Chemistry & Biology*, *11*(12), 1729–1741.

Shapiro, B. A., Bengali, D., Kasprzak, W., & Wu, J. C. (2001). RNA folding pathway functional intermediates: Their prediction and analysis. *Journal of Molecular Biology*, *312*(1), 27–44.

Singh, P., Bandyopadhyay, P., Bhattacharya, S., Krishnamachari, A., & Sengupta, S. (2009). Riboswitch detection using profile hidden Markov models. *BMC Bioinformatics*, *10*(1), 325. Retrieved from, http://dx.doi.org/10.1186/1471-2105-10-325.

Sudarsan, N., Barrick, J. E., & Breaker, R. R. (2003). Metabolite-binding RNA domains are present in the genes of eukaryotes. *RNA*, *9*, 644–647.

Torarinsson, E., Yao, Z., Wiklund, E. D., Bramsen, J. B., Hansen, C., Kjems, J., et al. (2008). Comparative genomics beyond sequence-based alignments: RNA structures in the ENCODE regions. *Genome Research*, *18*(2), 242–251.

Tucker, B. J., & Breaker, R. R. (2005). Riboswitches as versatile gene control elements. *Current Opinion in Structural Biology*, *15*(3), 342–348.

Tuschl, T. (2003). Functional genomics: RNA sets the standard. *Nature*, *421*, 220–221.

Van Bakel, H., Nislow, C., Blencowe, B. J., & Hughes, T. R. (2010). Most "dark matter" transcripts are associated with known genes. *PLoS Biology*, *8*(5), e1000371.

Vapnik, V. (1998). *Statistical learning theory*. New York: John Wiley & Sons, Inc.

Voss, B., Giegerich, R., & Rehmsmeier, M. (2006). Complete probabilistic analysis of RNA shapes. *BMC Biology*, *4*(5), 1–23.

Voss, B., Meyer, C., & Giegerich, R. (2004). Evaluating the predictability of conformational switching in RNA. *Bioinformatics*, *20*(10), 1573–1582.

Wachter, A., Tunc-Ozdemir, M., Grove, B. C., Green, P. J., Shintani, D. K., & Breaker, R. R. (2007). Riboswitch control of gene expression in plants by splicing and alternative 3′ end processing of mRNAs. *Plant Cell*, *19*(11), 3437–3450.

Wakeman, C. A., Winkler, W. C., & Dann, C. (2007). Structural features of metabolite-sensing riboswitches. *Trends in Biochemical Sciences*, *32*(9), 415–424.

Washietl, S., Hofacker, I. L., & Stadler, P. F. (2005). Fast and reliable prediction of noncoding RNAs. *Proceedings of the National Academy of Sciences of the United States of America*, *102*(7), 2454–2459.

Weinberg, Z., Barrick, J. E., Yao, Z., Roth, A., Kim, J. N., Gore, J., et al. (2007). Identification of 22 candidate structured RNAs in bacteria using the CMfinder comparative genomics pipeline. *Nucleic Acids Research*, *35*(14), 4809–4819.

Weinberg, Z., & Ruzzo, W. L. (2004). Exploiting conserved structure for faster annotation of non-coding RNAs without loss of accuracy. *Bioinformatics*, *20*, 334–341.

Weinberg, Z., Wang, J. X., Bogue, J., Yang, J., Corbino, K., Moy, R. H., et al. (2010). Comparative genomics reveals 104 candidate structured RNAs from bacteria, archaea, and their metagenomes. *Genome Biology*, *11*(3), R31.

Wickiser, J. K., Cheah, M. T., Breaker, R. R., & Crothers, D. M. (2005). The kinetics of ligand binding by an adenine-sensing riboswitch. *Biochemistry*, *44*(40), 13404–13414.

Winkler, W. C., Cohen-Chalamish, S., & Breaker, R. R. (2002). An mRNA structure that controls gene expression by binding FMN. *Proceedings of the National Academy of Sciences of the United States of America*, *99*, 15908–15913.

Wuchty, S., Fontana, W., Hofacker, I. L., & Schuster, P. (1999). Complete suboptimal folding of RNA and the stability of secondary structures. *Biopolymers*, *49*, 145–165.

Xue, C., Li, F., He, T., Liu, G. P., Li, Y., & Zhang, X. (2005). Classification of real and pseudo microRNA precursors using local structure-sequence features and support vector machine. *BMC Bioinformatics*, *6*, 310.

Yao, Z., Barrick, J., Weinberg, Z., Neph, S., Breaker, R., Tompa, M., et al. (2007). A computational pipeline for high-throughput discovery of cis-regulatory noncoding RNA in prokaryotes. *PLoS Computational Biology*, *3*(7), e126.

Yao, Z., Weinberg, Z., & Ruzzo, W. L. (2006). CMfinder—A covariance model based RNA motif finding algorithm. *Bioinformatics*, *22*, 445–452.

Zhang, S., Borovok, I., Aharonowitz, Y., Sharan, R., & Bafna, V. (2006). A sequence-based filtering method for ncRNA identification and its application to searching for riboswitch elements. *Bioinformatics*, *22*, 557–565.

CHAPTER THIRTEEN

Computational and Experimental Studies of Reassociating RNA/DNA Hybrids Containing Split Functionalities

Kirill A. Afonin*[,1], Eckart Bindewald[†,1], Maria Kireeva[‡], Bruce A. Shapiro*[,2]

*Basic Research Laboratory, Center for Cancer Research, National Cancer Institute, Frederick, Maryland, USA

[†]Basic Science Program, Leidos Biomedical Research Inc., National Cancer Institute, National Institutes of Health, Frederick, Maryland, USA

[‡]Gene Regulation and Chromosome Biology Laboratory, Center for Cancer Research, NCI, National Cancer Institute, Frederick, Maryland, USA

[1]These authors contributed equally.

[2]Corresponding author: e-mail address: shapirbr@mail.nih.gov

Contents

Abstract

Recently, we developed a novel technique based on RNA/DNA hybrid reassociation that allows conditional activation of different split functionalities inside diseased cells and *in vivo*. We further expanded this idea to permit simultaneous activation of multiple different functions in a fully controllable fashion. In this chapter, we discuss some novel computational approaches and experimental techniques aimed at the characterization, design, and production of reassociating RNA/DNA hybrids containing split functionalities. We also briefly describe several experimental techniques that can be used to test these hybrids *in vitro* and *in vivo*.

Methods in Enzymology, Volume 553
ISSN 0076-6879
http://dx.doi.org/10.1016/bs.mie.2014.10.058

1. INTRODUCTION

RNA interference (RNAi) is a natural cellular posttranscriptional gene regulation process that involves small double-stranded RNAs directing homology-dependent silencing of target genes (Fire et al., 1998). One of the ways to activate RNAi is through the exogenous introduction of small-interfering RNAs or siRNAs (Elbashir, Lendeckel, & Tuschl, 2001; Elbashir, Martinez, Patkaniowska, Lendeckel, & Tuschl, 2001). The RNAi mechanism is increasingly employed for treatment and therapeutic gene modulation of various diseases and viral infections as illustrated by several clinical trials that are testing novel RNAi-based therapeutics (Bramsen & Kjems, 2012; Thompson, 2013; Zhou et al., 2013).

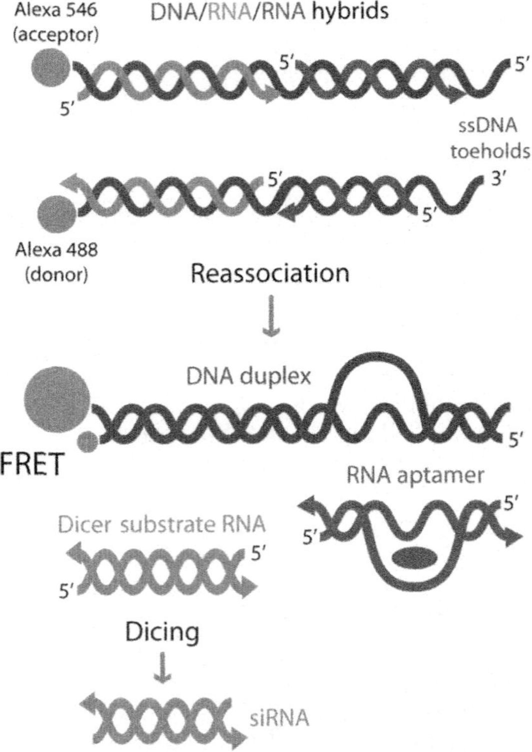

Figure 1 Schematic representation of reassociation for RNA/DNA hybrids carrying multiple split functionalities (FRET, Dicer Substrate RNA, and RNA aptamer such as malachite green aptamer). (See the color plate.)

We have developed a novel approach of split functionalities (schematically depicted in Fig. 1) based on RNA/DNA hybrids, which are activated only when two complementary hybrids are introduced into the same cell (Afonin, Viard, et al., 2013). This approach allows a greater degree of control over deliverable functionalities (such as siRNAs) and stabilities of RNA-based domains. Combining the properties of RNA and DNA molecules allows the hybrid constructs to have higher stability in blood serum, permits the attachment of fluorescent markers for tracking without interfering with RNA functionality, and permits the ability to split the components of functional elements inactivating them, but allowing later activation under the control of complementary toeholds by which the kinetics of reassociation can be fine-tuned. Thus, for example, a Dicer Substrate RNA (DS RNA) developed to enhance RNAi (Rose et al., 2005) could be split into two RNA/DNA hybrids, where the DNA contains a complementary single-stranded toehold to its counterpart found in a complementary hybrid. DS RNA has to be processed by Dicer first in order to induce RNAi. However, each RNA/DNA hybrid carrying one of the DS RNA strands cannot be diced and hence stays inactive. When transfected into cells, these two hybrids reassociate due to the presence of the single-stranded DNA toeholds and release DS RNA, thus activating RNAi. Extensive *in vitro* kinetics studies demonstrate that the average time of reassociation is hybrid toehold (length and composition) and concentration dependent (Afonin, Viard, et al., 2013). For example, for the hybrids with 12-nucleotide toeholds having 60% GC content, the limiting step of reassociation is the zipping of the toeholds at concentrations lower than ~30 n*M*, while at higher concentrations, the reassociation becomes the rate-determining step with $t_{1/2}$ not exceeding ~15 min. More detailed experimental and computational studies are currently being conducted utilizing various length and base compositions of DNA toeholds.

This concept has been expanded further to simultaneously release multiple split functionalities from two hybrid reassociations (Afonin, Desai, et al., 2014). As a proof of concept, we demonstrated the release of multiple split DS RNAs and RNA aptamers together with Förster resonance energy transfer (FRET) as shown in Fig. 1. Also, we were able to couple the hybrid concept with our multifunctional architectures such as nanocubes (Afonin et al., 2010, 2011; Afonin, Kasprzak, Bindewald, Kireeva, et al., 2014; Afonin, Kasprzak, Bindewald, Puppala, et al., 2014; Afonin, Viard, Kaglampakis, et al., In press). However, we demonstrated the use of

RNA-based nanoparticles (nanorings; Afonin et al., 2011; Grabow et al., 2011; Yingling & Shapiro, 2007) that simultaneously activate hybrid split functions in cancer cells (Afonin, Desai, et al., 2014). Due to the increasing complexity of the hybrid structures, there is a great demand for computer algorithms that aim to assist in the design as well as the process of simulating the reassociation of the RNA/DNA hybrids.

Currently, we are developing and improving the existing (Afonin, Desai, et al., 2014) computational algorithms that simulate both the kinetic and thermodynamic properties of multiple DNA and RNA hybrid assemblies. The computational characterization of RNA/DNA hybrids and their reassociation requires the ability to predict the folding properties of multiple nucleotide strands in solution. Several approaches for the computational prediction of nucleic acid secondary structures consisting of multiple strands have been described. RNAcofold (Lorenz et al., 2011) is a program that considers intrastrand and interstrand folding of two RNA strands. It can compute a predicted secondary structure corresponding to the minimum free energy as well as the concentrations of the resulting heterodimer structures and homodimers consisting only of RNAs. Pseudoknotted structures (Cao, Xu, & Chen, 2014) as well as higher order complexes consisting of more than two strands are not considered (Bernhart et al., 2006). NanoFolder can be used to predict potentially pseudoknotted RNA secondary structures consisting of multiple strands (Bindewald, Afonin, Jaeger, & Shapiro, 2011), but it does not consider DNA or RNA/DNA structures. The NUPACK software allows computing the secondary structure of multiple RNA or DNA strands (Dirks & Pierce, 2003; Zadeh et al., 2010) and reports predicted complex concentrations. Multifold can perform secondary structure predictions of multiple RNA strands based on a dynamic programming algorithm (Andronescu, Zhang, & Condon, 2005).

Recently, a program for the prediction of two-strand RNA/DNA hybrid heterodimers has been made available in the RNA Vienna package (Lorenz, Hofacker, & Bernhart, 2012). This program is based on dynamic programming and has algorithmic similarities with the RNAcofold program provided by the same package.

The motivation for the computational approach presented here is the development of software for computing the equilibrium thermodynamics of multiple RNA and DNA strands allowing for RNA/DNA hybrid interactions while also allowing for the formation of complex pseudoknots.

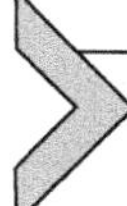

2. THERMODYNAMIC PREDICTION OF DIFFERENT COMPOSITIONS OF RNA AND DNA STRAND ASSOCIATIONS

2.1. Partition function of multiple nucleotide strands

Equilibrium thermodynamics is well understood via the statistical mechanics concept of a partition function. The probability of a state i in a system of constant volume and temperature is given through the canonical partition function Z:

$$Z = \sum_{\text{all microstates } i} e^{-E_i/RT} \tag{1}$$

where the terms E_i are the energies of the microstates of the system (a microstate is given through the positions and momenta of the involved atoms); R stands for the gas constant; and T is the temperature.

Different microstates can, however, have virtually identical energies. Examples of such occurrences are different backbone conformations in single-stranded regions of nucleotide strands (if single-stranded base–base stacking can be neglected) or translational degrees of freedom of molecules in a solvent. Such differences are not of much interest and lead to an unnecessary burden on attempts to estimate the partition function. To simplify the computation of the partition function, it is thus helpful to define a subset of distinct microstates by combining energetically degenerate microstates into combined states t using a statistical weight w_t that indicates how many different microstates have been combined:

$$Z = \sum_{\text{states } t} w_t e^{-E_t/RT} \tag{2}$$

Note that the weight w_t depends on the state t. This can be rewritten as

$$Z = \sum_{\text{states } t} e^{-((E_t - RT\log \mathrm{w}_t)/RT)} \tag{3}$$

We can thus define a free energy to a state t that consists of w_t microstates:

$$G_t = E_t - RT\log \mathrm{w}_t \tag{4}$$

Here, the different states correspond to the different base pairings (secondary structures) of the involved nucleotide strands. The underlying approximation is that the different translational and backbone conformation states of a set of nucleotides are energetically identical for the same nucleotide base pairing. Using a symbol s for a secondary structure that is an element of the set S of all secondary structures leads to (Dimitrov & Zuker, 2004; Dirks & Pierce, 2003):

$$Z = \sum_{s \in S} e^{-G_s/RT} \tag{5}$$

Note that the approach uses the Gibbs free energy and the Helmholtz free energy interchangeably, as is commonly the case in the secondary structure prediction field. This is defensible because the difference between those two quantities is a pressure–volume term (pV) that is important if a chemical reaction leads the system to perform work with respect to its environment in the form of an expansion or contraction. Since we are interested in base pairing of nucleotide strands in aqueous solution, this change in density is negligible.

One reason for the importance of the partition function is that the probability of observing a particular secondary structure s is given by

$$p_s = \frac{e^{-G_s/RT}}{Z} \tag{6}$$

In addition to the free energy G_s^0 of a particular secondary structure for a given set of positions of nucleotide strands (we utilize the helix and loop free energy estimates according to Mathews, Sabina, Zuker, & Turner (1999) in order to estimate G_s^0), there is a term G_s^{trans} that stands for the potentially different number of translational states of the set of simulated nucleotide strands in a given volume V:

$$G_s = G_s^0 + G_s^{\text{trans}} \tag{7}$$

The translational component G_s^{trans} is estimated as follows: the simulation volume V is given as the inverse of the concentration. The volume V is divided into small cubes each with volume v ($\sim$ the volume of each complex). The number of microstates for each complex is thus V/v. If m is the number of formed complexes, the free energy contribution is given through:

$$G_s^{\text{trans}} = -mRT \log \frac{V}{v} \tag{8}$$

Energy parameters for RNA/RNA, DNA/DNA, and RNA/DNA base pairing have been reported. To estimate the term G_s^0, we utilize the popular nearest neighbor model, in which a free energy contribution is assigned to two adjacent base pairs (Mathews et al., 1999; SantaLucia, 1998; Wu, Nakano, & Sugimoto, 2002). We also use a previously published approach to estimate loop entropies (Mathews et al., 1999). Note that electronic, vibrational, and rotational degrees of freedoms are not accounted for in our model.

This nearest neighbor model is a reflection of the fact that in addition to hydrogen bond formation of the base pairs, it is the stacking of the hydrophobic parts of the nucleobases that provide a major energetic contribution to the structure formation. A computational challenge for the computation of RNA/DNA complexes is that it involves the weighing of the three possible cases of RNA/RNA base pairing, DNA/DNA base pairing, and RNA/DNA base pairing. We model this by having a computational representation of eight different types of nucleotides.

2.2. Search algorithm

One common approach to estimate the partition function for multiple strands is to enumerate over all possible counts of involved molecular species (counting created complexes as a distinct molecular species). Because the partition function is a sum of a finite number of terms, there is, however, no requirement to follow one particular order of grouping the involved terms. For the sake of simplicity, we thus use an algorithm that generates distinct secondary structures, a way to estimate the free energy of a particular secondary structure (including a concentration-dependent term that accounts for translational entropy) and a way to store the interesting features of each structure (the estimated free energy as well as the types of resulting complexes it corresponds to). The found free energies are also used to estimate the partition function of the system. Note that each strand at the structure–enumeration stage is treated as a distinct molecular species, even if its sequence is identical to a different strand.

There is no known algorithm with polynomial complexity that would allow the exhaustive enumeration of all base-pairing states. One can cope with this computational challenge by restricting oneself to short sequences or reductions in the search space by (i) not searching conformations with a too complex topology (for example, not allowing nonnested, "pseudoknotted" structures), (ii) not searching energetically highly unfavorable states, or (iii) not searching states that are unlikely to be energetically substantially more stable compared to structures that are part of the search space (such as enforcing a minimal helix length of, for example, two base pairs or only considering maximally extended helices).

The challenge of the search algorithm approach is (i) to sample suboptimal structures in order to estimate the partition function, (ii) to identify the structures with the lowest free energy, (iii) to not sample any structure more than once, (iv) to (ideally) not omit any structures from the considered search space, and finally (v) to not sample structures that are not part of the considered search space.

Our approach involves the enumeration of states utilizing a "stem" approach: a list of all considered helices is generated initially. Next, a two-dimensional array of computational "containers" is generated. Each container is a computational data structure that contains an energetically sorted list of partially folded secondary structures. These sorted lists are henceforth referred to as queues. The two dimensions are the number of base pairs and the number of helices. Another data structure represents a secondary structure base pairing of all strands. This data structure represents the RNA and DNA strands whose folding is to be simulated. In other words, one secondary structure contains potentially several strands, which may potentially form one or several complexes that "live" within a simulation box of a defined volume corresponding to the concentrations.

The search algorithm proceeds as follows. Initially, the secondary structure object corresponding to the completely unfolded state is deposited into the queue corresponding to zero base pairs and zero helices. This structure is removed from its queue and then "expanded" by generating all possible structures that contain one additional helix. These expanded structures are placed into the appropriate queues. Now the algorithm performs the following steps, until all queues are empty: a score is utilized to decide which partially folded secondary structure is most promising to pursue further. Implemented are different heuristics for this score: it is, based on a flag, either set to the negative of the estimated free energy of the partially folded structure or to the number of base pairs minus three times the number of helices (the results presented below are based on the latter choice). The next "most promising" partially folded secondary structure that is on top of one of the queues is chosen, removed from its queue, and "expanded": all structures with one additional helix are placed into the appropriate queues. Note that there can be, depending on the user options, different constraints on the newly added helix: (i) it can be such that either nonnested base pairings are not allowed (thus prohibiting pseudoknots) or (ii) nonnested base pairings are not allowed when they correspond to the same strand interaction or (iii) no pseudoknot restrictions at all.

Associated with each queue is the data structure of a *set* of secondary structures. The data structure is utilized to ensure that no secondary structure is searched twice. A structure is placed into a queue only if it is not part of the set that keeps track of which structures have already been searched. A second criterion for placing a new partially folded secondary structure into a queue for further folding is that its estimated free energy is not less favorable than the so far best found structure with the same number of base pairs and helices (plus a "slop" term allowing for slightly unfavorable partially folded structures). This procedure tends to fill up initially empty queues during the search procedure. Near the end of the search when it is no longer possible to place additional helices, the queues are "emptied." The search terminates, once either all queues are empty or if a maximum number of search steps is reached. The free energy of each secondary structure that is encountered during the search is stored in a data structure representing the partition function.

2.3. Postprocessing of secondary structure predictions

Once the enumeration of strands is finished, the concentration of each molecular species (including complexes) is estimated by adding the estimated probabilities of occurrence of each examined secondary structure that leads to the formation of a molecular species in question. Note that at this stage, it *is* accounted for that strands with the same sequence are the same molecular species.

To compute the free energy of reassociation, one can take the difference between the free energy of the set of secondary structures that correspond to RNA/DNA complexes and the free energy of the secondary structures corresponding to the re-associated RNA/RNA + DNA/DNA complexes (indicated by arrows in Fig. 3).

The output of the program consists of the predicted concentrations of the encountered strands and complexes, a list of probability-sorted secondary structures as well as a probability-sorted list of combinations of simultaneously forming complexes. Each list of the combination of complexes corresponds to an ensemble of all found distinct base pairings in which these complexes form. The sum of the probabilities of the individual base pairing states corresponds to the estimated probability of the nucleotide strands forming this particular set of complexes (according to Eq. 6). This can be used to compute a free energy for this set of complexes. Subtracting the free energy of the reference state of a completely unfolded structure from the free energy of a folded structure leads to the free energy of folding. Subtracting instead the free energy of individually folded nucleotide strands (not allowing for interstrand interactions) leads to the free energy of binding. The program reports for each found complex the predicted concentration and the expected number of complexes in the simulated volume. A visual representation of the algorithm is shown in Fig. 2. Note that a previously utilized version of the program does not generate a list of possible stems but "expands" structures by placing one additional base pair at a time in all possible ways (Afonin, Desai, et al., 2014).

2.4. Implementation and example

We created a computer program in the C++ language that implements the described search strategy. The system allows us to specify a set of nucleotide

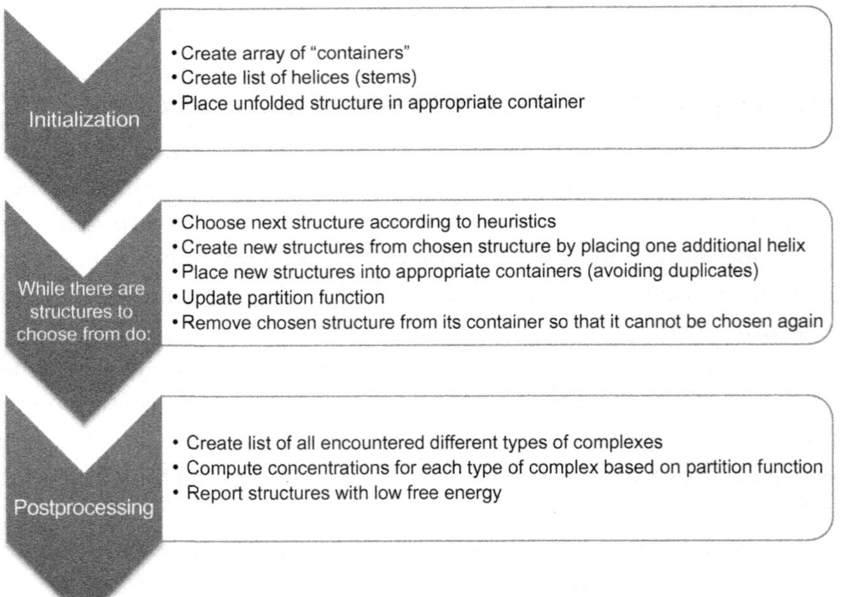

Figure 2 Schematic representation of the structure prediction algorithm. The algorithm consists of the three phases of initialization, structure search, and postprocessing. During the structure search, a partially folded structure is chosen according to a heuristic, and "expanded" by placing one additional helix in all possible ways. This process is repeated until there are no structures to choose from or until a maximum number of iterations have been reached.

strands, as well as a "multiplicity term" that indicates how many copies of each strand are being simulated. This potentially allows for considering higher order complexes such as homodimers and homotrimers.

An experimentally confirmed example of dual-release RNA/DNA hybrid complex reassociation was subjected to this algorithm as shown in Fig. 3. Shown are the input sequences (Fig. 3A, two copies of sense and antisense siRNAs, as well as two different DNA sequences) as well as the computational results for two different scenarios. One can see in Fig. 3B by the formation of AAB and CCD complexes that the dual-release hybrid structures are predicted to form. As expected, the siRNA duplexes (named "AC") and DNA duplexes (named "BD") are predicted to also form, albeit with a lower free energy (compare arrows in Fig. 3c).

3. SEQUENCE DESIGN OF RNA/DNA HYBRIDS

The sequence design of RNA/DNA hybrids is facilitated by the fact that most nucleotides are determined by the chosen siRNA target site on the mRNA. This target site specifies the siRNA, and by extension the cognate RNA and DNA sequences. Also, as previously described, it is beneficial to extend the 5′-end of the siRNA-antisense strand by ~4–8 nucleotides complementary with the mRNA target strand to make it a substrate for Dicer. We implemented an algorithm in the R programming language that for a given siRNA sequence and mRNA sequence finds a matching binding site on the mRNA sequence and performs the steps to extend the siRNA sequence and define the DNA strands accordingly. In addition to these "given" nucleotide positions, the DNA toehold sequences are, in principle, freely designable.

The chosen approach for designing DNA toeholds was such that a "criton" method was chosen to identify randomly generated sequences that do not have reverse complementary regions with respect to itself and with respect to the other RNA and DNA strand regions (Seeman, 1982; Bindewald et al., 2011). Another important aspect is the average G + C content of a nucleotide strand toehold. The R implementation generates, in a randomized fashion, for a given set of scaffold RNA or DNA strands, toehold sequence regions that do not contain undesired reverse complementary regions with respect to the remaining nucleotide strands. The toehold regions also have a target G + C content, that is within an range specified by the user. We computationally designed hybrids with toeholds having

A

```
A ( originally 1)
ACCCUGAAGUUCAUCUGCACCACCG
A ( originally 2)
ACCCUGAAGUUCAUCUGCACCACCG
B ( originally 3)
ggagaccgtgaccggtggtgcagatgaacttcagggtcacggtggtgcagatgaacttcagggtca
C ( originally 4)
CGGUGGUGCAGAUGAACUUCAGGGUCA
C ( originally 5)
CGGUGGUGCAGAUGAACUUCAGGGUCA
D ( originally 6)
tgaccctgaagttcatctgcaccaccgtgaccctgaagttcatctgcaccaccggtcacggtctcc
```

B

# Complex	Abs.conc.(mol/l)	Expected	Rel.concentration
AC	2e-06	2	0.333333
BD	1e-06	1	0.166667
A	1.27026e-29	1.27026e-23	2.1171e-24
C	1.27026e-29	1.27026e-23	2.1171e-24
ABCD	9.99639e-45	9.99639e-39	1.66606e-39
AACC	9.86713e-47	9.86713e-41	1.64452e-41
ACC	2.32837e-50	2.32837e-44	3.88062e-45
CC	8.08809e-52	8.08809e-46	1.34801e-46
AA	8.08505e-52	8.08505e-46	1.34751e-46
BCD	2.12716e-53	2.12716e-47	3.54527e-48
CD	1.31234e-56	1.31234e-50	2.18723e-51
AB	1.31234e-56	1.31234e-50	2.18723e-51
CCD	9.83161e-57	9.83161e-51	1.6386e-51
AAB	9.83161e-57	9.83161e-51	1.6386e-51
AAC	8.55041e-60	8.55041e-54	1.42507e-54
B	1.6953e-61	1.6953e-55	2.82549e-56
ACD	1.69329e-61	1.69329e-55	2.82215e-56

C

```
# List of Complexes: Probability, Free energy of folding (dGfold) and free energy of binding (dGbind):
→ AC|AC|BD     Prob:  1             dGfold  -163.792    dGbind  -154.762
A|AC|BD|C      Prob:  1.27026e-23   dGfold  -131.316    dGbind  -122.286
ABCD|AC        Prob:  9.99639e-39   dGfold  -109.893    dGbind  -100.863
AACC|BD        Prob:  9.86713e-41   dGfold  -107.048    dGbind  -98.0179
A|ACC|BD       Prob:  2.32837e-44   dGfold  -101.903    dGbind  -92.8732
AA|BD|CC       Prob:  8.08505e-46   dGfold  -99.8333    dGbind  -90.8032
A|AC|BCD       Prob:  2.12716e-47   dGfold  -97.5924    dGbind  -88.5623
A|A|BD|CC      Prob:  3.03735e-49   dGfold  -94.9751    dGbind  -85.945
AB|AC|CD       Prob:  1.31234e-50   dGfold  -93.0397    dGbind  -84.0096
→ AAB|CCD      Prob:  9.83161e-51   dGfold  -92.8619    dGbind  -83.8317
A|A|BD|C|C     Prob:  5.83422e-52   dGfold  -91.122     dGbind  -82.0919
```

Figure 3 Example of program output for given dual-release RNA (in red; light gray in the print version)/DNA (in blue; dark gray in the print version) hybrid sequences. (A) Input consisting of two copies of sense DS RNAs, two copies of antisense DS RNAs, and two different cognate DNA strands with toeholds. The program notices that sequences 1 and 2 as well as 4 and 5 are identical, and renames them as "A" and "C," respectively. (B) Part of the output of the program. The predicted complex formation includes DS RNAs (named by the program "AC") as well as DNA duplexes (named "BD"). Note the formation of complexes named AAB and CCD that indicate the formation of dual-release RNA/DNA hybrids. The computer output shows predictions for the absolute concentrations ("Abs conc."), the expected number of complexes in the simulation volume and the relative concentration (the expected number of complexes divided by the number of simulated strands). (C) Another list that is part of the output shows secondary structures and their probabilities ("Probability"), free energies of folding (dGfold), and free energies of binding (dGbind). The computer results are based on a list of stems with a minimum length of 3 base pairs; only stems that cannot be extended further are considered. There are no restrictions in terms of pseudoknot complexity. For clarity, symbols A and C indicating RNA strands have been colored red (light gray in the print version), and symbols B and D (indicating DNA strands) have been colored blue (dark

G + C contents of ~60% as well as ~25% that were further extensively tested experimentally (Unpublished data).

4. ENZYME-ASSISTED *IN VITRO* PRODUCTION OF RNA/DNA HYBRIDS

Currently, RNA/DNA hybrids carrying multiple split functionalities can be produced in several steps: individual RNAs and DNAs are synthesized using chemical synthesis, purified, and mixed in equimolar concentrations. The mixture is subjected to thermal denaturation and renaturation in order to assemble RNA/DNA hybrids as shown in Fig. 4. The current limitations on the chemical synthesis of RNA chains longer than 60–70 nucleotides emphasize the importance of enzymatic RNA synthesis by *in vitro* transcription in biotechnology and medicine. In this chapter, we summarize the current state and perspectives of the *in vitro* transcription methodology for pipeline production of RNA/DNA hybrids with split functionalities.

Recently, we developed a new methodology that facilitates the production of the individual hybrids carrying long RNAs during *in vitro* transcription with RNA polymerase II-dependent transcription of ssDNA templates. RNA polymerase II is mixed with short synthetic RNA primers annealed to ssDNAs (Fig. 5) followed by extension of the RNA to the end of the template, creating a construct with an RNA length close to 100 nucleotides. Interestingly, in the same experimental setup, *Escherichia coli* RNA polymerase failed to extend the RNA primer to the required length (Afonin, Desai, et al., 2014). Apparently, the subtle difference in the size and structure of the lid element (a loop-like structure), located near the RNA/DNA separation region at the upstream edge of the transcription bubble in the *Saccharomyces cerevisiae* RNAP II and bacterial RNA polymerase (Vassylyev, Vassylyeva, Perederina, Tahirov, & Artsimovitch, 2007; Westover, Bushnell, & Kornberg, 2004), accounts for this difference in the function. Indeed, deletion of the lid element in *E. coli* RNAP promotes formation of the extended RNA/DNA hybrids (Naryshkina, Kuznedelov, & Severinov, 2006; Toulokhonov & Landick, 2006),

gray in the print version); tab characters have been inserted into the computer output. The arrows indicate the desired hybrid state and product states. One possible explanation for the estimated low probabilities of higher order complexes could be the simplification that only secondary structure states are considered in which all helices consist of at least 3 base pairs. Also, the theoretical treatment does not consider the exchange of molecules with the environment.

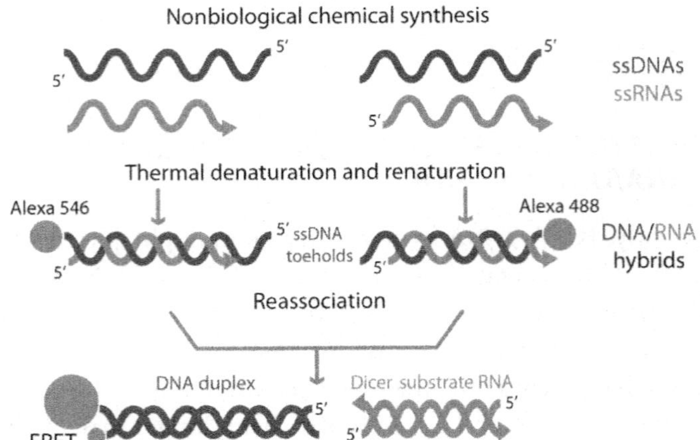

Figure 4 Schematic representation of RNA/DNA hybrid formation, reassociation, and release of split functionalities (Dicer Substrate RNA and FRET).

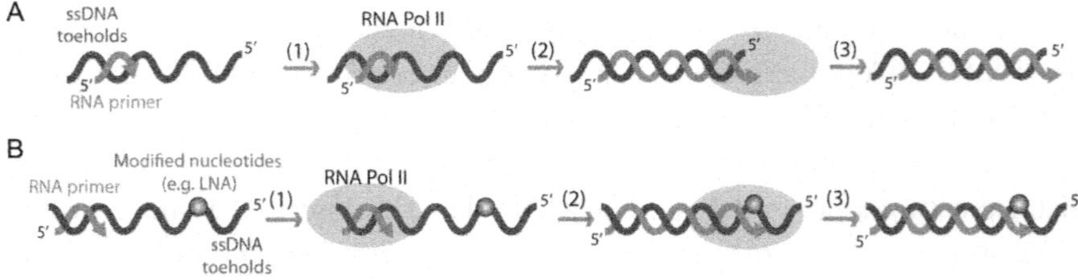

Figure 5 Schematic representation of RNA/DNA hybrids cotranscriptional production using RNA polymerase II. (A) RNA/DNA hybrids with upstream DNA toeholds are produced by run-off transcription. (B) RNA/DNA hybrids with downstream DNA toeholds are obtained by stopping transcription before RNA polymerase II runs off the template by incorporating at least two modified nucleotides (e.g., LNAs).

suggesting that this mutant might also be used for production of the hybrids. The T7 RNA polymerase, used before for cotranscriptional production of functional RNA nanoparticles (Afonin et al., 2010; Afonin, Kireeva, et al., 2012; Afonin, Lin, Calkins, & Jaeger, 2012), appears to be less suitable for this application. T7 RNA polymerase only partially transcribes the single-stranded DNA templates (Gopal, Brieba, Guajardo, McAllister, & Sousa, 1999; Milligan & Uhlenbeck, 1989), and therefore, production of the RNA/DNA hybrids with the proper ssDNA toeholds was not successful (Afonin, Desai, et al., 2014).

Preparative production of RNA molecules by *in vitro* transcription using multisubunit RNA polymerases is precluded by two main obstacles: these polymerases are not easy to purify and the purified protein complexes require extended promoters and specific protein factors for transcription initiation and termination; in addition, RNA elongation rates by multisubunit RNA polymerases are several times lower than those observed for bacteriophage RNA polymerases under similar *in vitro* transcription conditions. Bacterial RNA polymerase requires an approximately 40-base pair promoter and a single initiation factor that is σ^{70} for most *E. coli* RNA polymerase promoters, to initiate transcription as shown in Fig. 6. Termination may occur by ρ factor-dependent or factor-independent sequence-specific mechanisms; otherwise, *E. coli* RNA polymerase produces long continuous transcripts on a circular DNA template (Fried & Sokol, 1972). Initiation on eukaryotic promoters for RNA polymerase II is even more complex and requires at least five external transcription initiation factors (Roeder, 1996). Moreover, the initiation start-site selection in this system is not very precise (Sayre, Tschochner, & Kornberg, 1992). The efficiency of promoter-specific transcription initiation in purified systems is relatively low. That said, the possibility of using multisubunit RNA polymerases for preparative *in vitro* transcription has a few potentially important advantages. First, the high processivity of bacterial RNA polymerase compared to its single-subunit bacteriophage counterpart may be essential for synthesis of very long transcripts. Second, a slower transcription elongation rate may promote proper RNA folding. Third, the availability of a rapidly growing collection of *S. cerevisiae* mutants of RNA polymerase II that have increased elongation rates and/or relaxed substrate specificities (Kaplan, Larsson, & Kornberg, 2008; Kireeva et al., 2008, 2012; Strathern et al., 2013) opens new possibilities in using these mutants for preparative production of chemically modified transcripts. Use of yeast RNA polymerase II for *in vitro* transcription is attractive because *S. cerevisiae* is considered to be a safe and endotoxin-free organism, which facilitates therapeutic applications of the transcripts produced by this enzyme.

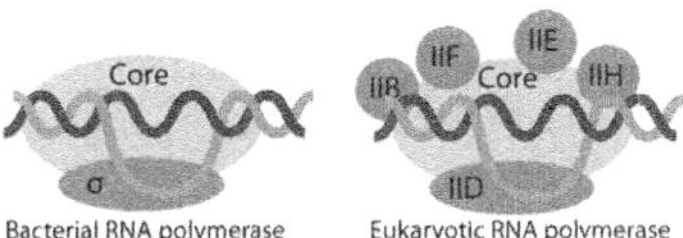

Figure 6 Promoter-dependent transcription initiation by multisubunit bacterial and eukaryotic RNA polymerases.

The essential methodology that circumvents the two main obstacles in the development of preparative *in vitro* transcription systems with multisubunit RNA polymerases has been developed in the course of investigations into the molecular mechanisms of transcription elongation in the past two decades. First, purification of *E. coli* RNA polymerase and *S. cerevisiae* RNA polymerase II has been greatly facilitated by the addition of hexahistidine tags to the C-terminus of the β' subunit of *E. coli* RNA polymerase (Kashlev et al., 1993, 1996) and N-terminus of Rpb3 (Kireeva, Komissarova, & Kashlev, 2000; Kireeva, Komissarova, Waugh, & Kashlev, 2000; Kireeva, Lubkowska, Komissarova, & Kashlev, 2003). Furthermore, especially important for RNA polymerase II, a promoter- and factor-independent system for the elongation complex assembly with core RNA polymerase and synthetic RNA and DNA oligonucleotides has been developed (Kireeva, Komissarova, & Kashlev, 2000; Kireeva, Komissarova, Waugh, et al., 2000; Sidorenkov, Komissarova, & Kashlev, 1998). This experimental approach was combined with the ligation of long PCR-derived downstream DNA fragments to the assembled elongation complexes (Kireeva et al., 2002). Immobilization of RNA polymerase on a Ni-NTA affinity resin not only promotes its purification but also allows for one-step pull-down of the active RNA polymerase from the crude cell lysate (Kireeva et al., 2003, 2009) and facilitates production and purification of the final product such as extended RNA/DNA hybrids. We reported the use of RNA polymerase II immobilized on Ni-NTA agarose cartridge for production of the extended RNA/DNA hybrids from a primer-template and NTP substrate mix (Afonin, Desai, et al., 2014). The low stability of an RNA polymerase II elongation complex carrying an RNA/DNA hybrid longer than 14 nucleotides (Kireeva, Komissarova, & Kashlev, 2000; Kireeva, Komissarova, Waugh, et al., 2000) promotes dissociation of the resulting hybrid from the immobilized RNA polymerase II, and the cycle of synthesis/dissociation is repeated multiple times until the desired amount of the RNA/DNA hybrid is obtained. Use of a fast RNA polymerase II mutant increased the yield of the full-length RNA/DNA hybrid and reduced contamination by the RNA species partially synthesized due to pausing or termination. Overall, solid-phase synthesis of RNA molecules hybridized to DNA emerges as a promising approach for preparative RNA/DNA hybrids synthesis *in vitro* (Afonin, Desai, et al., 2014).

5. EXPERIMENTAL TESTING OF RNA/DNA HYBRIDS

The RNA/DNA hybrids obtained by thermal annealing or during *in vitro* transcription can be used for the delivery and activation of functional RNAs *in vitro*, in various diseased cells, and *in vivo* (Afonin, Desai, et al., 2014; Afonin, Viard, et al., 2013). The experimental studies of RNA/DNA hybrids are outlined in Fig. 7. Nondenaturing native polyacrylamide gel electrophoresis can be employed for visualizing reassociation. Also, the fluorescently labeled DNAs or RNAs can be used to track reassociation through FRET in real time. When two RNA/DNA hybrids fluorescently labeled with Förster dye pairs (e.g., Alexa 488 as a donor and Alexa 546 as an acceptor) are mixed and incubated at 37 °C, their reassociation places the donor dye within the Förster distance of the acceptor dye. As a result, when the donor dye is excited, the emission of the acceptor dye tremendously increases and the signal of the donor dye drops. To track the reassociation inside living cells, fluorescently labeled hybrids can be cotransfected either on the same or on two different days. The FRET signal remaining upon bleed through correction can be calculated as detailed in Afonin, Viard, et al. (2013). The release of functional RNAs can be assessed either through

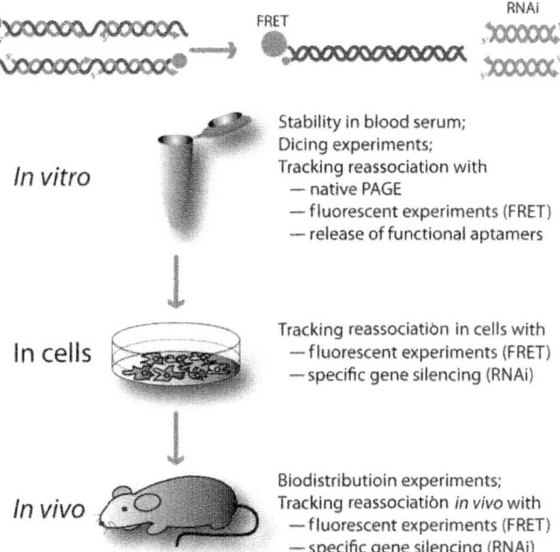

Figure 7 Experimental testing of reassociating RNA/DNA hybrids *in vitro*, in various cell cultures, and *in vivo* in murine models.

fluorescent experiments as in the case of malachite green aptamer release or through specific gene silencing experiments as in the case of RNAi activation.

6. CONCLUDING REMARKS

In this chapter, we described several computational and experimental techniques allowing design and production of RNA/DNA hybrids programmed to carry multiple split functionalities (FRET, RNAi, RNA aptamers). The computational approach allows one to estimate the equilibrium properties of multiple RNA and DNA strands in solution. The ability to computationally and experimentally characterize multiple RNA and DNA strands and their interactions could be an important step toward designing more complex nanoscale structures consisting of RNA and DNA strands. Altogether, it can tremendously benefit the expanding fields of RNA and DNA nanotechnologies (Afonin, Kasprzak, Bindewald, Kireeva, et al., 2014; Afonin, Kasprzak, Bindewald, Puppala, et al., 2014; Chworos et al., 2004; Douglas et al., 2009; Guo, 2010; Guo, Zhang, Chen, Garver, & Trottier, 1998; He et al., 2008; Jaeger & Chworos, 2006; Khisamutdinov, Jasinski, & Guo, 2014; Ko et al., 2010; Ohno et al., 2011; Osada et al., 2014; Pinheiro, Han, Shih, & Yan, 2011; Shu, Shu, Haque, Abdelmawla, & Guo, 2011; Shukla et al., 2011).

ACKNOWLEDGMENTS

This publication was funded in part with federal funds from the Frederick National Laboratory for Cancer Research, National Institutes of Health, under Contract HHSN261200800001E. This research was additionally supported in part by the Intramural Research Program of the National Institutes of Health, Center for Cancer Research. The content of this publication does not necessarily reflect the views or policies of the Department of Health and Human Services, nor does mention of trade names, commercial products, or organizations imply endorsement by the U.S. Government.

REFERENCES

Afonin, K. A., Bindewald, E., Yaghoubian, A. J., Voss, N., Jacovetty, E., Shapiro, B. A., et al. (2010). In vitro assembly of cubic RNA-based scaffolds designed in silico. *Nature Nanotechnology*, *5*(9), 676–682.

Afonin, K. A., Desai, R., Viard, M., Kireeva, M. L., Bindewald, E., Case, C. L., et al. (2014). Co-transcriptional production of RNA–DNA hybrids for simultaneous release of multiple split functionalities. *Nucleic Acids Research*, *42*(3), 2085–2097.

Afonin, K. A., Grabow, W. W., Walker, F. M., Bindewald, E., Dobrovolskaia, M. A., Shapiro, B. A., et al. (2011). Design and self-assembly of siRNA-functionalized RNA nanoparticles for use in automated nanomedicine. *Nature Protocols*, *6*(12), 2022–2034.

Afonin, K. A., Kasprzak, W. K., Bindewald, E., Kireeva, M., Viard, M., Kashlev, M., et al. (2014). In silico design and enzymatic synthesis of functional RNA nanoparticles. *Accounts of Chemical Research*, *47*(6), 1731–1741.

Afonin, K. A., Kasprzak, W., Bindewald, E., Puppala, P. S., Diehl, A. R., Hall, K. T., et al. (2014). Computational and experimental characterization of RNA cubic nanoscaffolds. *Methods*, *67*(2), 256–265.

Afonin, K. A., Kireeva, M., Grabow, W. W., Kashlev, M., Jaeger, L., & Shapiro, B. A. (2012). Co-transcriptional assembly of chemically modified RNA nanoparticles functionalized with siRNAs. *Nano Letters*, *12*(10), 5192–5195.

Afonin, K. A., Lin, Y. P., Calkins, E. R., & Jaeger, L. (2012). Attenuation of loop-receptor interactions with pseudoknot formation. *Nucleic Acids Research*, *40*(5), 2168–2180.

Afonin, K. A., Viard, M., Kagiampakis, I., Case, C. L., Dobrovolskaia, M. A., Hofmann, J., et al. (In Press). Triggering of RNA interference with RNA-RNA, RNA-DNA, and DNA-RNA nanoparticles. ACS Nano.

Afonin, K. A., Viard, M., Koyfman, A. Y., Martins, A. N., Kasprzak, W. K., Panigaj, M., et al. (2014). Multifunctional RNA nanoparticles. *Nano Letters*, *14*(10), 5662–5671.

Afonin, K. A., Viard, M., Martins, A. N., Lockett, S. J., Maciag, A. E., Freed, E. O., et al. (2013). Activation of different split functionalities on re-association of RNA–DNA hybrids. *Nature Nanotechnology*, *8*(4), 296–304.

Andronescu, M., Zhang, Z. C., & Condon, A. (2005). Secondary structure prediction of interacting RNA molecules. *Journal of Molecular Biology*, *345*(5), 987–1001.

Bernhart, S. H., Tafer, H., Mückstein, U., Flamm, C., Stadler, P. F., & Hofacker, I. L. (2006). Partition function and base pairing probabilities of RNA heterodimers. *Algorithms for Molecular Biology*, *1*(1), 3.

Bindewald, E., Afonin, K., Jaeger, L., & Shapiro, B. A. (2011). Multistrand RNA secondary structure prediction and nanostructure design including pseudoknots. *ACS Nano*, *5*(12), 9542–9551.

Bramsen, J. B., & Kjems, J. (2012). Development of therapeutic-grade small interfering RNAs by chemical engineering. *Frontiers in Genetics*, *3*(154).

Cao, S., Xu, X. J., & Chen, S. J. (2014). Predicting structure and stability for RNA complexes with intermolecular loop–loop base pairing. *RNA*, *20*, 835–845. http://dx.doi.org/10.1261/rna.043976.113.

Chworos, A., Severcan, I., Koyfman, A. Y., Weinkam, P., Oroudjev, E., Hansma, H. G., et al. (2004). Building programmable jigsaw puzzles with RNA. *Science*, *306*(5704), 2068–2072.

Dimitrov, R. A., & Zuker, M. (2004). Prediction of hybridization and melting for double-stranded nucleic acids. *Biophysical Journal*, *87*(1), 215–226.

Dirks, R. M., & Pierce, N. A. (2003). A partition function algorithm for nucleic acid secondary structure including pseudoknots. *Journal of Computational Chemistry*, *24*(13), 1664–1677.

Douglas, S. M., Dietz, H., Liedl, T., Högberg, B., Graf, F., & Shih, W. M. (2009). Self-assembly of DNA into nanoscale three-dimensional shapes. *Nature*, *459*(7245), 414–418.

Elbashir, S. M., Lendeckel, W., & Tuschl, T. (2001). RNA interference is mediated by 21- and 22-nucleotide RNAs. *Genes and Development*, *15*(2), 188–200.

Elbashir, S. M., Martinez, J., Patkaniowska, A., Lendeckel, W., & Tuschl, T. (2001). Functional anatomy of siRNAs for mediating efficient RNAi in Drosophila melanogaster embryo lysate. *EMBO Journal*, *20*(23), 6877–6888.

Fire, A., Xu, S., Montgomery, M. K., Kostas, S. A., Driver, S. E., & Mello, C. C. (1998). Potent and specific genetic interference by double-stranded RNA in Caenorhabditis elegans. *Nature*, *391*(6669), 806–811.

Fried, A. H., & Sokol, F. (1972). Synthesis in vitro by bacterial RNA-polymerase of simian virus 40-specific RNA: Multiple transcription of the DNA template into a continuous polyribonucleotide. *Journal of General Virology*, *17*(1), 69–79.

Gopal, V., Brieba, L. G., Guajardo, R., McAllister, W. T., & Sousa, R. (1999). Characterization of structural features important for T7 RNAP elongation complex stability reveals competing complex conformations and a role for the non-template strand in RNA displacement. *Journal of Molecular Biology*, *290*(2), 411–431.

Grabow, W. W., Zakrevsky, P., Afonin, K. A., Chworos, A., Shapiro, B. A., & Jaeger, L. (2011). Self-assembling RNA nanorings based on RNAI/II inverse kissing complexes. *Nano Letters*, *11*(2), 878–887.

Guo, P. (2010). The emerging field of RNA nanotechnology. *Nature Nanotechnology*, *5*(12), 833–842.

Guo, P., Zhang, C., Chen, C., Garver, K., & Trottier, M. (1998). Inter-RNA interaction of phage phi29 pRNA to form a hexameric complex for viral DNA transportation. *Molecular Cell*, *2*(1), 149–155.

He, Y., Ye, T., Su, M., Zhang, C., Ribbe, A. E., Jiang, W., et al. (2008). Hierarchical self-assembly of DNA into symmetric supramolecular polyhedra. *Nature*, *452*(7184), 198–201.

Jaeger, L., & Chworos, A. (2006). The architectonics of programmable RNA and DNA nanostructures. *Current Opinion in Structural Biology*, *16*(4), 531–543.

Kaplan, C. D., Larsson, K. M., & Kornberg, R. D. (2008). The RNA polymerase II trigger loop functions in substrate selection and is directly targeted by alpha-amanitin. *Molecular Cell*, *30*(5), 547–556.

Kashlev, M., Martin, E., Polyakov, A., Severinov, K., Nikiforov, V., & Goldfarb, A. (1993). Histidine-tagged RNA polymerase: Dissection of the transcription cycle using immobilized enzyme. *Gene*, *130*(1), 9–14.

Kashlev, M., Nudler, E., Severinov, K., Borukhov, S., Komissarova, N., & Goldfarb, A. (1996). Histidine-tagged RNA polymerase of Escherichia coli and transcription in solid phase. *Methods in Enzymology*, *274*, 326–334.

Khisamutdinov, E. F., Jasinski, D. L., & Guo, P. (2014). RNA as a boiling-resistant anionic polymer material to build robust structures with defined shape and stoichiometry. *ACS Nano*, *8*(5), 4771–4781.

Kireeva, M. L., Komissarova, N., & Kashlev, M. (2000). Overextended RNA:DNA hybrid as a negative regulator of RNA polymerase II processivity. *Journal of Molecular Biology*, *299*(2), 325–335.

Kireeva, M. L., Komissarova, N., Waugh, D. S., & Kashlev, M. (2000). The 8-nucleotide-long RNA:DNA hybrid is a primary stability determinant of the RNA polymerase II elongation complex. *Journal of Biological Chemistry*, *275*(9), 6530–6536.

Kireeva, M. L., Lubkowska, L., Komissarova, N., & Kashlev, M. (2003). Assays and affinity purification of biotinylated and nonbiotinylated forms of double-tagged core RNA polymerase II from Saccharomyces cerevisiae. *Methods in Enzymology*, *370*, 138–155.

Kireeva, M. L., Nedialkov, Y. A., Cremona, G. H., Purtov, Y. A., Lubkowska, L., Malagon, F., et al. (2008). Transient reversal of RNA polymerase II active site closing controls fidelity of transcription elongation. *Molecular Cell*, *30*(5), 557–566.

Kireeva, M., Nedialkov, Y. A., Gong, X. Q., Zhang, C., Xiong, Y., Moon, W., et al. (2009). Millisecond phase kinetic analysis of elongation catalyzed by human, yeast, and Escherichia coli RNA polymerase. *Methods*, *48*(4), 333–345.

Kireeva, M. L., Opron, K., Seibold, S. A., Domecq, C., Cukier, R. I., Coulombe, B., et al. (2012). Molecular dynamics and mutational analysis of the catalytic and translocation cycle of RNA polymerase. *BMC Biophysics*, *5*, 11.

Kireeva, M. L., Walter, W., Tchernajenko, V., Bondarenko, V., Kashlev, M., & Studitsky, V. M. (2002). Nucleosome remodeling induced by RNA polymerase II: Loss of the H2A/H2B dimer during transcription. *Molecular Cell*, *9*(3), 541–552.

Ko, S. H., Su, M., Zhang, C., Ribbe, A. E., Jiang, W., & Mao, C. (2010). Synergistic self-assembly of RNA and DNA molecules. *Nature Chemistry*, *2*(12), 1050–1055.

Lorenz, R., Bernhart, S. H., Höner Zu Siederdissen, C., Tafer, H., Flamm, C., Stadler, P. F., et al. (2011). ViennaRNA package 2.0. *Algorithms for Molecular Biology*, *6*, 26.

Lorenz, R., Hofacker, I. L., & Bernhart, S. H. (2012). Folding RNA/DNA hybrid duplexes. *Bioinformatics*, *28*(19), 2530–2531.

Mathews, D. H., Sabina, J., Zuker, M., & Turner, D. H. (1999). Expanded sequence dependence of thermodynamic parameters improves prediction of RNA secondary structure. *Journal of Molecular Biology*, *288*(5), 911–940.

Milligan, J. F., & Uhlenbeck, O. C. (1989). Synthesis of small RNAs using T7 RNA polymerase. *Methods in Enzymology*, *180*, 51–62.

Naryshkina, T., Kuznedelov, K., & Severinov, K. (2006). The role of the largest RNA polymerase subunit lid element in preventing the formation of extended RNA–DNA hybrid. *Journal of Molecular Biology*, *361*(4), 634–643.

Ohno, H., Kobayashi, T., Kabata, R., Endo, K., Iwasa, T., Yoshimura, S. H., et al. (2011). Synthetic RNA–protein complex shaped like an equilateral triangle. *Nature Nanotechnology*, *6*(2), 116–120.

Osada, E., Suzuki, Y., Hidaka, K., Ohno, H., Sugiyama, H., Sugiyama, M. E., et al. (2014). Engineering RNA–protein complexes with different shapes for imaging and therapeutic applications. *ACS Nano*, *8*, 8130–8140.

Pinheiro, A. V., Han, D., Shih, W. M., & Yan, H. (2011). Challenges and opportunities for structural DNA nanotechnology. *Nature Nanotechnology*, *6*(12), 763–772.

Roeder, R. G. (1996). The role of general initiation factors in transcription by RNA polymerase II. *Trends in Biochemical Sciences*, *21*(9), 327–335.

Rose, S. D., Kim, D. H., Amarzguioui, M., Heidel, J. D., Collingwood, M. A., Davis, M. E., et al. (2005). Functional polarity is introduced by Dicer processing of short substrate RNAs. *Nucleic Acids Research*, *33*(13), 4140–4156.

SantaLucia, J., Jr. (1998). A unified view of polymer, dumbbell, and oligonucleotide DNA nearest-neighbor thermodynamics. *Proceedings of the National Academy of Sciences of the United States of America*, *95*(4), 1460–1465.

Sayre, M. H., Tschochner, H., & Kornberg, R. D. (1992). Reconstitution of transcription with five purified initiation factors and RNA polymerase II from Saccharomyces cerevisiae. *Journal of Biological Chemistry*, *267*(32), 23376–23382.

Seeman, N. C. (1982). Nucleic acid junctions and lattices. *Journal of Theoretical Biology*, *99*(2), 237–247.

Shu, D., Shu, Y., Haque, F., Abdelmawla, S., & Guo, P. (2011). Thermodynamically stable RNA three-way junction for constructing multifunctional nanoparticles for delivery of therapeutics. *Nature Nanotechnology*, *6*(10), 658–667.

Shukla, G. C., Haque, F., Tor, Y., Wilhelmsson, L. M., Toulmé, J.-J., Isambert, H., et al. (2011). A boost for the emerging field of RNA nanotechnology. *ACS Nano*, *5*(5), 3405–3418.

Sidorenkov, I., Komissarova, N., & Kashlev, M. (1998). Crucial role of the RNA:DNA hybrid in the processivity of transcription. *Molecular Cell*, *2*(1), 55–64.

Strathern, J., Malagon, F., Irvin, J., Gotte, D., Shafer, B., Kireeva, M., et al. (2013). The fidelity of transcription: RPB1 (RPO21) mutations that increase transcriptional slippage in S. cerevisiae. *Journal of Biological Chemistry*, *288*(4), 2689–2699.

Thompson, J. D. (2013). Clinical development of synthetic siRNA therapeutics. *Drug Discovery Today: Therapeutic Strategies*. http://dx.doi.org/10.1016/j.ddstr.2013.03.002.

Toulokhonov, I., & Landick, R. (2006). The role of the lid element in transcription by E. coli RNA polymerase. *Journal of Molecular Biology*, *361*(4), 644–658.

Vassylyev, D. G., Vassylyeva, M. N., Perederina, A., Tahirov, T. H., & Artsimovitch, I. (2007). Structural basis for transcription elongation by bacterial RNA polymerase. *Nature*, *448*(7150), 157–162.

Westover, K. D., Bushnell, D. A., & Kornberg, R. D. (2004). Structural basis of transcription: Separation of RNA from DNA by RNA polymerase II. *Science*, *303*(5660), 1014–1016.

Wu, P., Nakano, S., & Sugimoto, N. (2002). Temperature dependence of thermodynamic properties for DNA/DNA and RNA/DNA duplex formation. *European Journal of Biochemistry*, *269*(12), 2821–2830.

Yingling, Y., & Shapiro, B. A. (2007). Computational design of an RNA hexagonal nanoring and an RNA nanotube. *Nano Letters*, 7(8), 2328–2334.

Zadeh, J. N., Steenberg, C. D., Bois, J. S., Wolfe, B. R., Pierce, M. B., Khan, A. R., et al. (2010). NUPACK: Analysis and design of nucleic acid systems. *Journal of Computational Chemistry*, *32*(1), 170–173.

Zhou, J., Shum, K. T., Miele, E., Di Fabrizio, E., Ferretti, E., Tomao, S., et al. (2013). Nanoparticle-based delivery of RNAi therapeutics: Progress and challenges. *Pharmaceuticals (Basel)*, *6*(1), 85–107.

CHAPTER FOURTEEN

Multiscale Methods for Computational RNA Enzymology

Maria T. Panteva*, Thakshila Dissanayake*, Haoyuan Chen*, Brian K. Radak*, Erich R. Kuechler*, George M. Giambaşu*, Tai-Sung Lee*, Darrin M. York[1],*

*Center for Integrative Proteomics Research, BioMaPS Institute and Department of Chemistry and Chemical Biology, Rutgers University, Piscataway, New Jersey, USA

[1]Corresponding author: e-mail address: york@biomaps.rutgers.edu

Contents

Methods in Enzymology, Volume 553
ISSN 0076-6879
http://dx.doi.org/10.1016/bs.mie.2014.10.064

Abstract

RNA catalysis is of fundamental importance to biology and yet remains ill-understood due to its complex nature. The multidimensional "problem space" of RNA catalysis includes both local and global conformational rearrangements, changes in the ion atmosphere around nucleic acids and metal ion binding, dependence on potentially correlated protonation states of key residues, and bond breaking/forming in the chemical steps of the reaction. The goal of this chapter is to summarize and apply multiscale modeling methods in an effort to target the different parts of the RNA catalysis problem space while also addressing the limitations and pitfalls of these methods. Classical molecular dynamics simulations, reference interaction site model calculations, constant pH molecular dynamics (CpHMD) simulations, Hamiltonian replica exchange molecular dynamics, and quantum mechanical/molecular mechanical simulations will be discussed in the context of the study of RNA backbone cleavage transesterification. This reaction is catalyzed by both RNA and protein enzymes, and here we examine the different mechanistic strategies taken by the hepatitis delta virus ribozyme and RNase A.

1. INTRODUCTION

RNA plays several key roles in cellular function (Garst, Edwards, & Batey, 2011; Guttman & Rinn, 2012), ranging from the regulation of gene expression and signaling pathways to catalysis of important biochemical reactions, including protein synthesis (Schmeing & Ramakrishnan, 2009; D. N. Wilson & Cate, 2012) and pre-mRNA splicing (Hoskins & Moore, 2012; Valadkhan, 2010). Since the discovery that RNA molecules could act as enzymes, the study of the novel catalytic properties of RNA has been an area of intense interest and research (Fedor & Williamson, 2005; Sharp, 2009). These efforts have given birth to the field of ribozyme engineering (X. Chen & Ellington, 2009; Link & Breaker, 2009) and influenced theories into of the origin of life itself (X. Chen, Li, & Ellington, 2007; T. J. Wilson & Lilley, 2009). Ultimately, the elucidation of the mechanisms of RNA catalysis promises to yield a wealth of new insights that will extend our understanding of biological processes and facilitate the design of new RNA-based technologies including allosterically controlled ribozymes that may act as molecular switches (Fastrez, 2009; Suess & Weigand, 2008) in RNA chips and new ultra-sensitive biosensing devices (Penchovsky, 2014).

A detailed understanding of the general strategies whereby molecules of RNA can catalyze chemical reactions, including the factors that regulate their activity, provides guiding principles to aid in molecular design. A deep understanding of mechanism, however, requires an interdisciplinary approach that integrates both theory and experiment (Kellerman, York, Piccirilli, & Harris, 2014; Rhodes, Réblová, Sponer, & Walter, 2006). Experiments can be designed to probe mechanism, but due to the inherent difficulty of directly observing a transition state, primary experimental observables only indirectly report on mechanism and do not provide an atomic-level picture of catalysis (Al-Hashimi & Walter, 2008; Fedor, 2009; Walter, 2007). Computational approaches, and in particular molecular simulation methods, provide a wealth of molecular-level detail, but are derived from approximate models that, without validation from experiments, are not meaningful (Hashem & Auffinger, 2009; McDowell, Špačková, Šponer, & Walter, 2006). Thus, while experimental structural biology, molecular biology, and molecular biophysics approaches provide critical data about ribozyme mechanisms, the translation of these data into knowledge requires interpretation through theoretical models.

In this chapter, we provide a discussion of the multiscale methods used for computational RNA enzymology. These methods have particular relevance to the study of catalytic riboswitches that combine molecular recognition and binding of an effector molecule with conformational changes in the expression platform that lead to catalysis. Each of these elements is sensitive to environmental conditions, including the specific content of the ionic atmosphere as well as pH. Recently, there have been advances in both computational tools as well as experimental methods that have advanced our understanding of RNA catalysis. The goal of this chapter is to present an overview of the multiscale modeling tools used in computational enzymology, what they are used for, and how they can be connected with experiments. Ultimately, a goal of these efforts is to develop robust computational models that not only aid in the interpretation of experimental data but also provide predictive insight.

This chapter is organized as follows. In the first section, an overview of the multiscale modeling approach is given, along with a discussion of why RNA poses particular computational challenges. The second section describes methods for modeling the ionic environment around RNA as a function of ionic conditions. The third section addresses issues of protonation states as a function of pH. The fourth section describes examination of conformational events required to reach a catalytically active structure, and how this structure evolves along the reaction path. The fifth section examines methods to study the catalytic chemical steps of the reaction once a catalytically active state has been achieved. The sixth section will focus on the calculation of kinetic isotope effects (KIEs) to verify the rate-controlling transition state structure. Along the way, we will use examples of the catalytic mechanism of the hepatitis delta virus ribozyme (HDVr) and RNase A to illustrate key points.

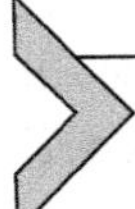

2. THE "PROBLEM SPACE" OF COMPUTATIONAL RNA ENZYMOLOGY

The goal of computational RNA enzymology is to provide a detailed atomic-level description of catalysis that is able to unify the interpretations of a wide range of experiments such that a consensus view of mechanism emerges. Ultimately, the elucidation of mechanism requires consideration of specific chemical reaction pathways through a multidimensional free energy landscape that connects reactants to products while departing from a presumed catalytically competent *active state* (Ensing, De Vivo, Liu, Moore, & Klein, 2006; T.-S. Lee, Radak, Pabis, & York, 2013; Vanden-Eijnden, 2009; Wojtas-Niziurski, Meng, Roux, & Bernèche, 2013). However, the mapping of any given pathway is not meaningful unless one also characterizes the free energy associated with formation of the active state itself, i.e., the probability of finding the system in the active state as a function of experimentally tunable environmental variables such as pH and ionic conditions. The active state will be a function of the RNA conformation, protonation state of key residues, and metal ion binding modes. Together with the catalytic chemical steps, these dimensions define the scope of the "problem space" (Fig. 1) that needs to be explored and characterized.

RNA molecules are highly charged and exhibit a high degree of structural variation that is sensitive to the protonation state of nucleobase residues as well as the binding of metal ions and, in the case of riboswitches, small molecules (Bevilacqua, Brown, Nakano, & Yajima, 2004; Draper, 2008; Roth & Breaker, 2009). Other factors involving greater complexity, such as the binding of proteins and intermolecular interactions that occur in macromolecular assemblies, also influence RNA structure and function, but are beyond the scope of the present discussion.

In the case of ribozymes, protonation state and metal ion binding, in addition to being important to achieve a catalytically active conformation, may also play a role in the chemical steps of the reaction. Divalent metal ions (usually Mg^{2+}) are universally important for folding under physiological conditions (Grilley, Soto, & Draper, 2006; Misra & Draper, 2002) and, in ribozymes, have been implicated in many instances as directly participating in catalysis (J. Chen et al., 2013; Wong & York, 2012). This makes their roles in folding and catalysis difficult to untangle (Gong, Chen, Bevilacqua, Golden, & Carey, 2009;

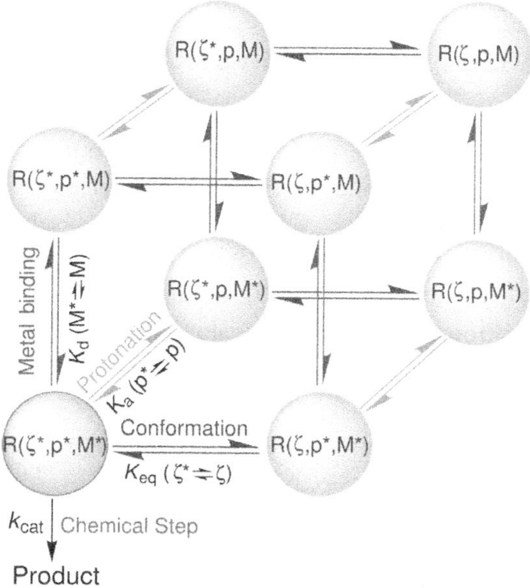

Figure 1 Complexity of the ribozyme (R) "problem space" involving metal ion interactions/metal binding (M), protonation state (p), and conformational state (ζ) coordinates that lead to a catalytically "active state" (green (dark gray in the print version)). Other variables that represent binding of small molecules and interaction with proteins or other macromolecules are not illustrated. Substrate binding is also not shown, as the primary application focus here is on small self-cleaving RNA enzymes.

Gong, Chen, Yajima, et al., 2009; Misra & Draper, 1998). Similarly, the structure and energetics of RNA, in particular tertiary interactions, are sensitive to pH (Murray, Dunham, & Scott, 2002; Nixon & Giedroc, 2000; Wadkins, Shih, Perrotta, & Been, 2001), as are the catalytic protonation state requirements of key active site residues in ribozymes (Butcher & Pyle, 2011; Doudna & Cech, 2002). The duality of influence of protonation state and metal ion binding on RNA structure and catalytic activity is further complicated by the fact that they are often coupled. Metal ion binding can perturb the pK_a values of active site residues such that they can be more effective as general acids or bases under physiological conditions, and conversely, ionization events can adjust the occupation of nearby metal binding sites.

In what follows, we outline a multiscale modeling strategy that takes a "divide-and-conquer" approach to deconstructing the problem space for RNA catalysis using different computational methods. We then go on to give examples of the use of these methods in application to RNA backbone cleavage transesterification reactions catalyzed by a protein and an RNA enzyme that employ two different mechanisms.

3. MULTISCALE MODELING STRATEGY

Multiscale simulations involve the integration of a hierarchy of models that, used together, can solve problems that span multiple spatial and/or temporal scales (Dama et al., 2013; Feig, Karanicolas, & Brooks, III, 2004; Lodola & Mulholland, 2013; Meier-Schellersheim, Fraser, & Klauschen, 2009; Sherwood, Brooks, & Sansom, 2008; van der Kamp & Mulholland, 2013; York & Lee, 2009). In the context of biocatalysis, this implies consideration of the enzyme, along with its substrate and any required cofactors, acting in a realistic condensed-phase environment. The active site, where chemical bond formation or cleavage occurs, generally requires a computationally intensive quantum mechanical (QM) description (Garcia-Viloca, Gao, Karplus, & Truhlar, 2004; Hou & Cui, 2013; Lodola & Mulholland, 2013; van der Kamp & Mulholland, 2013; Xie, Orozco, Truhlar, & Gao, 2009), and this problem is amplified by the need for different levels of phase space sampling to capture events that occur on vastly different timescales. Consequently, biocatalysis applications demand the use of multiscale models that can overcome these challenges in a fashion that is both reliable and computationally tractable.

Multiscale modeling simulations of RNA catalysis are laden with challenges that are more pronounced relative to their protein enzyme counterparts (A. A. Chen, Marucho, Baker, & Pappu, 2009; T.-S. Lee, Giambaşu, & York, 2010). The high degree of charge in RNA systems demands careful treatment of electrostatic interactions in the solvated ionic environment, accurate models for ion interactions, and long equilibration times. The increased conformational heterogeneity of RNA relative to typical proteins demands special consideration of sampling in order to get converged structural and thermodynamic properties. The methods used to address these challenges will be described within the context of the example applications that apply them.

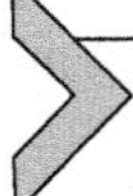

4. CATALYTIC STRATEGIES FOR CLEAVAGE OF THE RNA BACKBONE

We will focus on cleavage of the phosphodiester backbone of RNA, a fundamental reaction in biology that is catalyzed by protein and RNA enzymes. Studying protein and RNA enzymes together has been a long-standing strategy for better understanding biocatalysis at a fundamental level

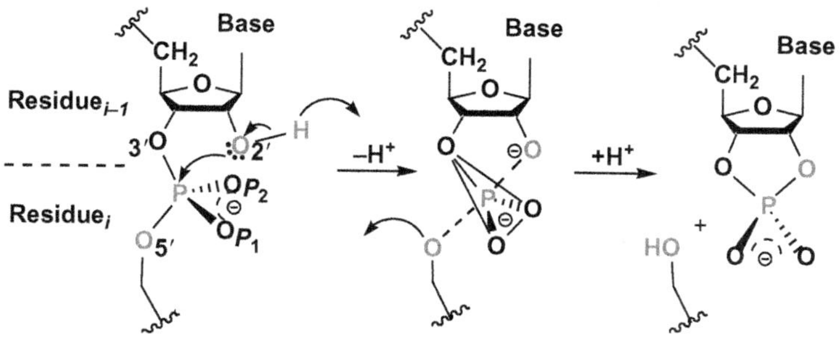

Figure 2 Reaction schematic of RNA 2′-*O*-transesterification. The $O_{2'}$ nucleophile is activated via loss of a proton to a general base and undergoes inline attack on phosphorous to form a pentacoordinate transition structure, followed by departure of the $O_{5'}$ leaving group which is facilitated by proton donation from a general acid.

(Doudna & Lorsch, 2005) since protein catalysts are composed of a diverse set of amino acid side chains while ribozymes only have four similar nucleobase building blocks to work with. Here, we apply a multiscale modeling approach in the context of a prototype protein enzyme, RNase A, and HDVr which belongs to the class of small self-cleaving nucleolytic ribozymes. RNase A and HDVr catalyze the same phosphoryl transfer reaction via general acid/base mechanisms (Lilley, 2011a, 2011b) (Fig. 2) and require one or more nucleobase or amino acid residues to be in a nonstandard protonation state for catalytic activity (Gong et al., 2007; Wilcox, Ahluwalia, & Bevilacqua, 2011). Enzymes belonging to the RNase superfamily degrade single-stranded RNA specifically, and bovine pancreatic RNase A has been a model enzyme system for studying protein structure and catalytic mechanism for decades (Herschlag, 1994). The HDVr is involved in, and essential for, the rolling-circle replication of the human pathogen, the hepatitis D virus (HDV) where its role is to self-cleave the replicated RNAs into unit length (Ferré-D'Amaré, Zhou, & Doudna, 1998). For RNase A, two histidine residues, His12 and His119, are the proposed general acid and base, respectively (Raines, 1998), while in the case of the HDVr, a Mg^{2+} ion has been directly implicated in catalysis as the general base, with C75 playing the role as the general acid (Das & Piccirilli, 2005; Gong, Chen, Bevilacqua, et al., 2009; Nakano, Chadalavada, & Bevilacqua, 2000).

5. MODELING ION AND NUCLEIC ACID INTERACTIONS

The first dimension in the computational RNA enzymology "problem space" (Fig. 1) refers to metal ion interactions with RNA. The structure and function of RNA are strongly influenced by interactions with metal ions in solution. These interactions range from tight, specific site binding of divalent metal ions to diffuse territorial binding of monovalent ions in the ion atmosphere around the RNA. Here we provide a brief overview of the most commonly applied ion models used in biomolecular simulations and highlight recent developments in the application of integral equation theories and their relation to recent "ion counting" experiments.

5.1. Ion models used in biomolecular simulations

The most commonly used force fields for simulating nucleic acid systems are based on nonpolarizable pairwise potential models (Case et al., 2014; Cornell et al., 1995; Foloppe & MacKerell, Jr., 2000; Jorgensen, Maxwell, & Tirado-Rives, 1996; Kaminski, Friesner, Tirado-Rives, & Jorgensen, 2001; MacKerell, Jr. & Banavali, 2000; Oostenbrink, Villa, Mark, & van Gunsteren, 2004; Pérez et al., 2007; Wang, Cieplak, & Kollman, 2000; Zgarbová et al., 2011) which gain tremendous advantage in computational efficiency at the expense of neglecting explicit many-body quantum effects that are known to be important for many problems (Anisimov, Lamoureux, Vorobyov, Roux, & MacKerell, Jr., 2005; Ponder et al., 2010). Ion parameters are typically parameterized to reproduce bulk properties such as solvation free energies, first shell ion–water distances, and water exchange barriers. For monatomic ions that lack internal conformational degrees of freedom, the main consideration for obtaining correct bulk properties involves balancing the ion–water and ion–ion interaction parameters.

Early monovalent ion parameters that were not properly balanced were found to form aggregated clusters in simulations of various salts in aqueous solution (Auffinger, Cheatham III, & Vaiana, 2007; A. A. Chen & Pappu, 2007). Joung and Cheatham subsequently derived new monovalent ion parameters that considered multiple experimental properties that included structure, dynamics, and solvation, and in addition, salt crystal lattice energies that were sensitive to the cation–anion interactions (Joung & Cheatham III, 2008). This led to a new set of alkali and halide monovalent ion parameters that corrected the "salting out" artifacts of some previous models, but because the ion–ion and ion–water interactions needed to be balanced, it was necessary that separate sets of ion parameters be developed for specific water models. Nonetheless, these parameter sets provided a necessary advance that allowed simulations of nucleic acids to be more reliably extended to longer time domains.

Progress in the development of divalent metal ion models is more challenging, but has nonetheless been the focus of considerable recent effort (Babu & Lim, 2006; Li & Merz, Jr., 2014; Li, Roberts, Chakravorty, & Merz, Jr., 2013; Martínez, Pappalardo, & Marcos, 1999). Recently, a Mg^{2+} ion model has been developed for biomolecular simulations that correctly reproduces the ion–water coordination and inner sphere water exchange barrier in solution (Allnér, Nilsson, & Villa, 2012), but that does not necessarily accurately predict absolute solvation thermodynamics (Panteva, Giambaşu, & York, n.d.). More recently, Li et al. (2013) have developed a series of water model-dependent "12-6" models for divalent metal ions that primarily target a single experimental observable. Unlike the monovalent ion parameters, however, the 12-6 divalent metal ion parameters cannot simultaneously reproduce both structural and thermodynamic properties at the same time, owing largely to the neglect of the electronic polarization energy of waters in the first coordination sphere. Follow-up work by the same authors (Li & Merz, Jr., 2014) then introduced "12-6-4" divalent metal ion parameters that make use of a pairwise potential that includes the contribution of the charge-induced dipole interaction in the form of an additional r^{-4} term added to the traditional Lennard-Jones potential. These divalent ion models have been shown to simultaneously reproduce multiple different properties (Li & Merz, Jr., 2014; Panteva et al., n.d.).

The result of these efforts clearly illustrates the need to create models for metal ions with properly balanced ion–ion and ion–water interactions in order to accurately model bulk properties. In the case of biomolecular simulations involving RNA, these models need to be extended so that the ion–RNA interactions are similarly balanced. The effort to create new models for metal ion interactions with RNA is still in its infancy, owing largely to the fact that there currently is a paucity of quantitative experimental binding and competition data that is amenable to robust force field parameterization efforts. Nonetheless, there has been some preliminary progress in the modeling of the ion atmosphere around nucleic acids, and our recent contributions to this area are described in the next section.

5.2. Modeling the ion atmosphere around nucleic acids

The most common approaches to study the distribution of ions around nucleic acids include explicit solvent MD simulations, the three-dimensional reference interaction site model (3D-RISM) (Beglov & Roux, 1997; Kovalenko & Hirata, 2000; Kovalenko, Ten-no, & Hirata, 1999), or through solving the nonlinear Poisson–Boltzmann (NLPB) equation (Bai et al., 2007; Bond, Anderson, & Record, Jr., 1994; Chu, Bai, Lipfert, Herschlag, & Doniach, 2007; Draper, 2008; Kirmizialtin, Silalahi, Elber, & Fenley, 2012; Pabit et al., 2009). Until recently, solving the NLPB equation was the most common way reported in the literature to study the ion atmosphere surrounding nucleic acids, providing solvation thermodynamics as well as three-dimensional ion distributions. NLPB calculations are simple and computationally efficient, but are limited in the treatment of water as a uniform dielectric and neglect explicit ion–ion correlation. Thus, there is compelling evidence that conventional NLPB does not accurately model the ion atmosphere around nucleic acids (Giambaşu, Luchko, Herschlag, York, & Case, 2014). Methods that consider explicitly the role of water and the correlations between ions, such as 3D-RISM and molecular dynamics simulations with explicit solvent, have only recently become practical to study such problems (A. A. Chen, Draper, & Pappu, 2009; Giambaşu et al., 2014; Luchko et al., 2010; Yoo & Aksimentiev, 2012).

Molecular dynamics simulations, from a theoretical perspective, offer the most rigorous description of solvent structure and dynamics. These simulations, however, are extremely costly and require consideration of very large system sizes (A. A. Chen, Draper, & Pappu, 2009; Giambaşu et al., 2014; Yoo & Aksimentiev, 2012) and long timescales for ion equilibration (Rueda, Cubero, Laughton, & Orozco, 2004; Thomas & Elcock, 2006). Further, solvation thermodynamics is extremely difficult to extract from these calculations. 3D-RISM calculations, on the other hand, integrate out the solvent degrees of freedom and thus allow direct access into solvation thermodynamics, and are sufficiently fast (for fixed solute configurations) that a wide range of ionic conditions can be examined. 3D-RISM calculations can also efficiently explore low salt concentrations (e.g., μM–mM range) where MD suffers from convergence issues that require very long equilibration times and more sophisticated enhanced sampling methods. 3D-RISM calculations are thus potentially very powerful as tools to study the ion atmosphere for nucleic acid systems that can be represented by a relatively small ensemble of rigid conformations. Both MD and 3D-RISM use molecular force fields and, unlike NLPB, yield similar layered solvent and ion distributions (Giambaşu et al., 2014; Howard, Lynch, & Pettitt, 2011; Maruyama, Yoshida, & Hirata, 2010; Yonetani, Maruyama, Hirata, & Kono, 2008) (Fig. 3). A challenge for both 3D-RISM calculations and MD simulations that has been fully recognized only recently (A. A. Chen, Draper, & Pappu, 2009; Giambaşu et al., 2014;

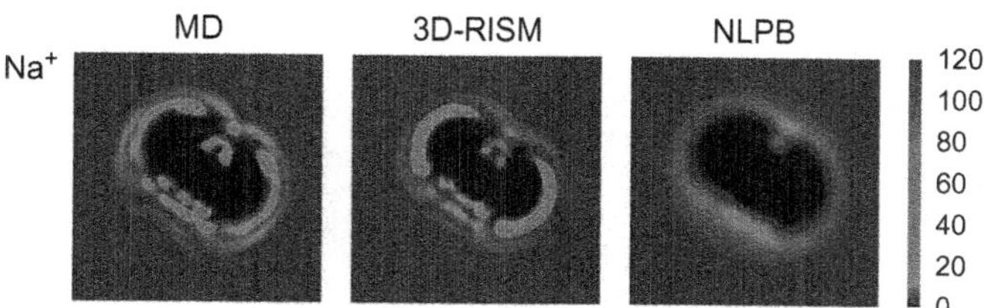

Figure 3 Comparison of ion distributions around a 24-mer of duplex B-DNA from MD simulation, 3D-RISM, and conventional NLPB. Distributions are shown along a rotating "untwisted" coordinate frame along the DNA axis as described in Giambaşu et al. (2014). Shown are the untwisted Na^+ densities from MD, 3D-RISM, and NLPB for 0.17 *M* bulk NaCl concentration. MD and 3D-RISM predict a layered Na^+ density, whereas NLPB is unstructured. (See the color plate.)

Yoo & Aksimentiev, 2012) is the need to consider a sufficiently large amount of solvent such that regions far from the solute exhibit bulk behavior. This is particularly important when comparisons are to be made with experiments that are being performed under different ionic conditions, and thus it is necessary to know what bulk salt concentration the simulations are in equilibrium with.

Recently, several experimental methods have been developed to examine the nature of ion atmosphere around nucleic acids through *ion counting* (IC) experiments that rely on anomalous small-angle X-ray scattering (Andresen et al., 2004, 2008; Pabit et al., 2010; Pollack, 2011), buffer equilibration atomic absorption spectroscopy (Bai et al., 2007; Greenfeld & Herschlag, 2009), and titration with fluorescent dyes (Grilley et al., 2006). In previous decades, ^{23}Na or ^{39}Co NMR relaxation rates have also been employed (Bleam, Anderson, & Record, Jr., 1980; Braunlin, Anderson, & Record, Jr., 1987). IC experiments are important as they quantitatively report directly on the contents of the ion atmosphere around nucleic acids and therefore can be used to facilitate the development of new models. The key observable that allows comparison with experiment is the preferential interaction parameter (Γ) that is, at the microscopic level, an integral measure of the perturbation of the local density of solution components by the highly charged nucleic acid. We have recently reproduced a series of IC measurements using MD simulations, 3D-RISM, and NLPB calculations (Giambaşu et al., 2014) (Fig. 4).

5.3. Current challenges

The difficulties in modeling ion–nucleic acid interactions molecular mechanically are more pronounced when divalent metal ions are involved.

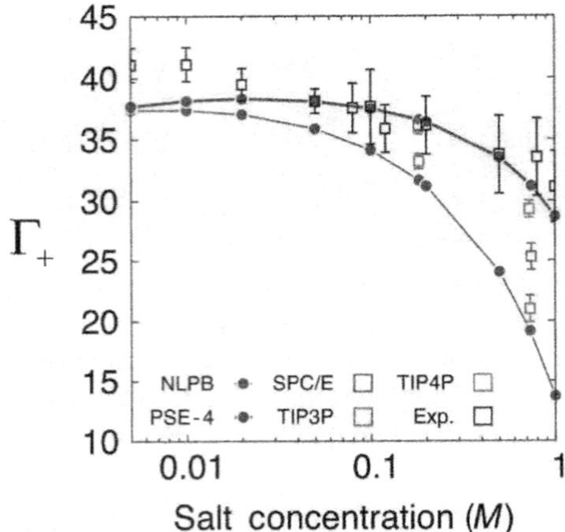

Figure 4 Ion counting profiles for Na^+ from MD, 3D-RISM, and NLPB. Data from Bai et al. (2007).

Although experimental IC data are available for divalent metal ions, MD simulations are challenging due to the very slow exchange rates of waters and ligands in the first coordination sphere, and generally slower equilibration timescales. It is likely that enhanced sampling methods will have to be developed in order to extend the scope of MD simulations that can be directly compared with IC experiments. 3D-RISM calculations, on the other hand, are made challenging with applications to divalent metal ions due to convergence issues that arise from the nonlinear equations that need to be solved. These are also influenced by the specific "closure relation" that is applied, and can lead to quantitatively different predicted preferential interaction parameters (Giambaşu et al., 2014). Consequently, there are technical challenges that need to be met before MD simulations and 3D-RISM calculations can be widely tested and applied to divalent metal ion interactions with nucleic acids, and new models to emerge.

6. MODELING pH-RATE PROFILES FOR ENZYMES

The second dimension in the computational RNA enzymology "problem space" (Fig. 1) involves examination of protonation states. Within the context of catalysis, the specific protonation states of key residues are a critical requirement of the catalytically active state. For general acid/base catalysis, such as in RNase A and HDVr, the general base must be in the deprotonated form (able to accept a proton) whereas the general acid must be in the protonated form (able to donate a proton) for catalytic activity.

In the analysis of kinetic data as a function of pH, the simplest mechanistic assumption is that the catalytic rate is directly proportional to the probability of the general acid and base being in the active state (Herschlag, 1994). If this assumption is true, and if other protonation events within the pH range of interest are benign with respect to affecting the rate, then general acid/base catalysts will give a classic bell-shaped profile that can be fit to a simple kinetic model where the parameters are the "apparent pK_a" values of the general acid and base (Bevilacqua, 2003). Agreement between the kinetic "apparent pK_a" values and macroscopic pK_a values from direct measurements (such as NMR) of the postulated general acid and base is usually considered as evidence in support of their roles in acid/base catalysis. An underlying assumption in this interpretation, however, is that the protonation states of the general acid and base are not correlated (Klingen, Bombarda, & Ullmann, 2006; Ullmann, 2003), i.e., the pK_a of the general acid is independent of the protonation state of the general base and vice versa. The validity of this assumption, which is system dependent, can be tested computationally in order to gain a more fundamental understanding of catalytic mechanism for a particular system (Dissanayake, Swails, Harris, Roitberg, & York, n.d.).

CpHMD and pH replica exchange molecular dynamics (pH-REMD) have emerged as powerful computational tools for deriving pH-rate profiles for general acid–base catalysts. CpHMD is a method where protonation states are sampled dynamically from a Boltzmann distribution at a fixed pH (Baptista, Martel, & Petersen, 1997; Baptista, Teixeira, & Soares, 2002; Khandogin & Brooks III, 2005; M. S. Lee, Salsbury, Jr., & Brooks III, 2004; Mongan, Case, & McCammon, 2004). We adopt a discrete protonation state model that employs Metropolis Monte Carlo (MC) exchange attempts between different protonation states throughout the course of the MD simulation, (Baptista et al., 2002; Mongan et al., 2004) which has been recently implemented in the AMBER software suite (Mongan et al., 2004) for proteins. Unlike free energy perturbation and thermodynamic integration calculations, CpHMD intrinsically takes into account the correlated effects of residue protonation states for a fixed value of pH. The use of pH-REMD allows multiple simulations to be performed over a range of discrete pH values, and is used to enhance sampling and ensure that simulations are in equilibrium with one another. The result is that complete atomic-level simulation data, including conditional probabilities for different protonation states (including tautomers), are generated over a range of pH values. From these data, titration and pH-activity curves can be predicted and used to aid in the interpretation of experimental data.

6.1. Application to apo and cCMP-bound RNase A

In this section, we apply the CpHMD/pH-REMD method in explicit solvent to RNase A, both in the apo form, and bound to a 2′3′-cyclic phosphate (cCMP) complex. The simulation data are used to predict the macroscopic and microscopic pK_a values for the general acid and base, as well as the shape of the pH-rate profile. These results allow us to examine the validity of assumptions about "apparent pK_a" values commonly used to interpret experimental pH-rate profiles.

The kinetic model illustrated in Fig. 5 used to interpret pH-rate data in which it is assumed that the functional forms of the general base and acid are B^- and AH^+, respectively, and only the active state ${}_{AH^+}E_{B^-}$ goes on to give products with a first-order rate constant k_{cat}. Based on the acid–base equilibrium, there are four different microstates whose probabilities (fractions) are denoted as $f_{(AH^+/B^-)}$, $f_{(AH^+/BH^-)}$, $f_{(A/BH)}$, and $f_{(A/B^-)}$. The most common and simplest assumption that can be made is that the protonation states of the general acid and base are uncorrelated and can be modeled by "apparent pK_a" values for the general acid (p$K_{a,A}$) and base (p$K_{a,B}$), i.e., $pK_{a,A}=pK_{a,A}^{BH}=pK_{a,A}^{B-}$ and $pK_{a,B}=pK_{a,B}^{A}=pK_{a,B}^{AH+}$. The "apparent p$K_a$" values are determined through fitting to the active fraction $f_{(AH^+/B^-)}$ and can be compared to the simulated macroscopic pK_a values obtained from the Hill equation. This procedure is analogous to what is typically done experimentally. Alternately, the full microscopic model in Fig. 5 can be used to fit all the fractions determined from the simulation data. This allows one to assess the degree to which the assumptions inherent in the "apparent pK_a"

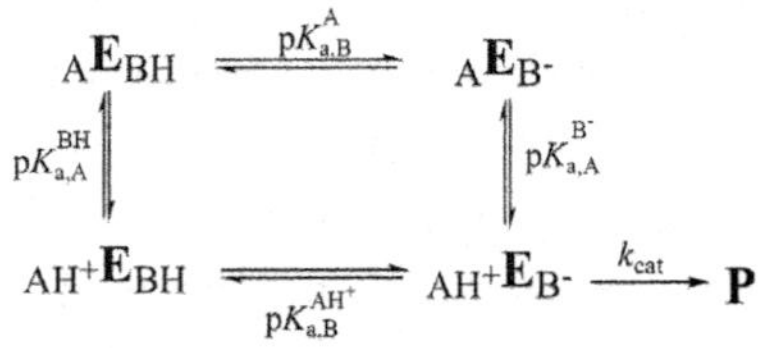

$$
\begin{array}{ccccc}
{}_{A}\mathbf{E}_{BH} & \underset{}{\overset{pK_{a,B}^{A}}{\rightleftharpoons}} & {}_{A}\mathbf{E}_{B^-} & & \\
pK_{a,A}^{BH}\updownarrow & & \updownarrow pK_{a,A}^{B^-} & & \\
{}_{AH^+}\mathbf{E}_{BH} & \underset{pK_{a,B}^{AH^+}}{\rightleftharpoons} & {}_{AH^+}\mathbf{E}_{B^-} & \xrightarrow{k_{cat}} & \mathbf{P}
\end{array}
$$

Figure 5 The microscopic model used in interpretation of pH-rate data. The protonation equilibria between pairwise protonation states are defined in the thermodynamic cycle. The active fraction $f_{(AH^+/B^-)}$ leads to the products.

model are valid, and to provide a more detailed and direct interpretation of experimental pH-rate data.

The CpHMD/pH-REMD simulations predict macroscopic pK_a values for His12 and His119 (6.0 and 6.2, respectively) in the apo structure that are quite close to the experimental values (5.8 and 6.2, respectively). The corresponding calculated macroscopic pK_a values for the cCMP complex (8.5 and 7.3, respectively) are in reasonable agreement with experimental data on 3′-CMP (8.0 and 7.4, respectively). Correspondence of these values with the "apparent pK_a" values derived from pH-activity profiles would suggest a mechanistic role for His12 and His119 as the general acid and base.

Figure 6 plots the predicted pH-activity curves (as fractions, or probabilities, for each microstate) from the pH-REMD simulations for RNase A in the apo form and complexed with cCMP. Plotted are the probabilities of the active fractions (red (gray in the print version)) points are from the simulations, and red lines are fitted with the full microscopic pK_a model illustrated in Fig. 5). Also shown are the fractions for the nonactive states. It is clear

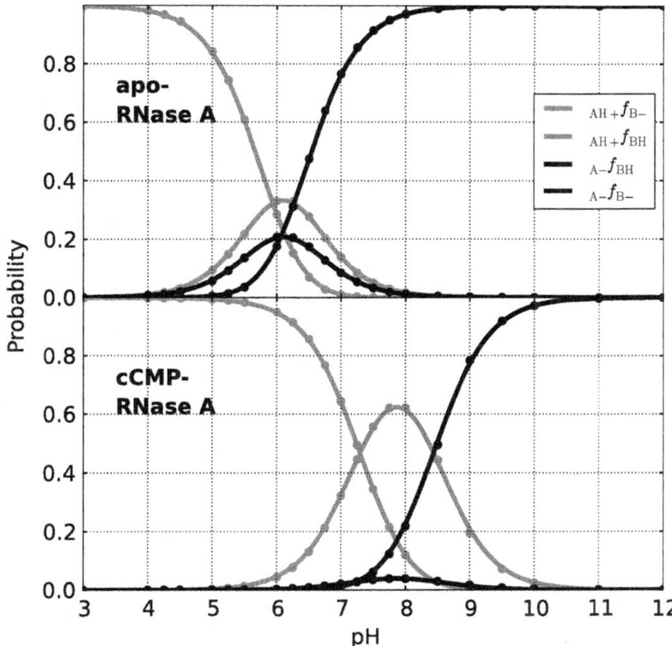

Figure 6 The pH-activity curves for top: apo-RNase A and bottom: cCMP-bound RNase A. The titratable residues are His12 and His119. The four curves represent the fractions (probabilities) for each of the four possible protonation states, with the red (gray in the print version) curve being the fraction of the catalytically active microstate, $f_{(AH^+/B^-)}$.

Table 1 Experimental and calculated microscopic pK_as for apo and cCMP-bound RNase A

	$pK_{a,B}^{AH+}$	$pK_{a,B}^{A}$	$\Delta pK_{a,B}$	$pK_{a,A}^{BH}$	$pK_{a,A}^{B-}$	$\Delta pK_{a,A}$
apo-RNase A						
Expt.[a]	5.87	6.18	0.31	6.03	6.34	−0.31
Microscopic model	5.94	6.06	0.12	6.15	6.26	−0.12
cCMP-RNase A						
Expt. (3′-UMP)[a]	7.95	7.85	−0.1	6.45	6.35	0.1
Microscopic model	7.30	7.24	−0.06	8.50	8.44	0.06

The calculated microscopic pK_a values are derived from the thermodynamic cycle illustrated in Fig. 5. The ΔpK_a values are the differences in the microscopic pK_a values and indicate coupling between protonation states (zero ΔpK_a values indicate no coupling, as in the "apparent pK_a" model).
[a]From Quirk et al. (1999) in order to validate the model.

from the figure that the microscopic pK_a model fits the simulated data extremely well.

Table 1 compares parameters derived from the microscopic model to relevant NMR experiments (Quirk & Raines, 1999). Overall, the calculated and experimental results are quite similar. The general acid and base microscopic pK_a values (i.e., small ΔpK_a values, i.e., less than 0.12 for the simulation results) indicate that the protonation states, in these two examples, are not strongly coupled. This supports the idea that the "apparent pK_a" model may be used. Indeed, the "apparent pK_a" model gives values of 6.0 and 6.3, respectively, for the apo enzyme, and 8.5 and 7.3, respectively, for the cCMP complex. These are within 0.05 pK_a units of one another. Applying the CpHMD/pH-REMD method in conjunction with fitting the computed pH-activity data with the microscopic model will be particularly useful when general acid and base residues are oppositely charged and in closer proximity in the active site, as is the case with many ribozyme systems.

6.2. Current challenges

The ability to accurately compute macroscopic and microscopic pK_a values of residues involved in acid–base catalysis while at the same time treating effects of correlated protonation states using the CpHMD/REMD method will be an invaluable tool to aid in the interpretation of pH-activity data for catalytic RNAs. This has not yet been demonstrated for RNA systems. It has only been very recently (Goh, Knight, & Brooks III, 2012, 2013) that models for simulating nucleic acids using CpHMD have been extended to nucleic acids, and they have yet to be thoroughly tested. Challenges that will need to be overcome for application to RNA include consideration of nucleobase tautomers, and coupling of protonation state with divalent metal ion binding (such as has been implicated for HDVr). Finally, at this point proton exchange attempts are accepted or rejected based on energies obtained using a Generalized Born (GB) implicit solvation model (Onufriev, Bashford, & Case, 2004), even though the conformational ensembles are generated through MD simulation in explicit solvent. The degree to which this is sufficiently robust for RNA applications has yet to be determined and may need further development.

7. MODELING CONFORMATIONAL STATES

The third dimension in the computational RNA enzymology "problem space" (Fig. 1) considers the exploration of thermally accessible conformational states. The characterization of relevant states requires exploration of a vast conformational landscape using accurate models and often specialized methods to enhance sampling. The past few decades have witnessed significant maturation of molecular mechanical (MM) force fields for nucleic acids based on relatively simple fixed charge models and pairwise potentials for nonbonded interactions (Brooks et al., 2009; Pérez et al., 2007; Wang et al., 2000; Zgarbová et al., 2011). The computational efficiency of these models allows MD simulations to routinely access μs timescales (Dror, Dirks, Grossman, Xu, & Shaw, 2012; Salomon-Ferrer, Götz, Poole, Le Grand, & Walker, 2013), making it a viable method for capturing large-scale conformational changes in catalytic riboswitches (Giambaşu, Lee, Scott, & York, 2012; Giambaşu et al., 2010). Taken together, these developments have provided insight into the condensed-phase structure and dynamics of ribozymes both in their precleaved ground state and at various points along a reaction path (T.-S. Lee, Giambaşu, Harris, & York, 2011; T.-S. Lee et al., 2010; T.-S. Lee, Wong, Giambasu, & York, 2013). Of key importance to the understanding of ribozyme function is to understand what conformational event leads to the catalytically active precleaved ground state, and how does the ribozyme environment respond so as to preferentially stabilize high-energy transition states and intermediates as the reaction progresses. Computational mutagenesis provides insights into the origin of experimental mutational effects on the catalytic rate (T.-S. Lee & York, 2008) and may lead to experimentally testable predictions such as chemical modifications that test a specific mechanistic hypothesis or correlated mutations that exhibit a rescue effect (T.-S. Lee & York, 2010). In this way, molecular simulations serve as a tool to aid in the interpretation of experimental functional studies and may guide the design of new experiments.

7.1. Catalytic strategies of ribozymes

Nucleolytic ribozymes employ a broad range of catalytic strategies for RNA backbone cleavage transesterification (Cochrane & Strobel, 2008; Lilley, 2011a). These may include activation of the nucleophile by a general base, promotion of leaving group departure by a general acid, electrostatic stabilization of the transition state by hydrogen bonding or cationic interactions, and facilitation of proton transfer by Lewis acid activation. However, these are merely general mechanistic considerations that are applicable to many different phosphoryl transfer enzymes (Golden, 2011; Ji & Zhang, 2011). The key question that remains at the heart of our understanding of RNA catalysis is how do certain molecules of RNA, with their relatively limited repertoire of reactive functional groups, adopt three-dimensional conformations that convey catalytic activity that rivals many protein enzymes. Insight into some of these questions may be gleaned from molecular simulations. To date, some general guiding principles have begun to emerge. Current molecular simulation evidence suggests that ribozymes are able to engineer electrostatically strained active sites that can cause shifts of the pK_a values of key residues or recruit solvent components, including solvent and in some cases divalent metal ions, to assist in catalysis. In the case of the hairpin ribozyme, electrostatic effects in the active site (Nam, Gao, & York, 2008a) account for a large part of the observed rate acceleration and cause a shift of the pK_a of an adenine nucleobase which acts as a general acid catalyst to facilitate leaving group departure (Nam, Gao, & York, 2008b). The hammerhead ribozyme, on the other hand, has engineered a highly electronegative active site that can recruit a threshold occupation of cationic charge (T.-S. Lee et al., 2009) (a Mg^{2+} ion under physiological conditions, or multiple monovalent cations under high salt conditions) that facilitates formation of an active in-line attack conformation (T.-S. Lee et al., 2008; T.-S. Lee, Wong, et al., 2013), stabilizes accumulating charge in the transition state, and increases the acidity of the 2′OH group of a conserved guanine residue in order to facilitate proton transfer to the leaving group (Wong, Lee, & York, 2011). In both the hairpin and hammerhead ribozymes, as well as other ribozymes such as the *glmS* riboswitch (Klein, Been, & Ferré-D'Amaré, 2007; Viladoms, Scott, & Fedor, 2011) and Varkud satellite ribozyme (T. J. Wilson et al., 2010), a guanine is positioned near to the nucleophile and possibly acts as a general base, but its role is still actively debated. In the case of HDVr, the situation appears to be more complex and controversial (Golden, 2011; Nakano, Proctor, & Bevilacqua, 2001). While the HDVr requires divalent metal ions for catalytic activity under physiological conditions, the role of the metal ion, its catalytically active binding mode, and its correlation with other protonation events in the active site are yet to be resolved (Golden, Hammes-Schiffer, Carey, & Bevilacqua, 2013; Lévesque, Reymond, & Perreault, 2012; Wadkins et al., 2001).

7.2. General considerations when starting MD simulations from inactive structures

Structural characterization of ribozymes by X-ray crystallography, NMR, and small-angle X-ray scattering have often implicated the involvement of key residues in catalysis based on close proximity to the cleavage site. On departing from these ribozyme structures to run molecular dynamics simulations, however, there are several considerations to take into account. Structures that are not highly resolved, have fractional occupations, or that exhibit a large degree of conformational variation pose difficulties for the computational chemist. Often is becomes necessary to perform many independent MD simulations departing from different starting structures in order to eliminate bias from using a single starting structure.

At the current point in time, the most abundant structural data for ribozymes have been derived from X-ray crystallography (Lilley, 2005; Scott, 2007). These data have been crucial for the field of computational RNA enzymology (Lodola & Mulholland, 2013). Crystal structures must be trapped in a particular state along the reaction path in order to be resolved. Oftentimes, this means deactivating the ribozyme by blocking the nucleophilic $O_{2'}$ group by either methylating it or else removing it completely, or by mutating other residues that are known to be critical for activity at different stages along the reaction coordinate. Invariably, these lead to structures that, with respect to the degree to which they represent an active state, are artificial. Ribozymes can also be trapped in transition state mimic structures, such as vanadate (Davies & Hol, 2004) or $2',5'$-phosphodiester linkage (Klein et al., 2007; Torelli, Krucinska, & Wedekind, 2007), at the cleavage site. One must further consider the effect of the crystalline environment as RNA structures can be influenced by crystal packing artifacts that are sensitive to crystallization conditions (Auffinger, Bielecki, & Westhof, 2004; Ennifar, Walter, & Dumas, 2003). In general, one must remember that, although X-ray crystal structures provide invaluable data and critical starting points for molecular simulations, they nonetheless represent static pictures of deactivated enzymes in a crystal environment, whereas meaningful biological interpretation of mechanism requires a dynamical picture of active enzymes in solution. In many cases, molecular dynamics crystal simulations are performed in order to aid in the interpretation of crystallographic data and help to separate the effects of chemical modifications and crystal packing environment on the structure and dynamics of the active enzyme in solution (Heldenbrand et al., 2014; Martick, Lee, York, & Scott, 2008).

7.3. Application to HDVr

In the case of the HDVr, several different experimental results are available which highlight the importance and challenge of computational modeling. First, to date, no precleavage wild-type HDVr crystallographic structure exists that includes a fully resolved active site. Until recently, the only precleavage HDVr structure available was that of a catalytically inactive C75U mutant, which has a distinctly different active site architecture compared to the wild-type product state (Ferré-D'Amaré et al., 1998; Ke, Zhou, Ding, Cate, & Doudna, 2004). Specifically, in the inactivated mutant ribozyme, U75 is poised to act as the general base while in the postcleavage structure C75 is located in close proximity to the leaving group, suggesting a general acid role (Ferré-D'Amaré et al., 1998). This role is now strongly supported by biochemical studies (Das & Piccirilli, 2005).

Furthermore, in the C75U structures, no resolved divalent ion shows apparently catalytic importance, although biochemical studies support a direct catalytic role for Mg^{2+}. It has also been suggested that the pK_a of C75 in the ribozyme environment is anticorrelated with the presence of Mg^{2+} (Gong et al., 2007) and that C75 protonation is linked to changes in Mg^{2+} inner-sphere coordination and binding mode (Gong, Chen, Bevilacqua, et al., 2009; Gong et al., 2008). More recently, a wild-type precleavage HDVr structure (via deoxy mutation of U-1) has been resolved to 1.9 Å. This structure includes a resolved Mg^{2+} ion with water-mediated contacts to a previously unobserved G-U reverse wobble pair in the active site. Unfortunately, much of the scissile phosphate (and all of the upstream U-1 nucleotide) in this structure is disordered and could not be resolved; instead, it was modeled using analogy to the active site of the HHR (J.-H. Chen et al., 2010). In this modeled active site structure, U-1 is directly coordinated to Mg^{2+} via both the $O_{2'}$ nucleophile and a nonbridging oxygen. These structural data, along with recent kinetic studies of G-U reverse wobble mutants (J. Chen et al., 2013; Lévesque et al., 2012), suggest that a Mg^{2+} ion may be responsible for activating the nucleophile.

We have previously performed a series of MD simulations of HDVr along the reaction path by using the precleaved C75U mutant structure as a starting point(T.-S. Lee et al., 2011). U75 was modeled as C75, and simulations of the reactant state, precursor state (nucleophile deprotonated), early transition state, late transition state, and product state were conducted for 350 ns each. Representative snapshots from the simulations can be found in Fig. 7.

In the inactive mutant structure, a Mg^{2+} ion was directly coordinating U75:$O_{4'}$ while in our simulations a Mg^{2+} starting in this position would not remain stably bound. Instead, the Mg^{2+} was restrained to be within

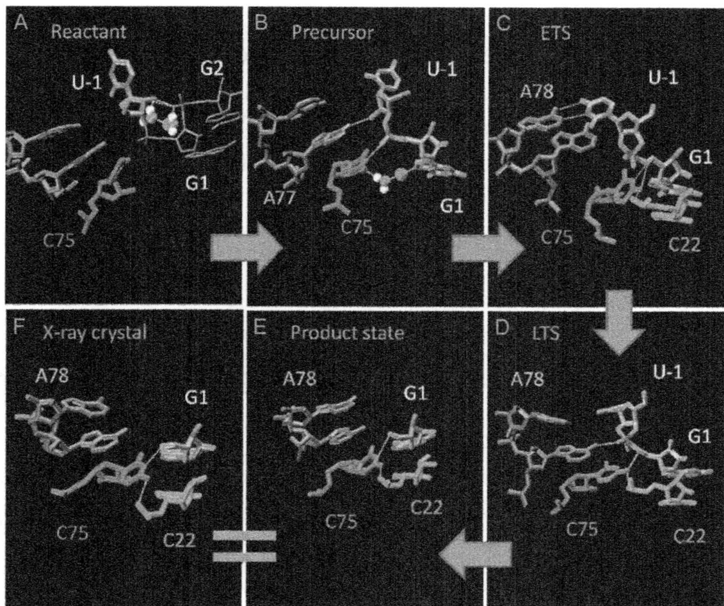

Figure 7 Representative snapshots taken from simulations of HDVr along the reaction path starting from an inactive C75U mutant crystal structure and compared with the crystal structure of the product state (PDB ID: 1CX0). (See the color plate.)

2Å of G1:N7 as suggested by J.-H. Chen, Gong, Bevilacqua, Carey, and Golden (2009), and along the reaction path the coordination of this ion in the active site changed. After about 100 ns, the active site of our starting structure for the product state simulation began to converge to the product state crystal structure. This illustrates the importance of running extensive MD simulations to correct for local and/or global conformational changes due to inactive or inaccurate experimental starting structures. Within the context of computational RNA enzymology, one of the main objectives of these MD simulations is to identify a set of plausible catalytically active structure, along with the conformational events that lead to their formation, in order to proceed to the final stage of investigation: to explore the chemical steps of the reaction by testing specific mechanistic hypotheses using QM simulations. This topic will be discussed in the following section.

7.4. Current challenges

Although MD simulation is an increasingly powerful computational tool to study ribozyme mechanism, several limitations exist. The timescales accessible to MD simulations (now on the order of μs) still restricts the types of motions that can be investigated computationally (Dror et al., 2012; Zwier & Chong, 2010). Comprehensive nucleic acid force fields, especially for divalent metal ions, are still being tested and developed and are not as robust as their protein counterparts (Pérez, Luque, & Orozco, 2011). Finally, as mentioned earlier, exploration of the "problem space" of RNA catalysis is complicated by the fact that metal ion interactions, protonation states, and conformational events are intimately coupled, and more progress needs to be made in the development of methods that can efficiently and reliably model this coupling.

8. MODELING THE CHEMICAL STEPS OF CATALYSIS

Thus far, we have discussed the use of computational methods to explore the three dimensions of the computational RNA enzymology "problem space" (Fig. 1) in order to arrive at an active state that is competent to proceed on to the catalytic chemical steps of the reaction. It may be the case that more than one plausible active state is identified, and each may have a different probability of being realized under different pH and ionic conditions. To complete the mechanistic picture, it remains to characterize the free energy landscape corresponding to the chemical steps of the reaction. The individual pathways through this landscape correspond to specific mechanisms, and elucidation of the free energy barriers for each pathway allows prediction of those paths that are most probable. In order to predict reaction kinetics, it is necessary to know the probability of observing the catalytically active state (or more specifically, the precatalytic "reactant" ground state), the pathway, and free energy barrier(s) that connect the reactant state with the product state, as well as other factors (Garcia-Viloca et al., 2004) such as barrier re-crossings and quantum tunneling contributions. In the remainder of this section, we discuss in detail the issue of computing the free energy landscapes and identifying mechanistic pathways for the chemical reaction. This sets the stage for the last section, which is to validate the rate-controlling transition state of a predicted pathway by analyzing KIEs.

Exploration of the free energy landscape for the chemical steps of catalysis where bond formation and cleavage are occurring requires a QM model to describe the changes in electronic structure and energetics. Most enzyme systems are far too large to treat with a fully QM method, although recently advances in so-called linear-scaling quantum force fields may alter that paradigm (Giese, Chen, Huang, & York, 2014; Giese, Huang, Chen, & York, 2014). An attractive alternative that has been widely applied is to use so-called combined QM/MM models (Field et al., 1990; Warshel & Levitt, 1976). These models typically treat a relatively small localized region of the system, such as the key residues in the enzyme active site, with a QM model, whereas the vast remainder of the system is treated with a classical MM force field. QM/MM methods have been widely applied to simulations of enzyme reactions (Acevedo & Jorgensen, 2010; Garcia-Viloca et al., 2004; Senn & Thiel, 2009; van der Kamp & Mulholland, 2013).

8.1. General considerations

In QM/MM methods, the first important factor to be considered is the choice of the QM and MM models. The QM method must be accurate to model the reactive chemistry of interest, while also being sufficiently fast to be applied with the required amount of sampling that the application demands. The MM method should be able to reliably model the electrostatic environment surrounding the QM region as well as the relevant conformational events that occur. With a choice of QM and MM models, it then becomes necessary that the QM/MM interaction parameters, and in particular the nonbonded Lennard-Jones potentials, are appropriately balanced so as to give correct energetics. Other specialized terms are required when the boundary between the QM and MM systems occurs across a chemical bond, and have been described in detail elsewhere (Gao, Amara, Alhambra, & Field, 1998; Reuter, Dejaegere, Maigret, & Karplus, 2000; Y. Zhang, Lee, & Yang, 1999).

Here we will use a fast, approximate semi-empirical quantum model that has been especially designed to accurately model phosphoryl transfer reactions such as those considered in the present work (Nam, Cui, Gao, & York, 2007). This model has been applied previously to examine cleavage transesterification in the hairpin (Nam et al., 2008a, 2008b) and hammerhead (Wong et al., 2011) ribozymes.

Reliable determination of free energy landscapes for enzyme reactions requires sufficient sampling of the generalized coordinates used to define the landscape, in addition to the degrees of freedom orthogonal to the reaction coordinates. A wide range of sampling methods have been developed to overcome these challenges (Zuckerman, 2011). Some of the most widespread include multistage/stratified sampling (Valleau & Card, 1972), statically (Hamelberg, Mongan, & McCammon, 2004; Torrie & Valleau, 1974, 1977) and adaptively (Babin, Roland, & Sagui, 2008; Darve, Rodríguez-Gómez, & Pohorille, 2008; Laio & Parrinello, 2002) biased sampling, self-guided dynamics (Wu & Brooks, 2012), constrained dynamics (Darve & Pohorille, 2001; den Otter, 2000), as well as multicanonical (Berg & Neuhaus, 1992; Nakajima, Nakamura, & Kidera, 1997) and replica exchange (Chodera & Shirts, 2011; Sugita, Kitao, & Okamoto, 2000) algorithms.

Free energy analysis methods need to be incorporated with all simulation data to construct the free energy profile of the reaction coordinates. The weighted histogram analysis method (Kumar, Bouzida, Swendsen, Kollman, & Rosenberg, 1992; Souaille & Roux, 2001) is widely used but requires highly overlapped data and the results are often noisy. The multistate Bennett acceptance ratio (Bennett, 1976; Shirts & Chodera, 2008; Tan, Gallicchio, Lapelosa, & Levy, 2012) methods are broadly applicable but can also suffer from statistical error in the estimation of free energy surfaces when there is low sampling coverage. The recently developed vFEP method uses a general maximum likelihood framework to provide robust analytical estimates to the free energy surface, and offers some advantage over alternative methods (T.-S. Lee, Radak, Huang, Wong, & York, 2014; T.-S. Lee, Radak, et al., 2013).

8.2. Constructing free energy profiles of HDVr

In this section, we provide a demonstration application of the calculation of the 2D free energy profile for the general acid step in HDVr catalysis. The goal is to examine the feasibility of a previously proposed mechanistic hypothesis (J.-H. Chen et al., 2010; Golden, 2011) whereby the catalytic precursor state involves a Mg^{2+} ion bound in a bridging position between the scissile phosphate and a phosphate from a neighboring strand. In this position, the divalent ion is able to coordinate the nucleophile and facilitates its activation. The presumed general acid in this mechanism is a protonated cytosine residue (C75).

Figure 8 shows the 2D free energy surfaces, departing from the activated nucleophile, for the reaction in the presence and absence of a bound Mg^{2+}

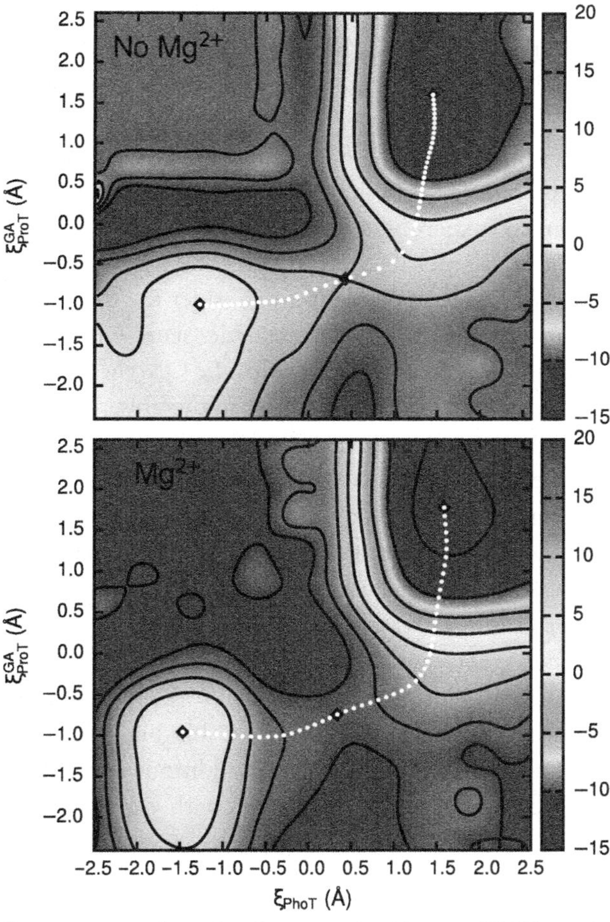

Figure 8 2D free energy surfaces for the general acid step of HDVr catalysis departing from a state where the nucleophile has been activated in a prior step. Shown are simulations in the absence (top) and in the presence (bottom) of a Mg^{2+} bound in the active site. Minima and saddle points (diamonds) and the minimum free energy path (white points) are also indicated. (See the color plate.)

ion, henceforth referred to as simply "Mg^{2+}" and "no Mg^{2+}" simulations. Two reaction coordinates are used to describe the reaction progression: ξ^{GA}_{ProT} is a "general acid proton transfer" coordinate, defined as the difference in distances between the proton and the general acid, and the proton and leaving group; ξ_{PhoT} is a "phosphoryl transfer" progression coordinate defined as the difference in distances between the phosphorus and the leaving group, and the phosphorus and the nucleophile.

Analysis of the 2D free energy profiles in the presence and absence of a Mg^{2+} bound at the active site suggests the minimum free energy pathways are similar, with phosphoryl transfer leading to a late transition state with an almost fully cleaved bond to the leaving group, followed by asynchronous proton transfer from the general acid. Although the mechanistic pathways are similar, the free energy barrier in the absence of Mg^{2+} (10.0 kcal/mol) is approximately 8 kcal/mol lower than that for the model where Mg^{2+} is present (18.0 kcal/mol), and both of the barriers are considerably lower than the experimental catalytic barrier (estimated to be approximately 19.6–19.8 kcal/mol in the presence of Mg^{2+}). The reason for this apparent discrepancy is that one needs to consider the free energy associated with formation of the activated precatalytic state with the nucleophile deprotonated, which was the starting state for the QM/MM calculations. This activation is expected to be considerably less in the presence of the Mg^{2+} ion, but it is not yet clear as to whether this is enough to account for the experimental difference. Work to further reconcile these issues is in progress.

Hence, the simulation-derived free energy profiles suggest that based on the proposed mechanistic hypothesis (J.-H. Chen et al., 2010; Golden, 2011), these two cases should have similar mechanisms but the reaction barriers departing from the activated nucleophile are different due to the presence of the proposed active site Mg^{2+}. The QM/MM work presented here is therefore not yet conclusive, and ongoing work is needed to characterize the free energy associated with Mg^{2+} ion binding, and determine the resulting pK_a shift on the nucleophile. In addition, alternative competing mechanistic hypotheses that are consistent with experiments should also be explored. Finally, as will be discussed in the next section, once plausible pathways have been determined and the rate-controlling transition state identified, further validation of the pathway can be sought through the measurement and calculation of KIEs.

8.3. Current challenges

Computationally tractable methods are needed that allow accurate determination of free energy surfaces using high-level density-functional methods.

Currently, the QM/MM studies that have used density-functional methods for ribozymes have done so with small basis sets (which are notoriously problematic for anionic systems) and either neglected to do any simulation or else employed very short timescales. A promising research direction involves developing methods that allow free energy surfaces generated from exhaustive sampling with low level methods to be systematically corrected to higher levels with significantly reduced sampling requirements. Further, the current models treat the QM/MM interactions as decoupled from the electron density, e.g., they are independent of the local charge. This can lead to overstabilization of anions which are larger, and hence less solvated. Finally, the ability to calculate free energy profiles from linear-scaling quantum force fields is forthcoming and promises to advance the field.

9. COMPUTING KIEs TO VERIFY TRANSITION STATE STRUCTURE

KIEs are powerful experimental probes that report directly on properties of the rate-controlling transition state (Cleland & Hengge, 2006; Hengge, 2002). In these experiments, the isotopic mass of one or more atoms involved in the reaction is selectively altered, typically to a heavier isotope. Experiments are then devised to accurately measure the ratio of rate constants corresponding to reactions of the light and heavy isotope (KIE = k_{light}/k_{heavy}). KIE values that are greater than unity are referred to as "normal," whereas values less than unity are referred to as "inverse." Most importantly, KIEs are very sensitive to changes in transition state bonding environment and ultimately encode information about the transition state that allows validation of predicted pathways that pass through it, which in turn provides insight into enzyme mechanism (Harris & Cassano, 2008; Lassila, Zalatan, & Herschlag, 2011). Nonetheless, a detailed interpretation of KIE data in terms of structure and bonding in the transition state requires the use of computational QM models (H. Chen et al., 2014; Wong et al., 2012).

9.1. Application of KIE on RNase A and Zn^{2+} catalytic mechanisms

We have recently investigated the mechanistic details of RNase A using a joint experimental and theoretical approach through the determination of experimental KIEs and their interpretation using computational models (Gu et al., 2013). These results are placed into context of baseline nonenzymatic reaction models (Wong et al., 2012), and models where catalysis is affected by Zn^{2+} ions in solution (H. Chen, Harris, & York, n.d.). KIE values for the nucleophile $O_{2'}$, nonbridging phosphoryl oxygens, and leaving group $O_{5'}$, designated $^{18}k_{NUC}$, $^{18}k_{NPO}$ and $^{18}k_{LG}$, respectively, were calculated with density-functional QM models (H. Chen et al., n.d.) as well as measured. The results are summarized in Fig. 9.

The nonenzymatic model has a $^{18}k_{NUC}$ value near unity and a very large $^{18}k_{LG}$ value. The models indicate the transition state is very late (cleavage to the leaving group is almost complete). The reaction catalyzed by RNase A shows a $^{18}k_{NUC}$ value that is trending toward being slightly inverse, and the $^{18}k_{LG}$ value is significantly reduced. Overall, the models indicate that this corresponds to a late transition state that is overall more compact than

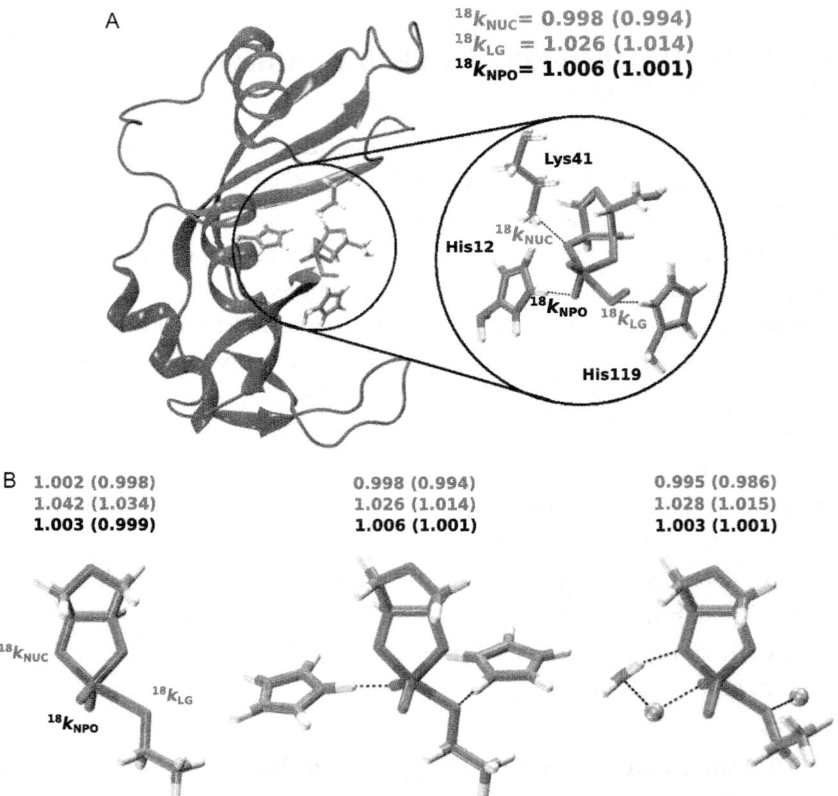

Figure 9 (A) Snapshot of RNase A transition state mimic structure from MD simulation and (B) transition state geometries and KIEs of nonenzymatic (left), RNase A-catalyzed (middle), and Zn^{2+}-catalyzed (right) RNA transphosphorylation model reactions obtained from QM calculations.

the nonenzymatic model reaction. More recently, the KIEs for RNA transphosphorylation catalyzed by Zn^{2+} ions in solution have been measured (S. Zhang et al., n.d.). It is of considerable interest that the experimental KIE values for the Zn^{2+}-catalyzed reaction are similar to those measured for the reaction catalyzed by RNase A. Further, the computational models for both these reactions that give the closest agreement with experiment are strikingly similar. The Zn^{2+} ion positions mimic closely those of the protonated histidine residues for RNase A. A slight difference is that, for the Zn^{2+}-catalyzed reaction, there is an additional hydrogen bond to the nucleophile donated from a Zn^{2+}-coordinated water molecule. This manifests itself in making the $^{18}k_{NUC}$ value slightly more inverse. Overall, these results suggest that the transition states for the catalyzed reactions are altered from that of the nonenzymatic model in a similar fashion by the RNase A enzyme environment or by Zn^{2+} ions in solution. Further, the agreement between the experimental and calculated KIE results provides support that the predicted mechanistic pathway for RNase A passes through a late transition state where the interactions illustrated in Fig. 9 are preserved.

9.2. Current challenges

At the moment, the calculation of KIEs for large enzyme or ribozyme systems is very tedious and time consuming. The development of computationally more efficient electronic structure methods that were made to be linear scaling and seamlessly integrated into a multiscale modeling framework for the calculation of KIEs would be extremely valuable.

10. CONCLUSIONS

In this chapter, we have applied a multiscale modeling strategy to the computational RNA enzymology "problem space" that, for the purposes here, consists of four major modeling components: metal ion–nucleic acid interactions, pH-rate profiles, catalytically active conformations, and the catalytic chemical steps in the reaction. Each of these components has a direct connection with experiment and can be integrated to form a detailed, atomic-level picture of ribozyme mechanism. The ultimate goal of computational RNA enzymology is to provide a unified interpretation of a wide range of experiments that leads to a consensus view of mechanism. Toward this end, a variety of computational methods have been brought to bear on different elements of the problem space.

Classical MD simulations have proven to be instrumental to probe structure and dynamics of ribozymes along their reaction path, as well as providing the most rigorous (although less practical) description of the solvated ionic atmosphere around nucleic acids. Other molecular solvation theory models, such as 3D-RISM, appear very promising as a practical tool to investigate a wide range of ionic conditions for a sufficiently small ensemble of structures. Molecular simulations under conditions of constant pH, together with pH-REMD, can be used to predict and interpret pH-rate profiles for general acid/base catalysts and account for coupling between protonation states that are difficult to probe experimentally. Departing from a presumed active state, QM/MM simulations can be used with enhanced sampling methods such as Hamiltonian replica exchange methods to determine multidimensional free energy landscapes for catalysis. Minimum free energy pathways through these surfaces provide predictions of the specific mechanisms. Predicted mechanisms can be further tested by calculation of KIEs for the rate-controlling transition state along a given path, which can then be verified experimentally. Overall, this field is still rapidly maturing, and much progress is to be expected over the next decade in the development of integrated methods that will allow even closer connections between theory and experiment to be made, and models that provide a predictive understanding of ribozyme mechanism.

ACKNOWLEDGMENTS

This work was made possible by the National Institutes of Health (NIH) grant numbers P01GM066275 and GM62248 to D. M. Y. and by the National Science Foundation (NSF) CDI type-II grant #1125332 fund to D. M. Y. Computational resources utilized for this research include the Extreme Science and Engineering Discovery Environment (XSEDE), NSF grant number OCI-1053575, the Blue Waters super computer, NSF grant numbers ACI-0725070 and ACI-1238993, and the Minnesota Supercomputing Institute for Advanced Computational Research (MSI).

REFERENCES

Acevedo, O., & Jorgensen, W. L. (2010). Advances in quantum and molecular mechanical (QM/MM) simulations for organic and enzymatic reactions. *Accounts of Chemical Research*, *43*, 142–151.

Al-Hashimi, H. M., & Walter, N. G. (2008). RNA dynamics: It is about time. *Current Opinion in Structural Biology*, *18*(3), 321–329.

Allnér, O., Nilsson, L., & Villa, A. (2012). Magnesium ion-water coordination and exchange in biomolecular simulations. *Journal of Chemical Theory and Computation*, *8*(4), 1493–1502.

Andresen, K., Das, R., Park, H. Y., Smith, H., Kwok, L., Lamb, J., et al. (2004). Spatial distribution of competing ions around DNA in solution. *Physical Review Letters*, *93*(24), 248103.

Andresen, K., Qiu, X., Pabit, S. A., Lamb, J. S., Park, H. Y., Kwok, L. W., & Pollack, L. (2008). Mono- and trivalent ions around DNA: A small-angle scattering study of competition and interactions. *Biophysical Journal, 95*(1), 287–295.

Anisimov, V. M., Lamoureux, G., Vorobyov, I. V., Huang, N., Roux, B., & MacKerell, A. D., Jr. (2005). Determination of electrostatic parameters for a polarizable force field based on the classical drude oscillator. *Journal of Chemical Theory and Computation, 1*, 153–168.

Auffinger, P., Bielecki, L., & Westhof, E. (2004). Anion binding to nucleic acids. *Structure, 12*, 379–388.

Auffinger, P., Cheatham, T. E., III, & Vaiana, A. C. (2007). Spontaneous formation of KCl aggregates in biomolecular simulations: A force field issue? *Journal of Chemical Theory and Computation, 3*, 1851–1859.

Babin, V., Roland, C., & Sagui, C. (2008). Adaptively biased molecular dynamics for free energy calculations. *Journal of Chemical Physics, 128*, 134101.

Babu, C. S., & Lim, C. (2006). Empirical force fields for biologically active divalent metal cations in water. *Journal of Physical Chemistry A, 110*, 691–699.

Bai, Y., Greenfeld, M., Travers, K. J., Chu, V. B., Lipfert, J., Doniach, S., & Herschlag, D. (2007). Quantitative and comprehensive decomposition of the ion atmosphere around nucleic acids. *Journal of the American Chemical Society, 129*(48), 14981–14988.

Baptista, A. M., Martel, P. J., & Petersen, S. B. (1997). Simulation of protein conformational freedom as a function of pH: Constant-pH molecular dynamics using implicit titration. *Proteins, 27*, 523–544.

Baptista, A. M., Teixeira, V. H., & Soares, C. M. (2002). Constant-pH molecular dynamics using stochastic titration. *Journal of Chemical Physics, 117*, 4184–4200.

Beglov, D., & Roux, B. (1997). An integral equation to describe the solvation of polar molecules in liquid water. *Journal of Physical Chemistry. B, 101*, 7821–7826.

Bennett, C. H. (1976). Efficient estimation of free energy differences from Monte Carlo data. *Journal of Computational Physics, 22*, 245–268.

Berg, B. A., & Neuhaus, T. (1992). Multicanonical ensemble: A new approach to simulate first-order phase transitions. *Physical Review Letters, 68*, 9–12.

Bevilacqua, P. C. (2003). Mechanistic considerations for general acid-base catalysis by RNA: Revisiting the mechanism of the hairpin ribozyme. *Biochemistry, 42*, 2259–2265.

Bevilacqua, P. C., Brown, T. S., Nakano, S., & Yajima, R. (2004). Catalytic roles for proton transfer and protonation in ribozymes. *Biopolymers, 73*, 90–109.

Bleam, M. L., Anderson, C. F., & Record, T., Jr. (1980). Relative binding affinities of monovalent cations for double-stranded DNA. *Proceedings of the National Academy of Sciences of the United States of America*, 77(6), 3085–3089.

Bond, J. P., Anderson, C. F., & Record, M. T., Jr. (1994). Conformational transitions of duplex and triplex nucleic acid helices: Thermodynamic analysis of effects of salt concentration on stability using preferential interaction coefficients. *Biophysical Journal, 67*(2), 825–836.

Braunlin, W. H., Anderson, C. F., & Record, M. T., Jr. (1987). Competitive interactions of $Co(NH_3)_6{}^{3+}$) and Na^+ with helical B-DNA probed by ^{5}9Co and ^{2}3Na NMR. *Biochemistry, 26*(24), 7724–7731.

Brooks, B. R., Brooks, C. L., III, MacKerell, A. D., Jr., Nilsson, L., Petrella, R. J., Roux, B., et al. (2009). CHARMM: The biomolecular simulation program. *Journal of Computational Chemistry, 30*(10), 1545–1614.

Butcher, S. E., & Pyle, A. M. (2011). The molecular interactions that stabilize RNA tertiary structure: RNA motifs, patterns, and networks. *Accounts of Chemical Research, 44*, 1302–1311.

Case, D., Babin, V., Berryman, J., Betz, R., Cai, Q., Cerutti, D., et al. (2014). *AMBER 14*. San Francisco, CA: University of California, San Francisco.

Chen, A. A., Draper, D. E., & Pappu, R. V. (2009). Molecular simulation studies of monovalent counterion-mediated interactions in a model RNA kissing loop. *Journal of Molecular Biology*, *390*(4), 805–819.

Chen, A. A., Marucho, M., Baker, N. A., & Pappu, R. V. (2009). Simulations of RNA interactions with monovalent ions. *Methods in Enzymology*, *469*, 411–432.

Chen, A. A., & Pappu, R. V. (2007). Parameters of monovalent ions in the AMBER-99 forcefield: Assessment of inaccuracies and proposed improvements. *Journal of Physical Chemistry. B*, *111*, 11884–11887.

Chen, H., Giese, T. J., Huang, M., Wong, K.-Y., Harris, M. E., & York, D. M. (2014). Mechanistic insights into RNA transphosphorylation from kinetic isotope effects and linear free energy relationships of model reactions. *Chemistry: A European Journal*, *20*, 14336–14343.

Chen, H., Harris, M. E., & York, D. M. (n.d.). The effect of Zn^{2+} binding on the mechanism of RNA transphosphorylation interpreted through kinetic isotope effects. *Biochimica et Biophysica Acta,* in press.

Chen, J., Ganguly, A., Miswan, Z., Hammes-Schiffer, S., Bevilacqua, P. C., & Golden, B. L. (2013). Identification of the catalytic Mg^{2+} ion in the hepatitis delta virus ribozyme. *Biochemistry*, *52*(3), 557–567.

Chen, J.-H., Gong, B., Bevilacqua, P. C., Carey, P. R., & Golden, B. L. (2009). A catalytic metal ion interacts with the cleavage site GU wobble in the HDV ribozyme. *Biochemistry*, *48*, 1498–1507.

Chen, J.-H., Yajima, R., Chadalavada, D. M., Chase, E., Bevilacqua, P. C., & Golden, B. L. (2010). A 1.9 Å crystal structure of the HDV ribozyme precleavage suggests both Lewis acid and general acid mechanisms contribute to phosphodiester cleavage. *Biochemistry*, *49*(31), 6508–6518.

Chen, X., & Ellington, A. D. (2009). Design principles for ligand-sensing, conformation-switching ribozymes. *PLoS Computational Biology*, *5*, 1000620.

Chen, X., Li, N., & Ellington, A. D. (2007). Ribozyme catalysis of metabolism in the RNA world. *Chemistry & Biodiversity*, *4*, 633–655.

Chodera, J. D., & Shirts, M. R. (2011). Replica exchange and expanded ensemble simulations as Gibbs sampling: Simple improvements for enhanced mixing. *Journal of Chemical Physics*, *135*, 194110.

Chu, V. B., Bai, Y., Lipfert, J., Herschlag, D., & Doniach, S. (2007). Evaluation of ion binding to DNA duplexes using a size-modified Poisson-Boltzmann theory. *Biophysical Journal*, *93*(9), 3202–3209.

Cleland, W. W., & Hengge, A. C. (2006). Enzymatic mechanisms of phosphate and sulfate transfer. *Chemical Reviews*, *106*, 3252–3278.

Cochrane, J. C., & Strobel, S. A. (2008). Catalytic strategies of self-cleaving ribozymes. *Accounts of Chemical Research*, *41*, 1027–1035.

Cornell, W. D., Cieplak, P., Bayly, C. I., Gould, I. R., Merz, K. M., Jr., Ferguson, D. M., et al. (1995). A second generation force field for the simulation of proteins, nucleic acids and organic molecules. *Journal of the American Chemical Society*, *117*, 5179–5197.

Dama, J. F., Sinitskiy, A. V., McCullagh, M., Weare, J., Roux, B., Dinner, A. R., & Voth, G. A. (2013). The theory of ultra-coarse-graining. 1. General principles. *Journal of Chemical Theory and Computation*, *9*, 2466–2480.

Darve, E., & Pohorille, A. (2001). Calculating free energies using average force. *Journal of Chemical Physics*, *115*(20), 9169–9183.

Darve, E., Rodríguez-Gómez, D., & Pohorille, A. (2008). Adaptive biasing force method for scalar and vector free energy calculations. *Journal of Chemical Physics*, *128*(14), 144120.

Das, S., & Piccirilli, J. (2005). General acid catalysis by the hepatitis delta virus ribozyme. *Nature Chemical Biology*, *1*(1), 45–52.

Davies, D. R., & Hol, W. G. J. (2004). The power of vanadate in crystallographic investigations of phosphoryl transfer enzymes. *FEBS Letters*, *577*(3), 315–321.

den Otter, W. K. (2000). Thermodynamic integration of the free energy along a reaction coordinate in Cartesian coordinates. *Journal of Chemical Physics*, *112*(17), 7283–7292.

Dissanayake, T., Swails, J., Harris, M. E., Roitberg, A. E., & York, D. M. (n.d.). Interpretation of pH-rate profiles for acid-base catalysis from molecular simulations., *Biochemistry*, in press.

Doudna, J. A., & Cech, T. R. (2002). The chemical repertoire of natural ribozymes. *Nature*, *418*, 222–228.

Doudna, J. A., & Lorsch, J. R. (2005). Ribozyme catalysis: Not different, just worse. *Nature Structural and Molecular Biology*, *12*(5), 395–402.

Draper, D. E. (2008). RNA folding: Thermodynamic and molecular descriptions of the roles of ions. *Biophysical Journal*, *95*(12), 5489–5495.

Dror, R. O., Dirks, R. M., Grossman, J. P., Xu, H., & Shaw, D. E. (2012). Biomolecular simulation: A computational microscope for molecular biology. *Annual Review of Biophysics*, *41*, 429–452.

Ennifar, E., Walter, P., & Dumas, P. (2003). A crystallographic study of the binding of 13 metal ions to two related RNA duplexes. *Nucleic Acids Research*, *31*(10), 2671–2682.

Ensing, B., De Vivo, M., Liu, Z., Moore, P., & Klein, M. L. (2006). Metadynamics as a tool for exploring free energy landscapes of chemical reactions. *Accounts of Chemical Research*, *39*(2), 73–81.

Fastrez, J. (2009). Engineering allosteric regulation into biological catalysts. *Chembiochem*, *10*, 2824–2835.

Fedor, M. J. (2009). Comparative enzymology and structural biology of RNA self-cleavage. *Annual Review of Biophysics*, *38*, 271–299.

Fedor, M. J., & Williamson, J. R. (2005). The catalytic diversity of RNAs. Nature Reviews. *Molecular Cell Biology*, *6*, 399–412.

Feig, M., Karanicolas, J., & Brooks, C. L., III. (2004). MMTSB tool set: Enhanced sampling and multiscale modeling methods for applications in structure biology. *Journal of Molecular Graphics & Modelling*, *22*, 377–395.

Ferré-D'Amaré, A. R., Zhou, K., & Doudna, J. A. (1998). Crystal structure of a hepatitis delta virus ribozyme. *Nature*, *395*, 567–574.

Field, M. J., Bash, P. A., & Karplus, M. (1990). A combined quantum mechanical and molecular mechanical potential for molecular dynamics simulations. *Journal of Computational Chemistry*, *11*, 700–733.

Foloppe, N., & MacKerell, A. D., Jr. (2000). All-atom empirical force field for nucleic acids: I. *Parameter optimization based on small molecule and condensed phase macromolecular target data. Journal of Computational Chemistry*, *21*, 86–104.

Gao, J., Amara, P., Alhambra, C., & Field, M. J. (1998). A generalized hybrid orbital (GHO) method for the treatment of boundary atoms in combined QM/MM calculations. *Journal of Physical Chemistry A*, *102*, 4714–4721.

Garcia-Viloca, M., Gao, J., Karplus, M., & Truhlar, D. G. (2004). How enzymes work: Analysis by modern rate theory and computer simulations. *Science*, *303*, 186–195.

Garst, A. D., Edwards, A. L., & Batey, R. T. (2011). Riboswitches: Structures and mechanisms. *Cold Spring Harbor Perspectives in Biology*, *3*(6), a003533.

Giambaşu, G. M., Lee, T.-S., Scott, W. G., & York, D. M. (2012). Mapping L1 ligase ribozyme conformational switch. *Journal of Molecular Biology*, *423*(1), 106–122.

Giambaşu, G. M., Lee, T.-S., Sosa, C. P., Robertson, M. P., Scott, W. G., & York, D. M. (2010). Identification of dynamical hinge points of the L1 ligase molecular switch. *RNA*, *16*(4), 769–780.

Giambaşu, G. M., Luchko, T., Herschlag, D., York, D. M., & Case, D. A. (2014). Ion counting from explicit-solvent simulations and 3D-RISM. *Biophysical Journal*, *106*, 883–894.

Giese, T. J., Chen, H., Huang, M., & York, D. M. (2014). Parametrization of an orbital-based linear-scaling quantum force field for noncovalent interactions. *Journal of Chemical Theory and Computation*, *10*, 1086–1098.

Giese, T. J., Huang, M., Chen, H., & York, D. M. (2014). Recent advances toward a general purpose linear-scaling quantum force field. *Accounts of Chemical Research, 47*, 2812–2820.

Goh, G. B., Knight, J. L., & Brooks, C. L., III. (2012). Constant pH molecular dynamics simulations of nucleic acids in explicit solvent. *Journal of Chemical Theory and Computation, 8*, 36–46.

Goh, G. B., Knight, J. L., & Brooks, C. L., III. (2013). Towards accurate prediction of protonation equilibrium of nucleic acids. *Journal of Physical Chemistry Letters, 4*(5), 760–766.

Golden, B. L. (2011). Two distinct catalytic strategies in the hepatitis delta virus ribozyme cleavage reaction. *Biochemistry, 50*(44), 9424–9433.

Golden, B. L., Hammes-Schiffer, S., Carey, P. R., & Bevilacqua, P. C. (2013). An integrated picture of HDV ribozyme catalysis. In R. Russell (Ed.), *Biophysics of RNA folding: Vol. 3* (pp. 135–167). New York: Springer.

Gong, B., Chen, J.-H., Bevilacqua, P. C., Golden, B. L., & Carey, P. R. (2009). Competition between $Co(NH_3)_6^{3+}$ and inner sphere Mg^{2+} ions in the HDV ribozyme. *Biochemistry, 48*, 11961–11970.

Gong, B., Chen, J.-H., Chase, E., Chadalavada, D. M., Yajima, R., Golden, B. L., et al. (2007). Direct measurement of a pK_a near neutrality for the catalytic cytosine in the genomic HDV ribozyme using Raman crystallography. *Journal of the American Chemical Society, 129*, 13335–13342.

Gong, B., Chen, J.-H., Yajima, R., Chen, Y., Chase, E., Chadalavada, D. M., et al. (2009). Raman crystallography of RNA. *Methods, 49*(2), 101–111.

Gong, B., Chen, Y., Christian, E. L., Chen, J.-H., Chase, E., Chadalavada, D. M., et al. (2008). Detection of innersphere interactions between magnesium hydrate and the phosphate backbone of the HDV ribozyme using Raman crystallography. *Journal of the American Chemical Society, 130*, 9670–9672.

Greenfeld, M., & Herschlag, D. (2009). Probing nucleic acid-ion interactions with buffer exchange-atomic emission spectroscopy. *Methods in Enzymology, 469*(10), 375–389.

Grilley, D., Soto, A. M., & Draper, D. E. (2006). Mg^{2+} - RNA interaction free energies and their relationship to the folding of RNA tertiary structures. *Proceedings of the National Academy of Sciences of the United States of America, 103*(38), 14003–14008.

Gu, H., Zhang, S., Wong, K.-Y., Radak, B. K., Dissanayake, T., Kellerman, D. L., et al. (2013). Experimental and computational analysis of the transition state for ribonuclease A-catalyzed RNA 2′-O-transphosphorylation. *Proceedings of the National Academy of Sciences of the United States of America, 110*, 13002–13007.

Guttman, M., & Rinn, J. L. (2012). Modular regulatory principles of large non-coding RNAs. *Nature, 482*, 339–346.

Hamelberg, D., Mongan, J., & McCammon, J. A. (2004). Accelerated molecular dynamics: A promising and efficient simulation method for biomolecules. *Journal of Chemical Physics, 120*, 11919–11929.

Harris, M. E., & Cassano, A. G. (2008). Experimental analyses of the chemical dynamics of ribozyme catalysis. *Current Opinion in Chemical Biology, 12*, 626–639.

Hashem, Y., & Auffinger, P. (2009). A short guide for molecular dynamics simulations of RNA systems. *Methods, 47*(3), 187–197.

Heldenbrand, H., Janowski, P. A., Giambaşu, G., Giese, T. J., Wedekind, J. E., & York, D. M. (2014). Evidence for the role of active site residues in the hairpin ribozyme from molecular simulations along the reaction path. *Journal of the American Chemical Society, 136*, 7789–7792.

Hengge, A. C. (2002). Isotope effects in the study of phosphoryl and sulfuryl transfer reactions. *Accounts of Chemical Research, 35*, 105–112.

Herschlag, D. (1994). Ribonuclease revisited: Catalysis via the classical general acid-base mechanism or a triester-like mechanism? *Journal of the American Chemical Society, 116*(26), 11631–11635.

Hoskins, A. A., & Moore, M. J. (2012). The spliceosome: A flexible, reversible macromolecular machine. *Trends in Biochemical Sciences*, *37*(5), 179–188.

Hou, G., & Cui, Q. (2013). Stabilization of different types of transition states in a single enzyme active site: QM/MM analysis of enzymes in the alkaline phosphatase superfamily. *Journal of the American Chemical Society*, *135*, 10457–10469.

Howard, J. J., Lynch, G. C., & Pettitt, B. M. (2011). Ion and solvent density distributions around canonical B-DNA from integral equations. *Journal of Physical Chemistry. B*, *115*(3), 547–556.

Ji, C. G., & Zhang, J. Z. H. (2011). Understanding the molecular mechanism of enzyme dynamics of ribonuclease A through protonation/deprotonation of HIS48. *Journal of the American Chemical Society*, *133*, 17727–17737.

Jorgensen, W. L., Maxwell, D. S., & Tirado-Rives, J. (1996). Development and testing of the OPLS all-atom force field on conformational energetics and properties of organic liquids. *Journal of the American Chemical Society*, *118*, 11225–11236.

Joung, I. S., & Cheatham, T. E., III. (2008). Determination of alkali and halide monovalent ion parameters for use in explicitly solvated biomolecular simulations. *Journal of Physical Chemistry. B*, *112*, 9020–9041.

Kaminski, G. A., Friesner, R. A., Tirado-Rives, J., & Jorgensen, W. L. (2001). Evaluation and reparametrization of the OPLS-AA force field for proteins via comparison with accurate quantum chemical calculations on peptides. *Journal of Physical Chemistry. B*, *105*, 6474–6487.

Ke, A., Zhou, K., Ding, F., Cate, J. H. D., & Doudna, J. A. (2004). A conformational switch controls hepatitis delta virus ribozyme catalysis. *Nature*, *429*, 201–205.

Kellerman, D. L., York, D. M., Piccirilli, J. A., & Harris, M. E. (2014). Altered (transition) states: mechanisms of solution and enzyme catalyzed RNA 2′-O-transphosphorylation. *Current Opinion in Chemical Biology*, *21*, 96–102.

Khandogin, J., & Brooks, C. L., III. (2005). Constant pH molecular dynamics with proton tautomerism. *Biophysical Journal*, *89*, 141–157.

Kirmizialtin, S., Silalahi, A. R. J., Elber, R., & Fenley, M. O. (2012). The ionic atmosphere around A-RNA: Poisson-Boltzmann and molecular dynamics simulations. *Biophysical Journal*, *102*(4), 829–838.

Klein, D. J., Been, M. D., & Ferré-D'Amaré, A. R. (2007). Essential role of an active-site guanine in glmS ribozyme catalysis. *Journal of the American Chemical Society*, *129*(48), 14858–14859.

Klingen, A. R., Bombarda, E., & Ullmann, G. M. (2006). Theoretical investigation of the behavior of titratable groups in proteins. *Photochemical & Photobiological Sciences*, *5*, 588–596.

Kovalenko, A., & Hirata, F. (2000). Potentials of mean force of simple ions in ambient aqueous solution. II. Solvation structure from the three-dimensional reference interaction site model approach, and comparison with simulations. *Journal of Chemical Physics*, *112*(23), 10403.

Kovalenko, A., Ten-no, S., & Hirata, F. (1999). Solution of three-dimensional reference interaction site model and hypernetted chain equations for simple point charge water by modified method of direct inversion in iterative subspace. *Journal of Computational Chemistry*, *20*(9), 928–936.

Kumar, S., Bouzida, D., Swendsen, R. H., Kollman, P. A., & Rosenberg, J. M. (1992). The weighted histogram analysis method for free-energy calculations on biomolecules. *I. The method. Journal of Computational Chemistry*, *13*, 1011–1021.

Laio, A., & Parrinello, M. (2002). Escaping free-energy minima. *Proceedings of the National Academy of Sciences of the United States of America*, *99*, 12562–12566.

Lassila, J. K., Zalatan, J. G., & Herschlag, D. (2011). Biological phosphoryl-transfer reactions: Understanding mechanism and catalysis. *Annual Review of Biochemistry*, *80*, 669–702.

Lee, M. S., Salsbury, F. R., Jr., & Brooks, C. L., III. (2004). Constant-pH molecular dynamics using continuous titration coordinates. *Proteins*, *56*, 738–752.

Lee, T.-S., Giambaşu, G. M., Harris, M. E., & York, D. M. (2011). Characterization of the structure and dynamics of the HDV ribozyme in different stages along the reaction path. *Journal of Physical Chemistry Letters*, *2*(20), 2538–2543.

Lee, T.-S., Giambaşu, G. M., Sosa, C. P., Martick, M., Scott, W. G., & York, D. M. (2009). Threshold occupancy and specific cation binding modes in the hammerhead ribozyme active site are required for active conformation. *Journal of Molecular Biology*, *388*, 195–206.

Lee, T.-S., Giambaşu, G. M., & York, D. M. (2010). Insights into the role of conformational transitions and metal ion binding in RNA catalysis from molecular simulations. In R. A. Wheeler (Ed.), *Annual reports in computational chemistry (Vol. 6, pp. 169–200)*: Amsterdam, The Netherlands: Elsevier.

Lee, T.-S., Radak, B. K., Huang, M., Wong, K.-Y., & York, D. M. (2014). Roadmaps through free energy landscapes calculated using the multidimensional vFEP approach. *Journal of Chemical Theory and Computation*, *10*, 24–34.

Lee, T.-S., Radak, B. K., Pabis, A., & York, D. M. (2013). A new maximum likelihood approach for free energy profile construction from molecular simulations. *Journal of Chemical Theory and Computation*, *9*, 153–164.

Lee, T.-S., Silva Lopez, C., Giambaşu, G. M., Martick, M., Scott, W. G., & York, D. M. (2008). Role of Mg^{2+} in hammerhead ribozyme catalysis from molecular simulation. *Journal of the American Chemical Society*, *130*(10), 3053–3064.

Lee, T.-S., Wong, K.-Y., Giambasu, G. M., & York, D. M. (2013). Bridging the gap between theory and experiment to derive a detailed understanding of hammerhead ribozyme catalysis. In G. A. Soukup (Ed.), *Progress in Molecular Biology and Translational Science: Vol. 120* (pp. 25–91). London, UK: Elsevier, Academic Press.

Lee, T.-S., & York, D. M. (2008). Origin of mutational effects at the C3 and G8 positions on hammerhead ribozyme catalysis from molecular dynamics simulations. *Journal of the American Chemical Society*, *130*(23), 7168–7169.

Lee, T.-S., & York, D. M. (2010). Computational mutagenesis studies of hammerhead ribozyme catalysis. *Journal of the American Chemical Society*, *132*(38), 13505–13518.

Lévesque, D., Reymond, C., & Perreault, J.-P. (2012). Characterization of the trans Watson-Crick GU base pair located in the catalytic core of the antigenomic HDV ribozyme. *PLoS One*, 7(6), 40309.

Li, P., & Merz, K. M., Jr. (2014). Taking into account the ion-induced dipole interaction in the nonbonded model of ions. *Journal of Chemical Theory and Computation*, *10*, 289–297.

Li, P., Roberts, B. P., Chakravorty, D. K., & Merz, K. M., Jr. (2013). Rational design of particle mesh Ewald compatible Lennard-Jones parameters for +2 metal cations in explicit solvent. *Journal of Chemical Theory and Computation*, *9*, 2733–2748.

Lilley, D. M. J. (2005). Structure, folding and mechanisms of ribozymes. *Current Opinion in Structural Biology*, *15*, 313–323.

Lilley, D. M. J. (2011). Catalysis by the nucleolytic ribozymes. *Biochemical Society Transactions*, *39*, 641–646.

Lilley, D. M. J. (2011). Mechanisms of RNA catalysis. *Philosophical Transactions of the Royal Society B*, *366*, 2910–2917.

Link, K. H., & Breaker, R. R. (2009). Engineering ligand-responsive gene-control elements: Lessons learned from natural riboswitches. *Gene Therapy*, *16*, 1189–1201.

Lodola, A., & Mulholland, A. J. (2013). Computational enzymology. In L. Monticelli & E. Salonen (Eds.), *Biomolecular simulations: Vol. 924* (pp. 67–89). New York, NY: Humana Press.

Luchko, T., Gusarov, S., Roe, D. R., Simmerling, C., Case, D. A., Tuszynski, J., & Kovalenko, A. (2010). Three-dimensional molecular theory of solvation coupled with molecular dynamics in AMBER. *Journal of Chemical Theory and Computation*, *6*, 607–624.

MacKerell, A. D., Jr., & Banavali, N. K. (2000). All-atom empirical force field for nucleic acids: II. *Application to molecular dynamics simulations of DNA and RNA in solution. Journal of Computational Chemistry, 21*, 105–120.

Martick, M., Lee, T.-S., York, D. M., & Scott, W. G. (2008). Solvent structure and hammerhead ribozyme catalysis. *Chemistry & Biology, 15*, 332–342.

Martínez, J. M., Pappalardo, R. R., & Marcos, E. S. (1999). First-principles ion-water interaction potentials for highly charged monatomic cations. Computer simulations of Al^{3+}, Mg^{2+}, and Be^{2+} in water. *Journal of the American Chemical Society, 121*, 3175–3184.

Maruyama, Y., Yoshida, N., & Hirata, F. (2010). Revisiting the salt-induced conformational change of DNA with 3D-RISM theory. *Journal of Chemical Physics B, 114*(19), 6464–6471.

McDowell, S. E., Špačková, N., Šponer, J., & Walter, N. G. (2006). Molecular dynamics simulations of RNA: An in silico single molecule approach. *Biopolymers, 85*, 169–184.

Meier-Schellersheim, M., Fraser, I. D. C., & Klauschen, F. (2009). Multiscale modeling for biologists. Wiley Interdisciplinary Reviews. *Systems Biology and Medicine, 1*(1), 4–14.

Misra, V. K., & Draper, D. E. (1998). On the role of magnesium ions in RNA stability. *Biopolymers, 48*, 113–135.

Misra, V. K., & Draper, D. E. (2002). The linkage between magnesium binding and RNA folding. *Journal of Molecular Biology, 317*, 507–521.

Mongan, J., Case, D. A., & McCammon, J. A. (2004). Constant pH molecular dynamics in generalized Born implicit solvent. *Journal of Computational Chemistry, 25*, 2038–2048.

Murray, J. B., Dunham, C. M., & Scott, W. G. (2002). A pH-dependent conformational change, rather than the chemical step, appears to be rate-limiting in the hammerhead ribozyme cleavage reaction. *Journal of Molecular Biology, 315*, 121–130.

Nakajima, N., Nakamura, H., & Kidera, A. (1997). Multicanonical ensemble generated by molecular dynamics simulation for enhanced conformational sampling of peptides. *Journal of Physical Chemistry. B, 101*, 817–824.

Nakano, S., Chadalavada, D. M., & Bevilacqua, P. C. (2000). General acid-base catalysis in the mechanism of a hepatitis delta virus ribozyme. *Science, 287*, 1493–1497.

Nakano, S., Proctor, D. J., & Bevilacqua, P. C. (2001). Mechanistic characterization of the HDV genomic ribozyme: Assessing the catalytic and structural contributions of divalent metal ions within a multichannel reaction mechanism. *Biochemistry, 40*, 12022–12038.

Nam, K., Cui, Q., Gao, J., & York, D. M. (2007). Specific reaction parametrization of the AM1/d Hamiltonian for phosphoryl transfer reactions: H, O, and P atoms. *Journal of Chemical Theory and Computation, 3*, 486–504.

Nam, K., Gao, J., & York, D. (2008). Electrostatic interactions in the hairpin ribozyme account for the majority of the rate acceleration without chemical participation by nucleobases. *RNA, 14*, 1501–1507.

Nam, K., Gao, J., & York, D. M. (2008). Quantum mechanical/molecular mechanical simulation study of the mechanism of hairpin ribozyme catalysis. *Journal of the American Chemical Society, 130*(14), 4680–4691.

Nixon, P. L., & Giedroc, D. P. (2000). Energetics of a strongly pH dependent RNA tertiary structure in a frameshifting pseudoknot. *Journal of Molecular Biology, 296*, 659–671.

Onufriev, A., Bashford, D., & Case, D. A. (2004). Exploring protein native states and large-scale conformational changes with a modified generalized Born model. *Proteins, 55*, 383–394.

Oostenbrink, C., Villa, A., Mark, A. E., & van Gunsteren, W. F. (2004). A biomolecular force field based on the free enthalpy of hydration and solvation: The GROMOS force-field parameter sets 53A5 and 53A6. *Journal of Computational Chemistry, 25*, 1656–1676.

Pabit, S. A., Meisburger, S. P., Li, L., Blose, J. M., Jones, C. D., & Pollack, L. (2010). Counting ions around DNA with anomalous small-angle X-ray scattering. *Journal of the American Chemical Society, 132*(46), 16334–16336.

Pabit, S. A., Qiu, X., Lamb, J. S., Li, L., Meisburger, S. P., & Pollack, L. (2009). Both helix topology and counterion distribution contribute to the more effective charge screening in dsRNA compared with dsDNA. *Nucleic Acids Research, 37*(12), 3887–3896.

Panteva, M. T., Giambaşu, G. M., & York, D. M. (n.d.). Comparison of structural, thermodynamic, kinetic and mass transport properties of Mg^{2+} ion models commonly used in biomolecular simulations. *Journal of Computational Chemistry*, in press.

Penchovsky, R. (2014). Computational design of allosteric ribozymes as molecular biosensors. *Biotechnology Advances, 32*, 1015–1027.

Pérez, A., Luque, F. J., & Orozco, M. (2011). Frontiers in molecular dynamics simulations of DNA. *Accounts of Chemical Research, 45*, 196–205.

Pérez, A., Marchán, I., Svozil, D., Sponer, J., Cheatham, T. E., III, Laughton, C. A., & Orozco, M. (2007). Refinement of the AMBER force field for nucleic acids: Improving the description of α/γ conformers. *Biophysical Journal, 92*, 3817–3829.

Pollack, L. (2011). SAXS studies of ion-nucleic acid interactions. *Annual Review of Biophysics, 40*, 225–242.

Ponder, J. W., Wu, C., Ren, P., Pande, V. S., Chodera, J. D., Schnieders, M. J., et al. (2010). Current status of the AMOEBA polarizable force field. *Journal of Physical Chemistry. B, 114*, 2549–2564.

Quirk, D. J., & Raines, R. T. (1999). His ... Asp catalytic dyad of ribonuclease A: Histidine pKa values in the wild-type, D121N, and D121A enzymes. *Biophysical Journal, 76*, 1571–1579.

Raines, R. T. (1998). Ribonuclease A. *Chemical Reviews, 98*, 1045–1065.

Reuter, N., Dejaegere, A., Maigret, B., & Karplus, M. (2000). Frontier bonds in QM/MM methods: A comparison of different approaches. *Journal of Physical Chemistry A, 104*, 1720–1735.

Rhodes, M. M., Réblová, K., Sponer, J., & Walter, N. G. (2006). Trapped water molecules are essential to structural dynamics and function of a ribozyme. *Proceedings of the National Academy of Sciences of the United States of America, 103*, 13380–13385.

Roth, A., & Breaker, R. R. (2009). The structural and functional diversity of metabolite-binding riboswitches. *Annual Review of Biochemistry, 78*, 305–334.

Rueda, M., Cubero, E., Laughton, C. A., & Orozco, M. (2004). Exploring the counterion atmosphere around DNA: What can be learned from molecular dynamics simulations? *Biophysical Journal, 87*, 800–811.

Salomon-Ferrer, R., Götz, A. W., Poole, D., Le Grand, S., & Walker, R. C. (2013). Routine microsecond molecular dynamics simulations with AMBER on GPUs. 2. Explicit solvent particle mesh Ewald. *Journal of Chemical Theory and Computation, 9*, 3878–3888.

Schmeing, T. M., & Ramakrishnan, V. (2009). What recent ribosome structures have revealed about the mechanism of translation. *Nature, 461*(7268), 1234–1242.

Scott, W. G. (2007). Ribozymes. *Current Opinion in Structural Biology, 17*, 280–286.

Senn, H. M., & Thiel, W. (2009). QM/MM methods for biomolecular systems. *Angewandte Chemie International Edition, 48*, 1198–1229.

Sharp, P. A. (2009). The centrality of RNA. *Cell, 136*(4), 577–580.

Sherwood, P., Brooks, B. R., & Sansom, M. S. P. (2008). Multiscale methods for macromolecular simulations. *Current Opinion in Structural Biology, 18*(5), 630–640.

Shirts, M. R., & Chodera, J. D. (2008). Statistically optimal analysis of samples from multiple equilibrium states. *Journal of Chemical Physics, 129*, 124105.

Souaille, M., & Roux, B. (2001). Extension to the weighted histogram analysis method: Combining umbrella sampling with free energy calculations. *Computer Physics Communications, 135*, 40–57.

Suess, B., & Weigand, J. E. (2008). Engineered riboswitches: Overview, problems and trends. *RNA Biology, 5*, 1–6.

Sugita, Y., Kitao, A., & Okamoto, Y. (2000). Multidimensional replica-exchange method for free-energy calculations. *Journal of Chemical Physics, 113*, 6042–6051.

Tan, Z., Gallicchio, E., Lapelosa, M., & Levy, R. M. (2012). Theory of binless multi-state free energy estimation with applications to protein-ligand binding. *Journal of Chemical Physics, 136*, 144102.

Thomas, A. S., & Elcock, A. H. (2006). Direct observation of salt effects on molecular interactions through explicit-solvent molecular dynamics simulations: Differential effects on electrostatic and hydrophobic interactions and comparisons to Poisson-Boltzmann theory. *Journal of the American Chemical Society, 128*, 7796–7806.

Torelli, A. T., Krucinska, J., & Wedekind, J. E. (2007). A comparison of vanadate to a 2′–5′ linkage at the active site of a small ribozyme suggests a role for water in transition-state stabilization. *RNA, 13*, 1052–1070.

Torrie, G. M., & Valleau, J. P. (1974). Monte Carlo free energy estimates using non-Boltzmann sampling: Application to the sub-critical Lennard-Jones fluid. *Chemical Physics Letters, 28*, 578–581.

Torrie, G. M., & Valleau, J. P. (1977). Nonphysical sampling distributions in Monte Carlo free-energy estimation: Umbrella sampling. *Journal of Computational Physics, 23*, 187–199.

Ullmann, G. M. (2003). Relations between protonation constants and titration curves in polyprotic acids: A critical view. *Journal of Physical Chemistry. B, 107*, 1263–1271.

Valadkhan, S. (2010). Role of the snRNAs in spliceosomal active site. *RNA Biology*, 7(3), 345–353.

Valleau, J. P., & Card, D. N. (1972). Monte Carlo estimation of the free energy by multistage sampling. *Journal of Chemical Physics, 57*, 5457–5462.

van der Kamp, M. W., & Mulholland, A. J. (2013). Combined quantum mechanics/molecular mechanics (QM/MM) methods in computational enzymology. *Biochemistry, 52*, 2708–2728.

Vanden-Eijnden, E. (2009). Some recent techniques for free energy calculations. *Journal of Computational Chemistry, 30*(11), 1737–1747.

Viladoms, J., Scott, L. G., & Fedor, M. J. (2011). An active-site guanine participates in glmS ribozyme catalysis in its protonated state. *Journal of the American Chemical Society, 133*(45), 18388–18396.

Wadkins, T. S., Shih, I., Perrotta, A. T., & Been, M. D. (2001). A pH-sensitive RNA tertiary interaction affects self-cleavage activity of the HDV ribozymes in the absence of added divalent metal ion. *Journal of Molecular Biology, 305*, 1045–1055.

Walter, N. G. (2007). Ribozyme catalysis revisited: Is water involved? *Molecular Cell, 28*, 923–929.

Wang, J., Cieplak, P., & Kollman, P. A. (2000). How well does a restrained electrostatic potential (RESP) model perform in calculating conformational energies of organic biological molecules. *Journal of Computational Chemistry, 21*(12), 1049–1074.

Warshel, A., & Levitt, M. (1976). Theoretical studies of enzymic reactions: Dielectric, electrostatic and steric stabilization of the carbonium ion in the reaction of lysozyme. *Journal of Molecular Biology, 103*, 227–249.

Wilcox, J. L., Ahluwalia, A. K., & Bevilacqua, P. C. (2011). Charged nucleobases and their potential for RNA catalysis. *Accounts of Chemical Research, 44*, 1270–1279.

Wilson, D. N., & Cate, J. H. D. (2012). The structure and function of the eukaryotic ribosome. *Cold Spring Harbor Perspectives in Biology*, *4*(5), a011536.

Wilson, T. J., Li, N.-S., Lu, J., Frederiksen, J. K., Piccirilli, J. A., & Lilley, D. M. J. (2010). Nucleobase-mediated general acid-base catalysis in the Varkud satellite ribozyme. *Proceedings of the National Academy of Sciences of the United States of America, 107*, 11751–11756.

Wilson, T. J., & Lilley, D. M. J. (2009). The evolution of ribozyme chemistry. *Science, 323*(5920), 1436–1438.

Wojtas-Niziurski, W., Meng, Y., Roux, B., & Bernèche, S. (2013). Self-learning adaptive umbrella sampling method for the determination of free energy landscapes in multiple dimensions. *Journal of Chemical Theory and Computation, 9*(4), 1885–1895.

Wong, K.-Y., Gu, H., Zhang, S., Piccirilli, J. A., Harris, M. E., & York, D. M. (2012). Characterization of the reaction path and transition states for RNA transphosphorylation models from theory and experiment. *Angewandte Chemie International Edition, 51*, 647–651.

Wong, K.-Y., Lee, T.-S., & York, D. M. (2011). Active participation of the Mg^{2+} ion in the reaction coordinate of RNA self-cleavage catalyzed by the hammerhead ribozyme. *Journal of Chemical Theory and Computation*, 7(1), 1–3.

Wong, K.-Y., & York, D. M. (2012). Exact relation between potential of mean force and free-energy profile. *Journal of Chemical Theory and Computation, 8*(11), 3998–4003.

Wu, X., & Brooks, B. R. (2012). Efficient and unbiased sampling of biomolecular systems in the canonical ensemble: A review of self-guided Langevin dynamics. *Advances in Chemical Physics, 150*, 255–326.

Xie, W., Orozco, M., Truhlar, D. G., & Gao, J. (2009). X-Pol potential: An electronic structure-based force field for molecular dynamics simulation of a solvated protein in water. *Journal of Chemical Theory and Computation, 5*, 459–467.

Yonetani, Y., Maruyama, Y., Hirata, F., & Kono, H. (2008). Comparison of DNA hydration patterns obtained using two distinct computational methods, molecular dynamics simulation and three-dimensional reference interaction site model theory. *Journal of Chemical Physics, 128*, 185102.

Yoo, J., & Aksimentiev, A. (2012). Competitive binding of cations to duplex DNA revealed through molecular dynamics simulations. *Journal of Physical Chemistry. B, 116*(43), 12946–12954.

York, D. M., & Lee, T.-S. (Eds.), (2009). *Multiscale quantum models for biocatalysis: Modern techniques and applications*. New York: Springer.

Zgarbová, M., Otyepka, M., Šponer, J., Mládek, A., Banáš, P., Cheatham, T. E., III, & Jurečka, P. (2011). Refinement of the Cornell et al. nucleic acids force field based on reference quantum chemical calculations of glycosidic torsion profiles. *Journal of Chemical Theory and Computation*, 7, 2886–2902.

Zhang, S., Gu, H., Chen, H., Strong, E., Liang, D., Dai, Q., Harris, M. E. (n.d.). An associative two metal ion mechanism for non-enzymatic RNA 2′-O-transphosphorylation. Nature Chemistry, submitted.

Zhang, Y., Lee, T.-S., & Yang, W. (1999). A pseudobond approach to combining quantum mechanical and molecular mechanical methods. *Journal of Chemical Physics, 110*, 46–54.

Zuckerman, D. M. (2011). Equilibrium sampling in biomolecular simulations. *Annual Review of Biophysics, 40*, 41–62.

Zwier, M. C., & Chong, L. T. (2010). Reaching biological timescales with all-atom molecular dynamics simulations. *Current Opinion in Pharmacology, 10*, 745–752.

AUTHOR INDEX

Note: Page numbers followed by "*f*" indicate figures and "*t*" indicate tables and "*np*" indicate notes.

I

J

K

N

O

P

Q

R

S

V

W

SUBJECT INDEX

Note: Page numbers followed by "*f*" indicate figures and "*t*" indicate tables.

S

T

W

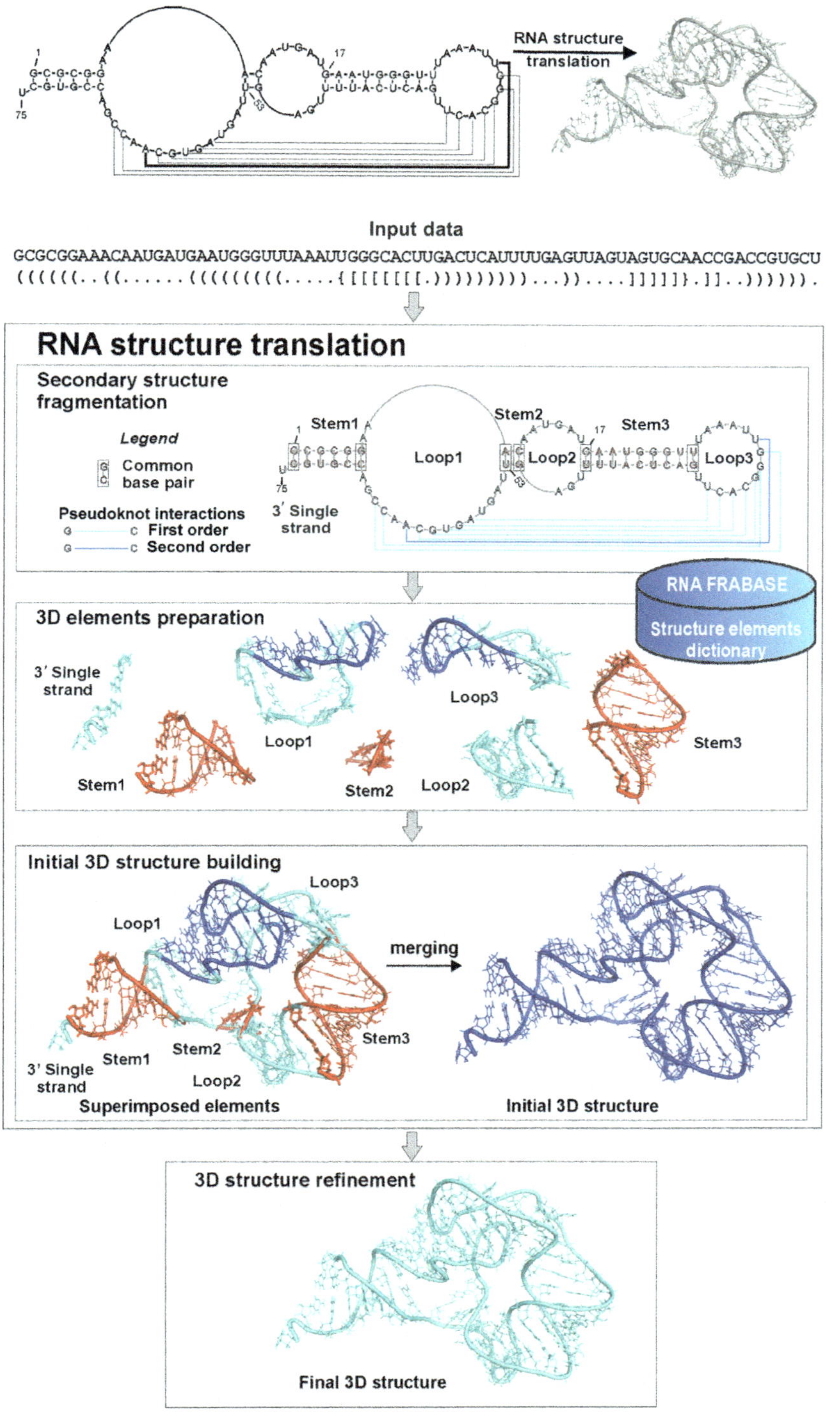

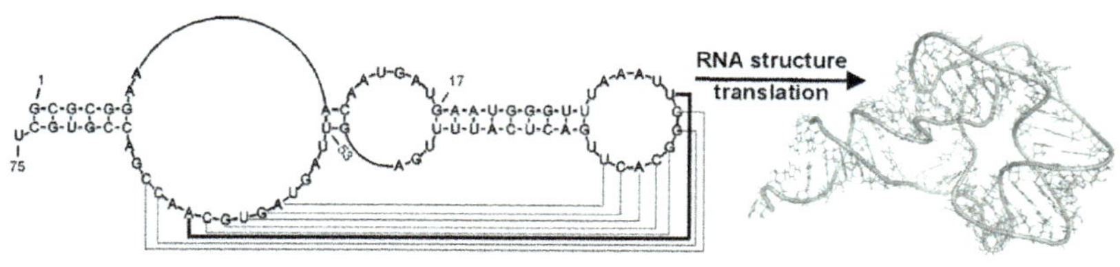

Input data

GCGCGGAAACAAUGAUGAAUGGGUUUAAAUUGGGCACUUGACUCAUUUUGAGUUAGUAGUGCAACCGACCGUGCU

((((((..((......(((((((((.....{[[[[[[[.))))))))))...))....]]]]]}.]]..))))))).

K.J. Purzycka *et al.*, Figure 1 The RNA secondary structure to the 3D structure translation. Basic steps of RNAComposer action exemplified on the cyclic di-GMP-II riboswitch structure from *C. acetobutylicum* (PDB ID 3Q3Z).

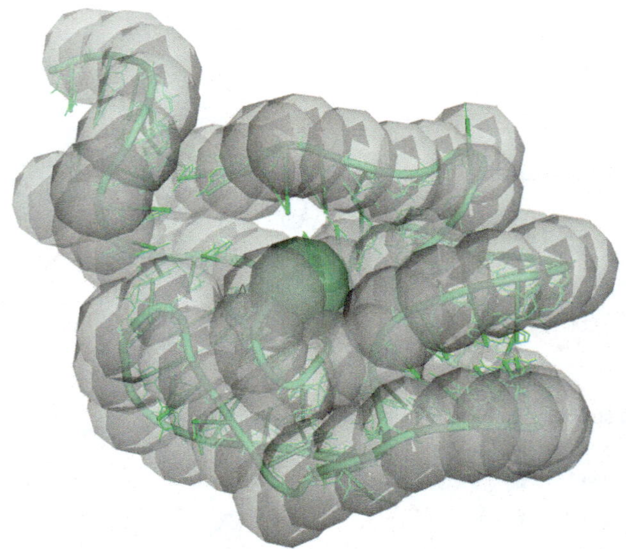

K.J. Purzycka *et al.*, Figure 2 RNAlyzer spheres defining range of the atoms in superimposition of 3D model and reference structure, as probing structural space of every nucleotide residue.

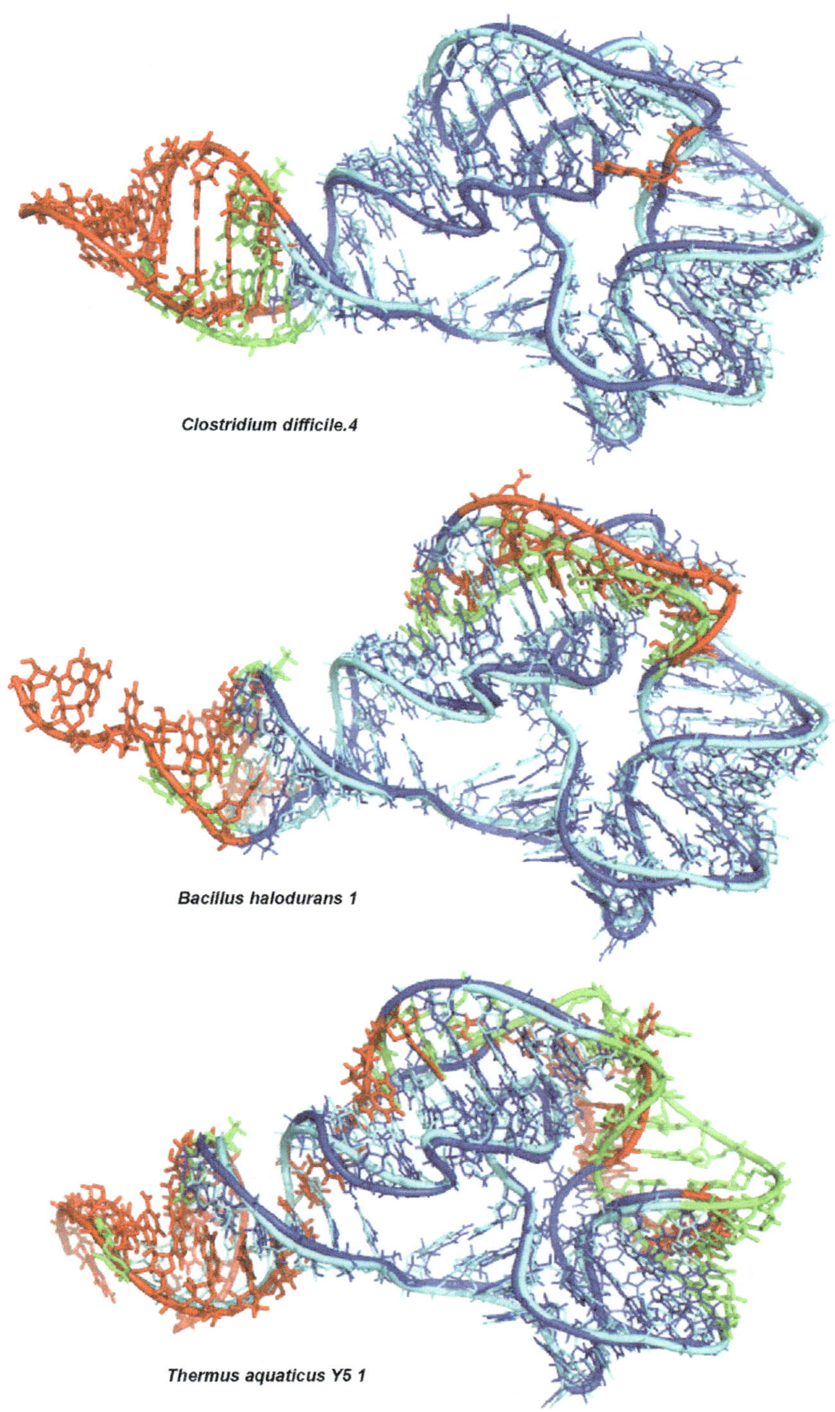

K.J. Purzycka *et al.*, Figure 5 Pairwise superimposition of the best 3D models of the cyclic di-GMP-II riboswitch from *C. difficile 4*, *B. halodurans 1*, and *T. aquaticus Y5.1* (red and blue-core), and the X-ray structure from *C. acetobutylicum* (PDB ID 3Q3Z) (green and cyan-core).

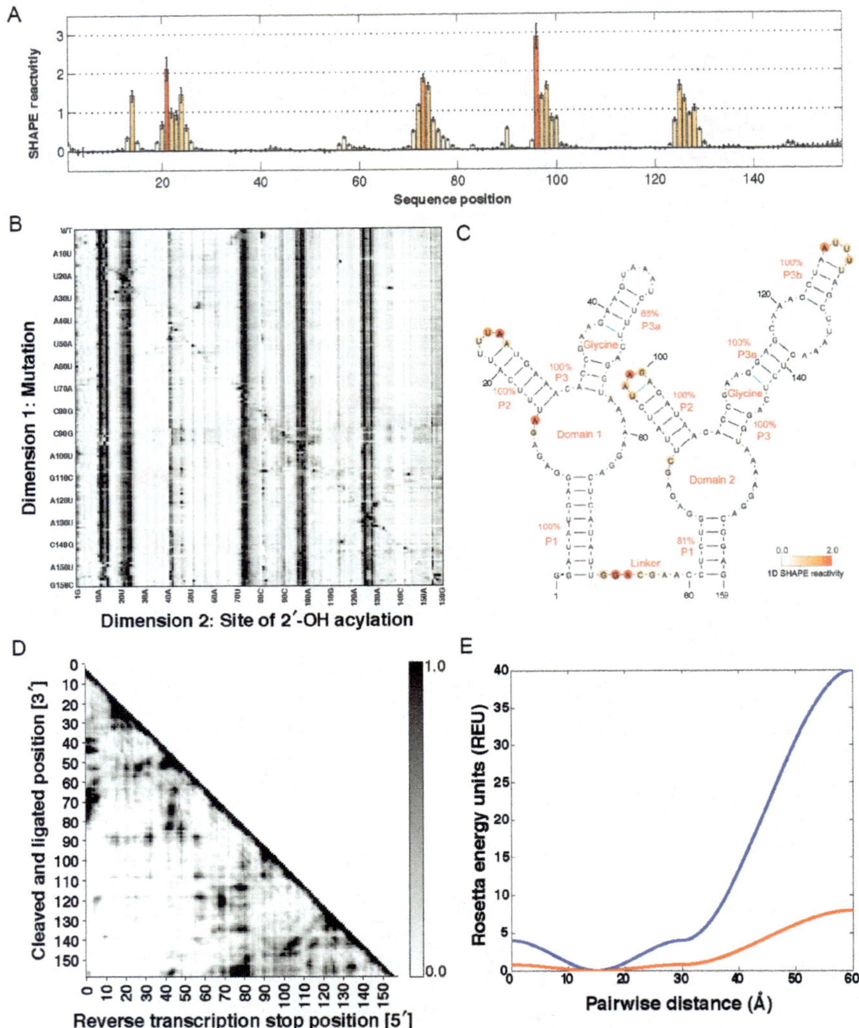

Clarence Yu Cheng *et al.*, Figure 2 Rapidly acquired chemical mapping data for modeling a complex RNA fold. (A) One-dimensional SHAPE chemical mapping data for the *F. nucleatum* glycine riboswitch double ligand-binding domain in the presence of 10 m*M* glycine. Reactivities are normalized to reference hairpins (not shown) (Kladwang et al., 2014). Data are available at the RNA Mapping Database (RMDB, http://rmdb.stanford.edu) under accession code GLYCFN_1M7_0005. (B) Mutate-and-map (M^2) chemical mapping data for the glycine riboswitch in the presence of 10 m*M* glycine. Data are available at the RMDB under accession code GLYCFN_SHP_0002. (C) M^2-derived secondary structure model of the glycine riboswitch in the presence of 10 m*M* glycine, from Kladwang et al. (2011). Blue lines indicate Watson–Crick base pairs predicted in the model but not present in the crystallographic secondary structure. Red percentage values for each helix indicate confidence estimates from bootstrapping two-dimensional SHAPE chemical mapping data.

(Continued)

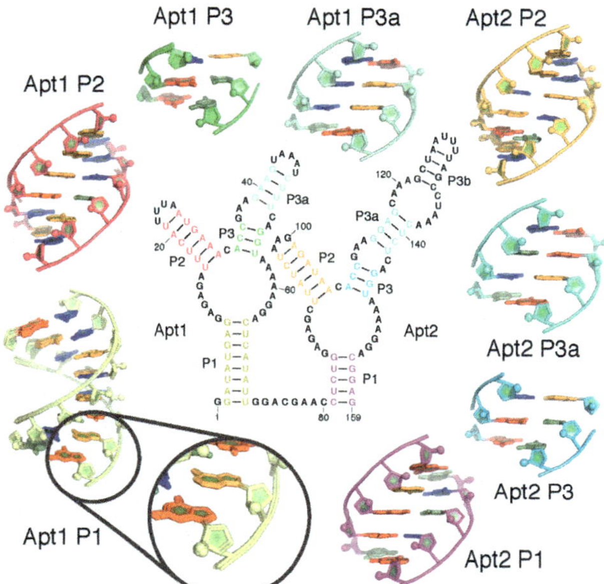

Clarence Yu Cheng *et al.*, Figure 3 Preassembled helices for *F. nucleatum* double glycine riboswitch ligand-binding domain. The secondary structure is shown at center with the residues used for helix preassembly highlighted in color. Ensembles of 10 models of each helix generated by the helix preassembly protocol in Rosetta are shown at the periphery, labeled with the aptamer and helix number (e.g., Apt1 P1 for the P1 helix of aptamer 1). The magnified view of the Apt1 P1 helix highlights the slight differences in conformation between the preassembled helix models.

Figure 2—Cont'd Nucleotides are colored according to SHAPE reactivity. (D) MOHCA-seq proximity map of the glycine riboswitch in the presence of 10 m*M* glycine, from Cheng et al. (2014). The *y*-axis represents positions that were cleaved by hydroxyl radicals, while the *x*-axis represents the locations of the radical sources from which the radicals originated. Pairwise positions are colored according to two-point correlation calculated by MAPseeker analysis (Seetin, Kladwang, Bida, & Das, 2014). Data are available at the RMDB under accession code GLYCFN_MCA_0000. (E) Pseudoenergy potential applied during modeling in Rosetta to constrain pairs of residues indicated to be in proximity by MOHCA-seq experimental data. Residue pairs showing strong MOHCA-seq signal are constrained with the blue potential and those with weaker signal are constrained with the red potential (1/5 of the blue potential).

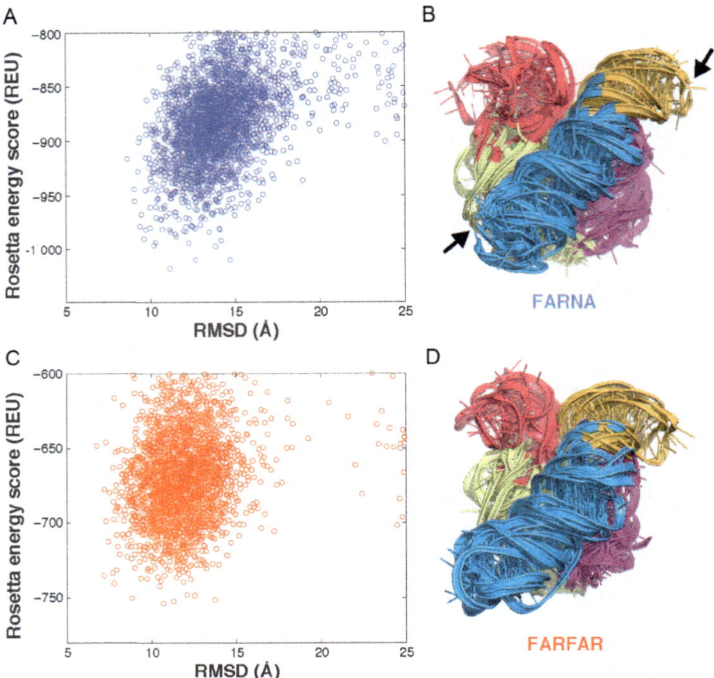

Clarence Yu Cheng *et al.*, Figure 4 Low-resolution modeling and full-atom refinement using FARNA and FARFAR. (A) Rosetta energy score versus RMSD plot after low-resolution modeling using FARNA. (B) Overlaid 10 lowest-energy models after low-resolution modeling using FARNA. Chain breaks are visible in many models (arrows), and residues commonly adopt unrealistic geometries. (C) Rosetta energy score versus RMSD plot after minimization using the FARFAR algorithm. (D) Overlaid 10 lowest-energy models after minimization using the FARFAR algorithm. The models do not show any chain breaks, and poor residue geometries are greatly reduced.

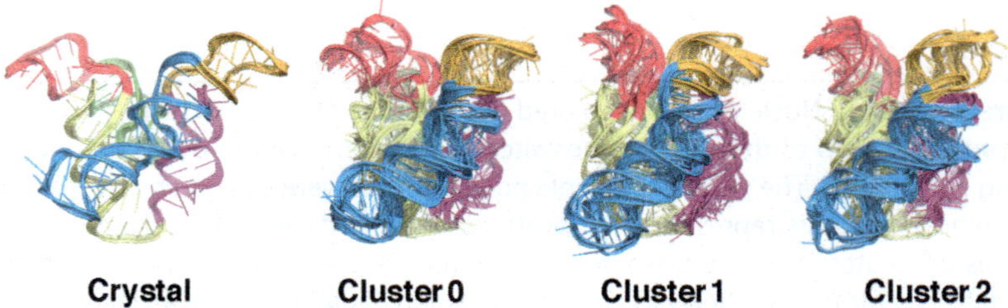

Clarence Yu Cheng *et al.*, Figure 5 Clustering of minimized models to select representative models. Comparison of models generated by the experimental/computational pipeline. The crystal structure (PDB ID 3P49) is shown at left. At right, four representative models are overlaid for each of the top three model clusters. The cluster center model of cluster0 has a 7.9 Å RMSD to the crystal structure.

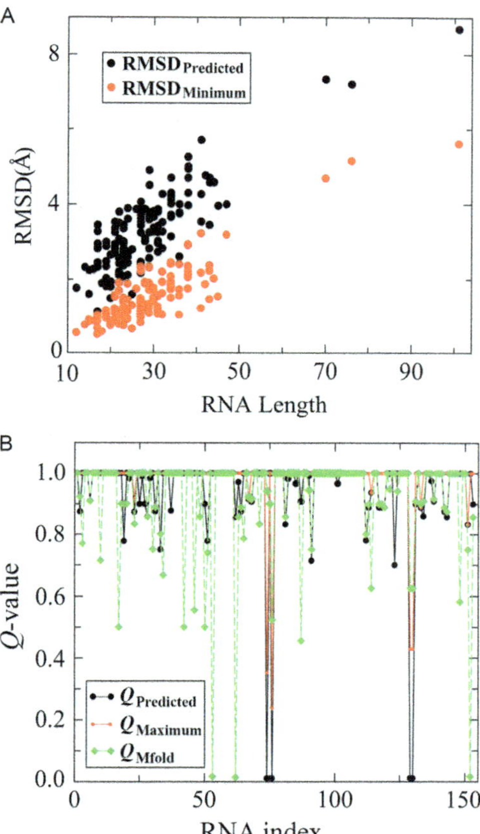

Andrey Krokhotin and Nikolay V. Dokholyan, Figure 3 (A) RMSD calculated between RNA structure obtained in the course of DMD simulations and experimental structure from the PDB. Black points correspond to the predicted structures with the lowest free energy and red points correspond to structures with minimum RMSD. (B) *Q*-values for the predicted structures with the lowest free energy (black dots), the maximum *Q*-values (red dots) and structures obtained using Mfold (green dots). *Adapted from Ding et al. (2008) with permission.*

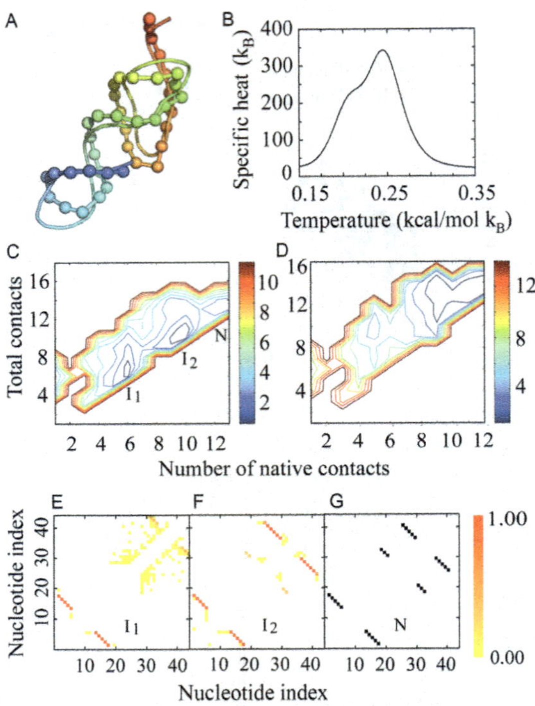

Andrey Krokhotin and Nikolay V. Dokholyan, Figure 4 Folding kinetics of pseudoknot RNA. (A) Structure obtained as a result of DMD simulation (ribbon backbone trace with backbone spheres) superimposed over experimentally determined RNA structure (PDB code: 1A60). Backbone ribbons are colored from blue (N terminus) to red (C-terminus). (B) Specific heat as a function of temperature. (C) Two-dimensional potential of mean force (2D-PMF) for pseudoknot folding at $T^* = 0.245$ (the major peak in the specific heat). I_1, I_2, and N denote two intermediate states and native state correspondingly. (D) 2D-PMF at $T^* = 0.21$. (E) Contact frequencies at the I_1 intermediate state, corresponding to the formation of 5′ hairpin. (F) Contact frequencies at the I_2 intermediate state, corresponding to the formation of the helix stem in the 3′ pseudoknot. (G) Contact map of the native state as observed in the experimental NMR structure (PDB code: 1A60). *Adapted from Ding et al. (2008) with permission.*

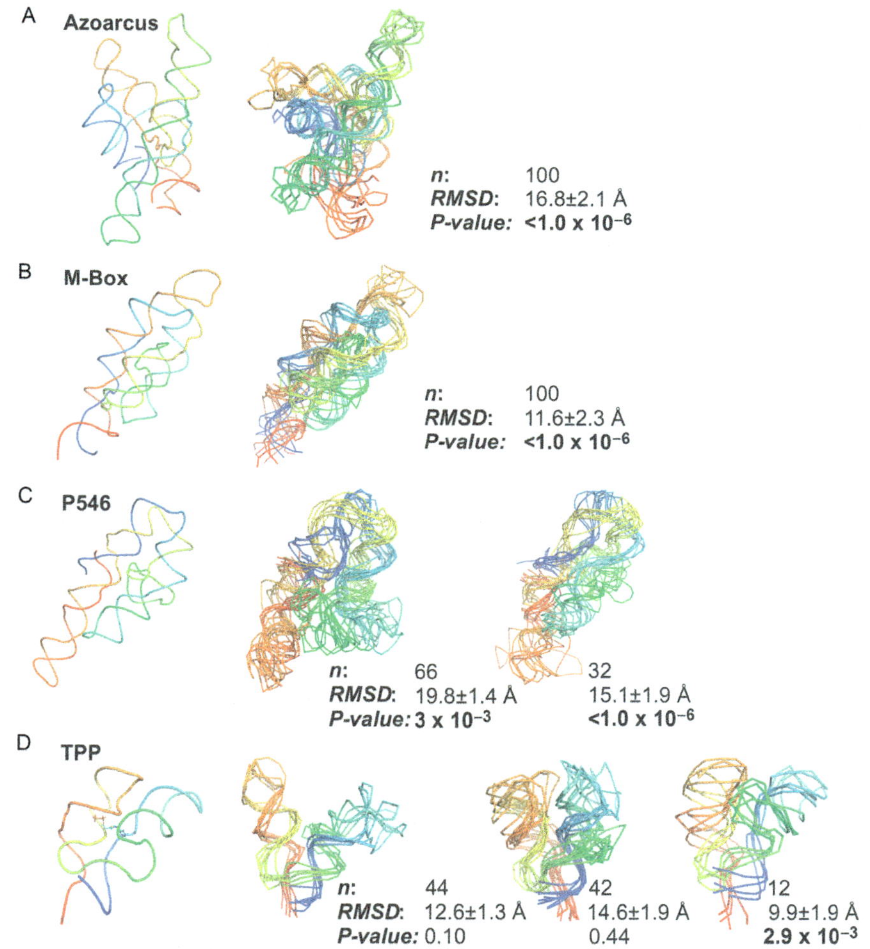

Andrey Krokhotin and Nikolay V. Dokholyan, Figure 9 Predicted structures of RNAs from the training set. RNAs are shown with backbone traces. The accepted structures are shown on the left. Representative structures for each of the highly populated cluster are shown on the right. Backbones are colored from blue to red in the direction from 5′ to 3′. The mean RMSD and *P*-values are shown, with significant *P*-values emphasized in bold. *Adapted from Ding et al. (2012) with permission.*

A

Probability >= 99%

99% > Probability >= 95%

95% > Probability >= 90%

90% > Probability >= 80%

80% > Probability >= 70%

70% > Probability >= 60%

60% > Probability >= 50%

50% > Probability

ENERGY = −51.5 07/31/14 01:02:24

Michael F. Sloma and David H. Mathews, Figure 2 Predicted *E. coli* 5S rRNA secondary structure without SHAPE data. Panel (A) shows the predicted structure, as displayed by the RNAstructure predict a secondary structure web server. The structure is color-annotated by estimated base-pairing probabilities.

(Continued)

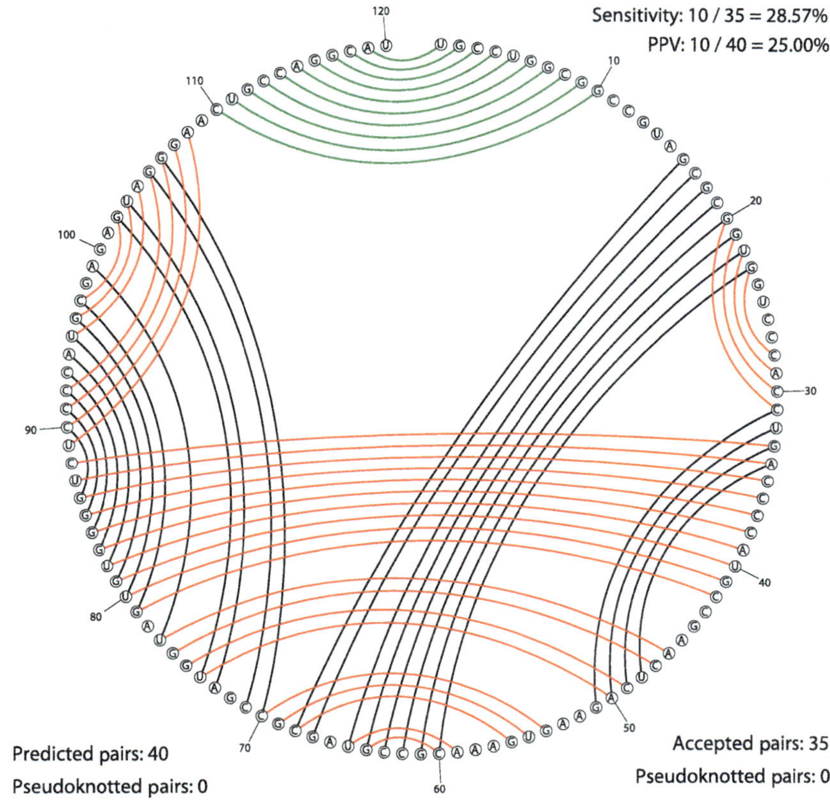

Pair present in both predicted and accepted structure (Green).

Pair present in predicted structure only (Red).

Pair present in accepted structure only (Black).

Figure 2—Cont'd Panel (B) shows a CircleCompare plot from the RNAstructure CircleCompare web server, which compares the predicted structure to the accepted structure. The sequence is drawn around a circle and lines indicate base pairs. Pairs present in both the predicted and accepted structure are green; pairs in the predicted structure only are red; and pairs in the accepted structure only are black. The upper-left shows that the sensitivity (fraction of accepted pairs correctly predicted) is 29% and the PPV (fraction of predicted pairs in the accepted structure) is 25%.

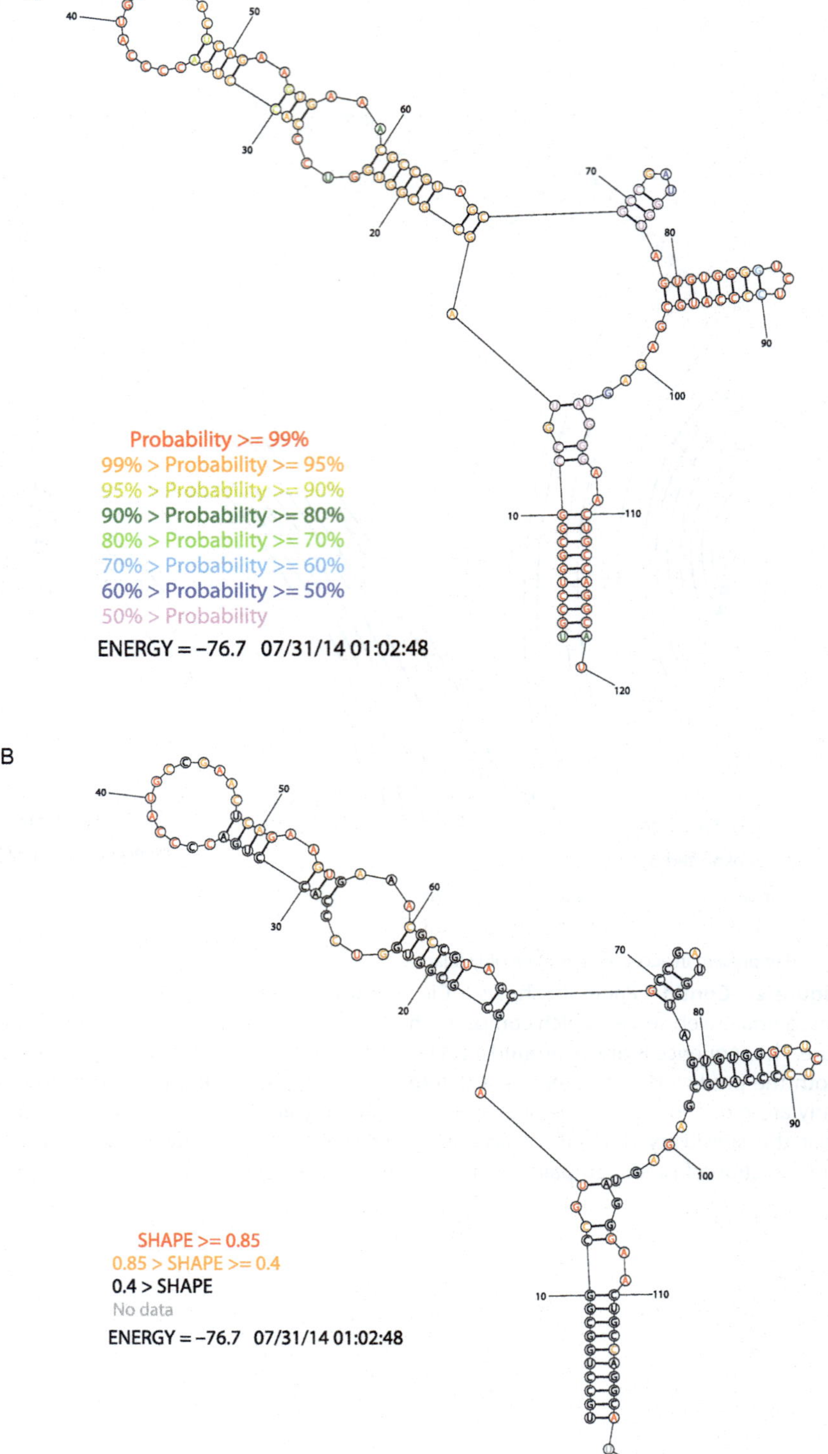

Figure 3 See legend on opposite page.

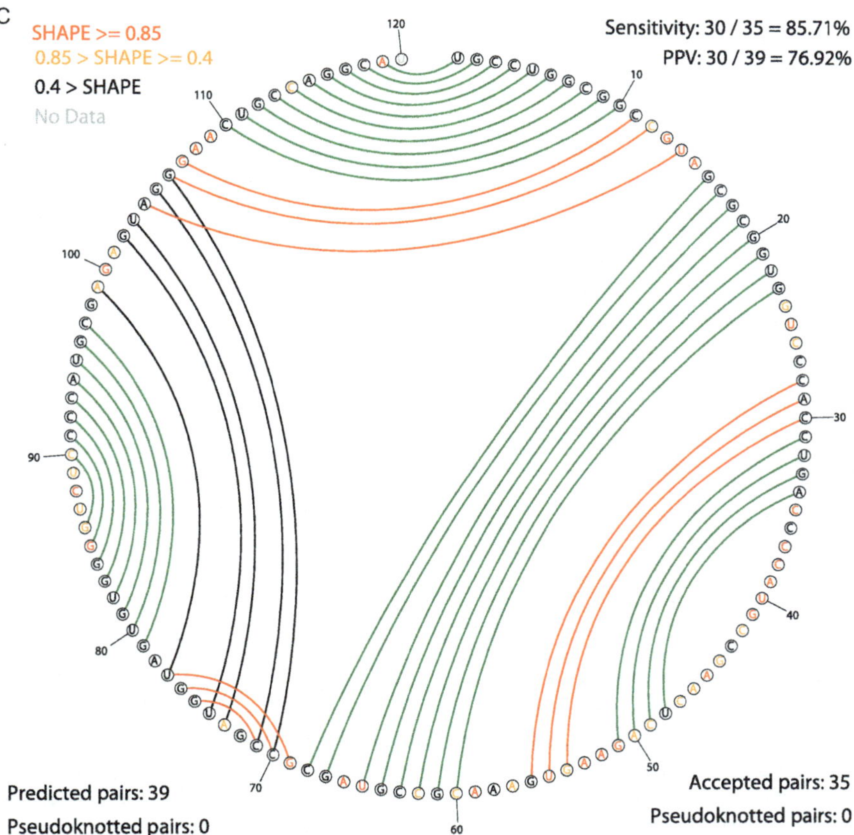

Pair present in both predicted and accepted structure (Green).

Pair present in predicted structure only (Red).

Pair present in accepted structure only (Black).

Michael F. Sloma and David H. Mathews, Figure 3 Predicted *E. coli* 5S rRNA secondary structure with restraints from SHAPE data. Panel (A) shows the predicted structure, as displayed by the RNAstructure predict a secondary structure web server. The structure is color-annotated by estimated base-pairing probabilities. Panel (B) shows the predicted structure with color-annotation by normalized SHAPE reactivities. These plots are both generated by default. Panel (C) shows a CircleCompare plot from the RNAstructure CircleCompare web server, which compares the predicted structure to the accepted structure. The sequence is drawn around a circle and lines indicate base pairs. Pairs present in both predicted and accepted structure are green; pairs in the predicted structure only are red; and pairs in the accepted structure only are black. The upper-left shows that the sensitivity (fraction of accepted pairs correctly predicted) is 86% and the PPV (fraction of predicted pairs in the accepted structure) is 77%. The sequence is color-annotated by SHAPE normalized SHAPE reactivities.

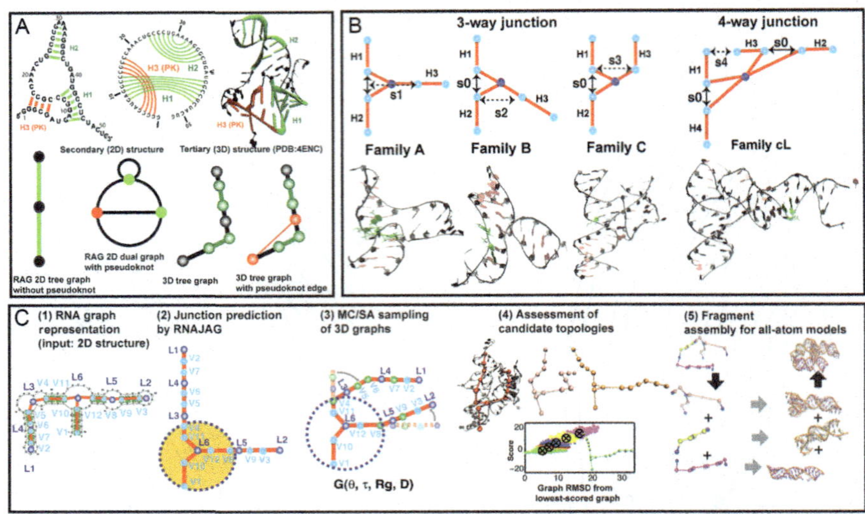

Namhee Kim *et al.*, Figure 2 Our hierarchical graph folding approach. (A) RNA-As-Graph (RAG) representation (e.g., fluoride riboswitch, PDB entry 4ENC): RNA 2D and 3D structure including a pseudoknot structure, with corresponding 2D tree graph, dual graph, 3D tree, and refined 3D tree graph; the pseudoknot structure is represented by an additional edge in the 3D tree graph (red). (B) Junction families of 3- and 4-way junctions in riboswitches (family types A, B, C for 3-way junction and cL for 4-way junction) where the helical arm has different helical arrangements (perpendicular, diagonal, or parallel) with respect to the coaxially stacked helices. (C) Our hierarchical graph folding approach RAGTOP, from 2D graph to all-atom models (e.g., TPP riboswitch, PDB entry 2GDI): (1) given a 2D structure, a 2D tree topology is annotated. (2) An initial planar graph is constructed by junction prediction by RNAJAG and edge size estimation. (3) Graphs are sampled by Monte Carlo sampling with two moves (random or restricted) and guided by knowledge-based statistical potentials for bending and torsion angles of internal loops, radii of gyration, and size of pseudoknot edge. (4) Candidate graphs are assessed by the lowest RMSD (P1), the lowest score (P2, for restricted moves), or the lowest cluster representatives (P3, for random moves), and compared with reference graphs translated from solved riboswitches based on local and global RMSD measures. (5) All-atom models are constructed by graph partitioning, fragment search, and assembly of corresponding all-atom modules in RAG-3D. See details in Kim, Laing, et al. (2014) and Laing et al. (2013).

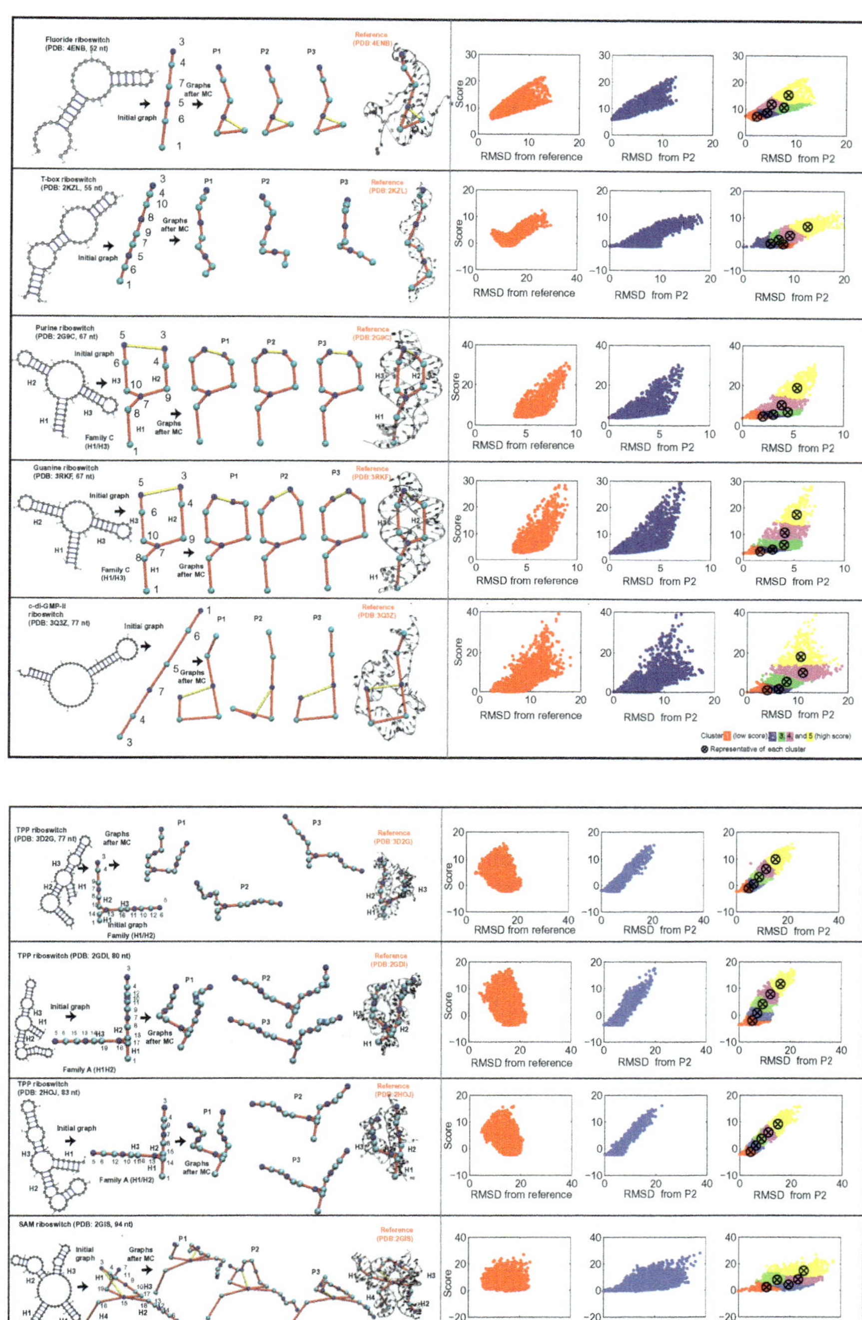

Namhee Kim *et al.*, Figure 3 See legend on the next page.

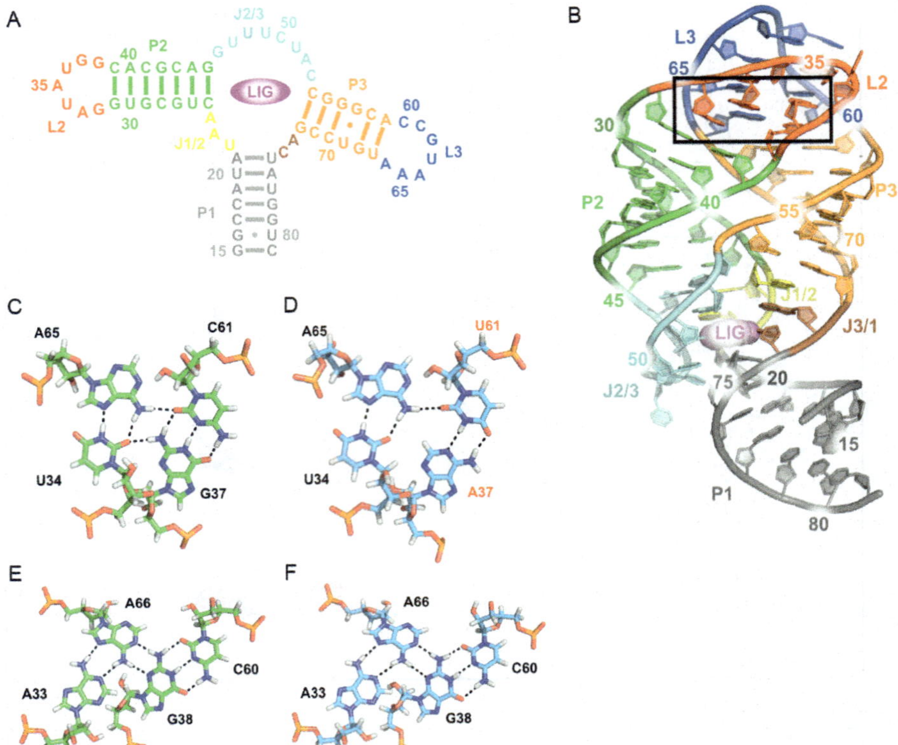

Christian A. Hanke and Holger Gohlke, Figure 1 (A) Secondary structure and sequence used in this study of the aptamer domain of Gsw^{apt}; the secondary structure elements are assigned according to Batey et al. (2004). In addition, the ligand-binding site is indicated by the purple oval and denoted by LIG. (B) Tertiary structure of the aptamer domain of the Gsw bound to hypoxanthine (LIG; PDB ID 4FE5, Stoddard et al., 2013). The secondary structure elements are assigned as in (A). The black rectangle indicates the location of the base quadruples. (C and D) Upper base quadruple formed by the L2 and L3 loops for Gsw^{apt} (C) and Gsw^{loop} (D). Here, the G37A/C61U mutation is located (highlighted by the red labels). (E and F) Lower base quadruple formed by the L2 and L3 loops for Gsw^{apt} (E) and Gsw^{loop} (F).

Namhee Kim *et al.,* Figure 3 Graph results for riboswitch structures—best graph with the lowest RMSD from reference graph based on random pivot moves (P1), the lowest-scored graph based on restricted pivot moves (P2), and the lowest cluster representative of landscapes with respect to the lowest-scored graph based on random pivot moves (P3). Pseudoknot edges in the graphs are indicated in yellow. Landscapes with respect to reference structure and the lowest-scored graph based on restricted and random pivot moves are also shown.

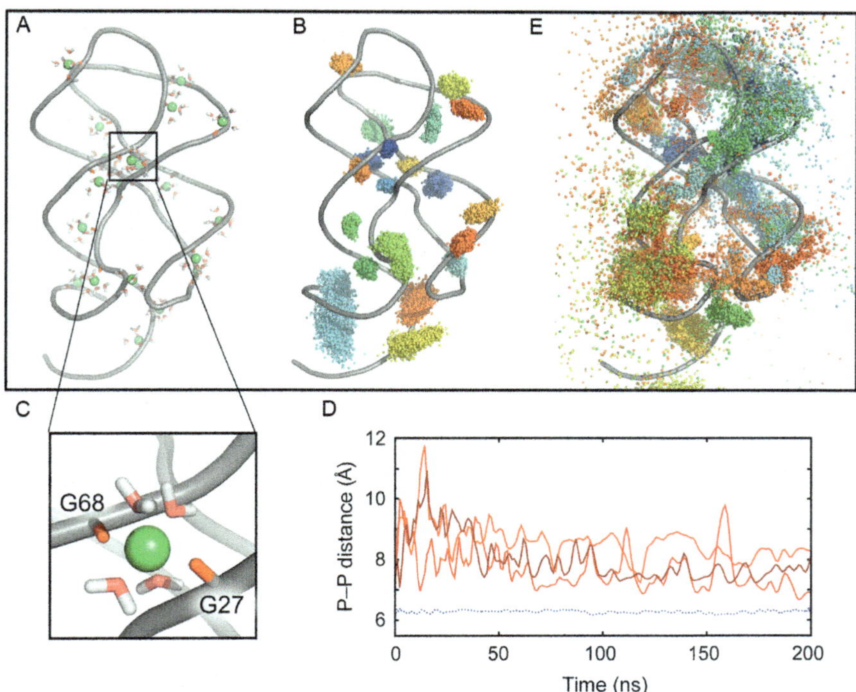

Christian A. Hanke and Holger Gohlke, Figure 2 (A) Initial placement of 20 Mg^{2+} ions (green spheres) and surrounding water molecules (sticks) around the structure of the aptamer domain of the Gsw (with the backbone displayed as gray line). The ions were placed prior to the water molecules at electrostatically favorable locations as implemented in *leap* (Case et al., 2005) using the default procedure and are thus located very closely to the RNA backbone. (B and E) Positions of 20 Mg^{2+} ions during 100 ns of MD simulation. Different colors represent different Mg^{2+} ions. The ions were initially placed by *leap* either using the default procedure (B) or using the improved placement procedure (see Section 2) considering a first hydration shell (E). (C) Close up of a Mg^{2+} ion connecting the OP1 atoms of nucleotides G27 and G68 (red sticks connected to the RNA backbone in dark gray). Four water molecules (red and white sticks) form the hydration shell of the Mg^{2+} ion. (D) Distance calculated between the P atoms of nucleotides G27 and G68 shown in (C) during MD simulations of 200 ns length. The blue dotted line represents the distance in a simulation with the default placement of the ions; the three red solid lines, which were smoothed for better visibility, represent the distances in simulations with the improved placement procedure.

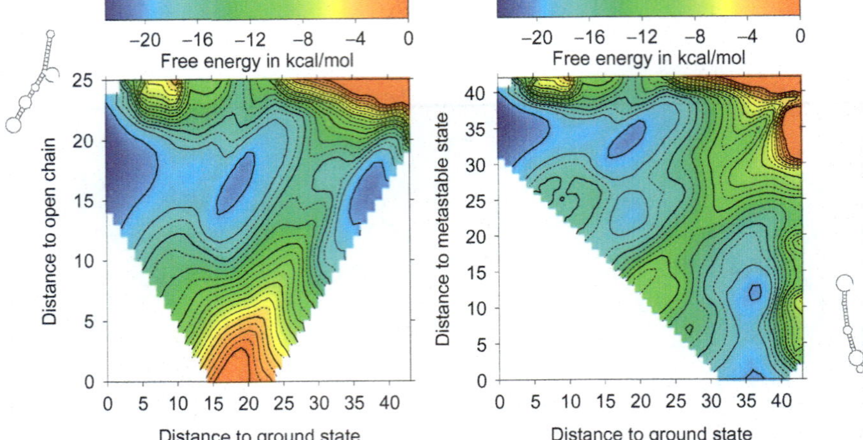

Stefan Badelt *et al.*, Figure 2 `RNA2Dfold` computed projection of the energy landscape for the leader sequence of the *E. coli* tRNA^{phe} synthetase operon. Left panel: Projection using the MFE structure (terminator hairpin, far left) and the open chain as references. Right panel: Projection using the MFE structure and the metastable structure found at position (36,17) of the first projection. The two structures are shown on the left and right of the landscapes.

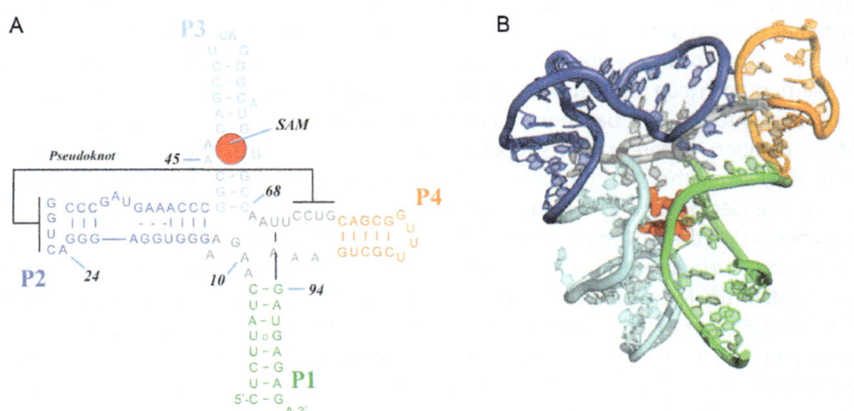

Serdal Kirmizialtin *et al.*, Figure 2 The *T. tengcongensis metF* SAM-I riboswitch aptamer domain in the off state. (A) Secondary structure of the aptamer domain with different colors representing secondary structure elements. (B) Tertiary structure of the sequence in the presence of *S*-adenosylmethionine (SAM) ligand.

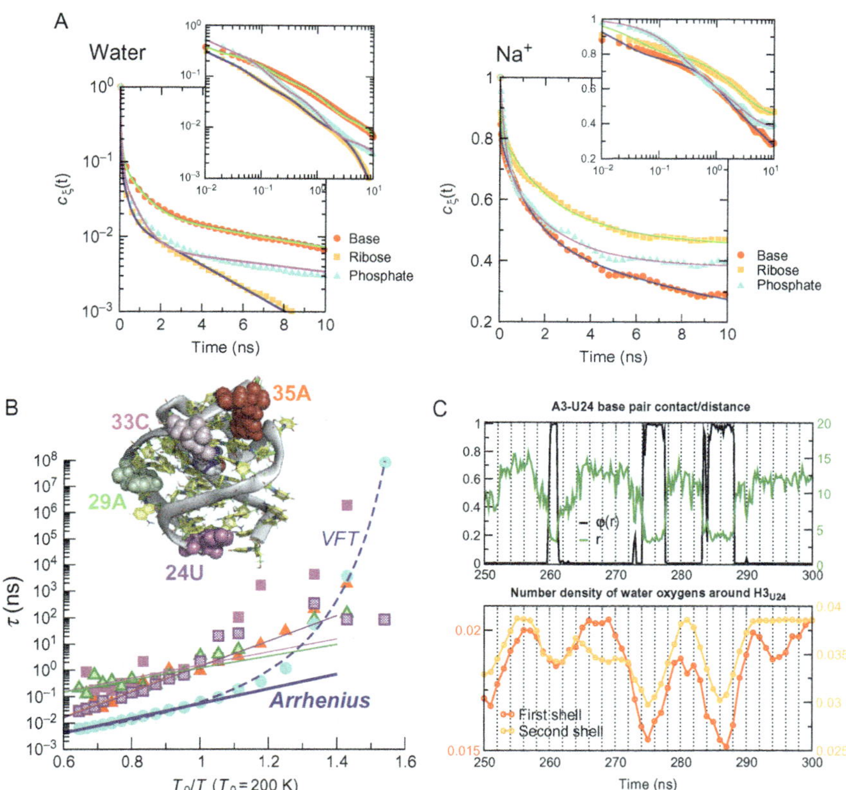

Jong-Chin Lin *et al.*, Figure 2 (A) The data for the dynamics of water HBs (left) and Na^+ (right) with base, ribose, and base (left) are described by the appropriate autocorrelation functions. (B) Temperature dependence of water HB relaxation time of bulk water and around four nucleotides, 24U, 29A, 33C, and 35A or $PreQ_1$. Arrhenius fits are made to $T > 200$ K for the four nucleotide and bulk water. Alternatively, the relaxation dynamics of bulk water HB can be fit using the Vogel–Fulcher–Tamman (VFT) equation over the full range of temperatures (dashed line) (Yoon et al., 2014). (C) Water dynamics induced fluctuations of a base pair. The status of base pair ($N1_{A3}$–$N3_{U24}$), quantified by calculating logistic function $\varphi(r) = (1 + e^{(r-r_0)/\sigma})^{-1}$ as well as distance r, shows apparent fluctuations, whose time scale is ~ 10 ns. *Figures adapted from Yoon et al. (2014, 2013).*

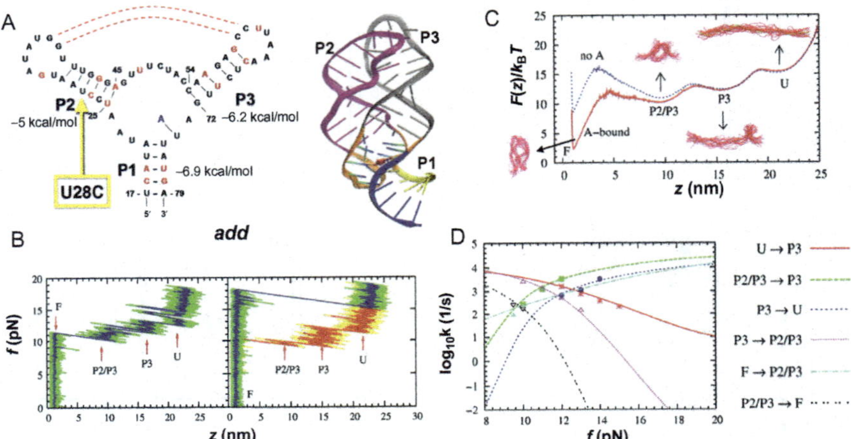

Jong-Chin Lin *et al.*, Figure 3 Force-induced dynamics of add A-riboswitch (RS). (A) Structure of the conserved domain of purine riboswitch containg a three way junction. On the left is the secondary structure map and on the right the three dimensional structure is shown. (B) Force-extension curves (FECs) obtained by pulling the RS at loading rate of 960 pN/s in the presence (left) and absence (right) of metabolite. The FEC in red on the right panel was obtained during the refolding of the RS while the exerted force is reduced. (C) Free energy profile *F*(*z*) with (red) and without (blue) the metabolite. (D) Force-dependent transition rates. The data points are directly from simulation; the lines were obtained by calculating mean first passage time using *F*(*z*) with a force-independent diffusion constant, which was calibrated by equating the theoretical and simulation rates. *Figure adapted from J. Lin and Thirumalai (2008).*

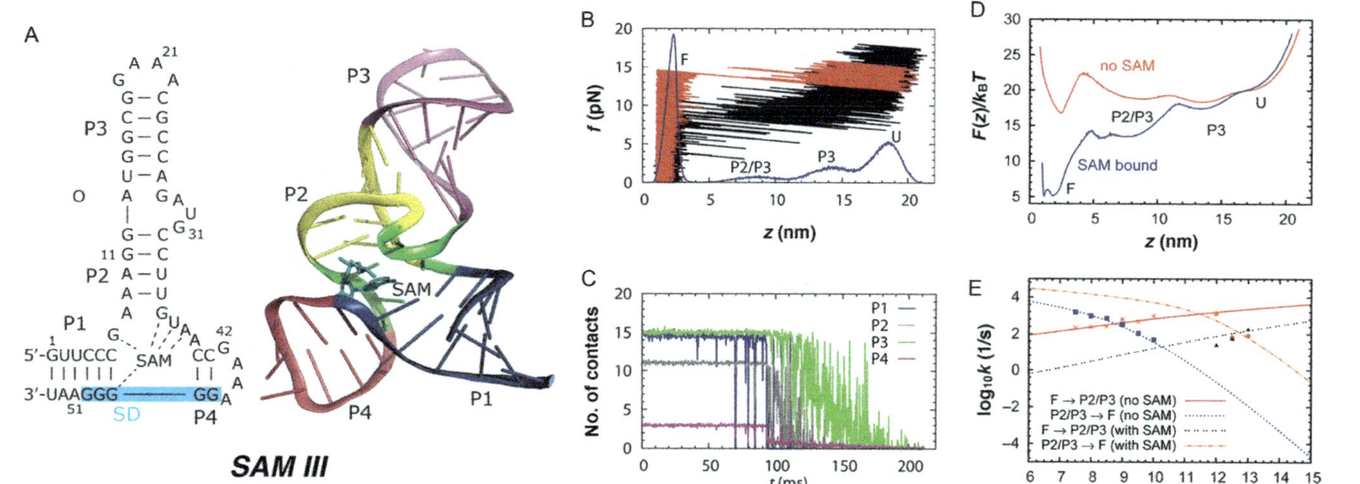

Jong-Chin Lin *et al.*, Figure 5 Dynamics of SAM-III riboswitch under force. (A) Structure of SAM-III RS. The blue shaded area on the left indicates the Shine–Dalgarno sequence recognized by the ribosome. (B) Simulated force-extension curve of SAM-III riboswitch in the absence of metabolite (black) produced at $r_f = 96$ pN/s. The distribution of molecular extension (z) during the pulling simulation is shown in blue at the bottom. FEC in red was produced in the presence of metabolite at the binding pocket. (C) Average number of contacts in each helix from P1 to P4. (D) Free energy profile at zero force calculated from streching simulation with and without metabolite (SAM) in the binding pocket. (E) Transition rates between F and P2/P3 states at varying forces. The data points are from explicit simulations. The lines were obtained from mean first passage time calculation on $F(z)$. *Figure adapted from J.-C. Lin and Thirumalai (2013).*

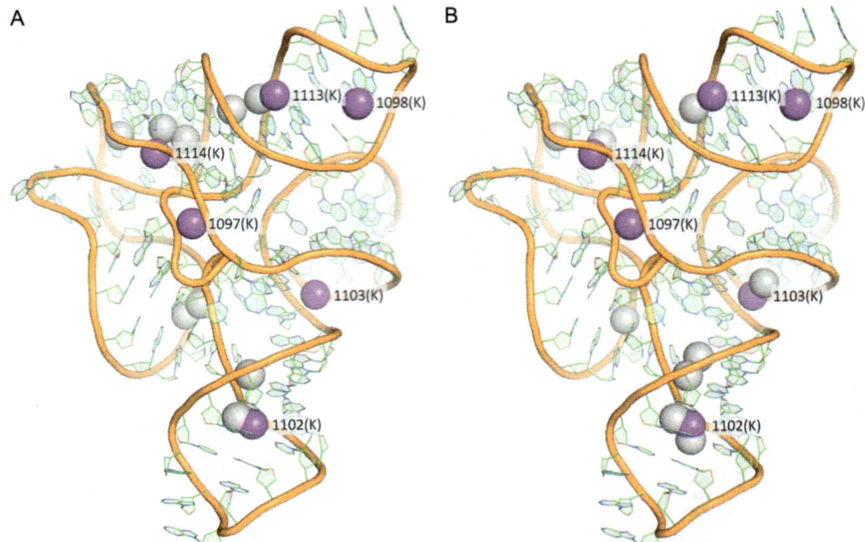

Anna Philips *et al.*, Figure 3 SAM-I riboswitch structure (PDB ID: 4B5R) with the experimentally determined positions of K^+ cations indicated by purple labeled balls. The top-scoring K^+ cations predicted by MetalionRNA are shown as gray balls (A) using the original potential and (B) using the updated potential; for more details, see Table 2.

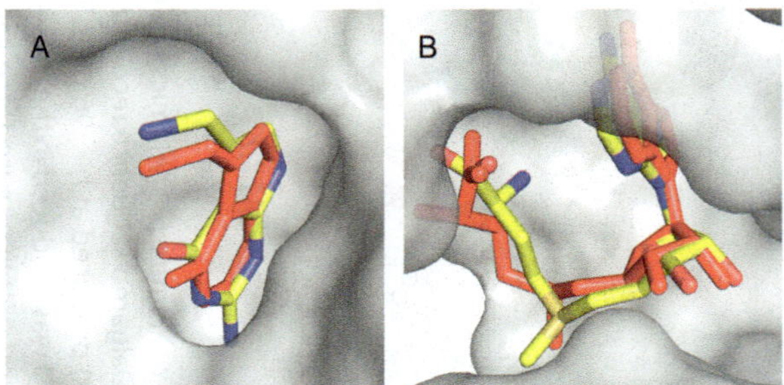

Anna Philips *et al.*, Figure 4 (A) Class II preQ1 riboswitch (PDB ID: 2MIY) in complex with 7-deaza-7-aminomethyl-guanine (PRF) and (B) SAM-I riboswitches (PDB ID: 4KQY) in complex with s-adenosylmethionine (SAM). The experimentally identified ligand poses are shown in light yellow, and the poses from computational docking identified as best-scored by LigandRNA are depicted in red.

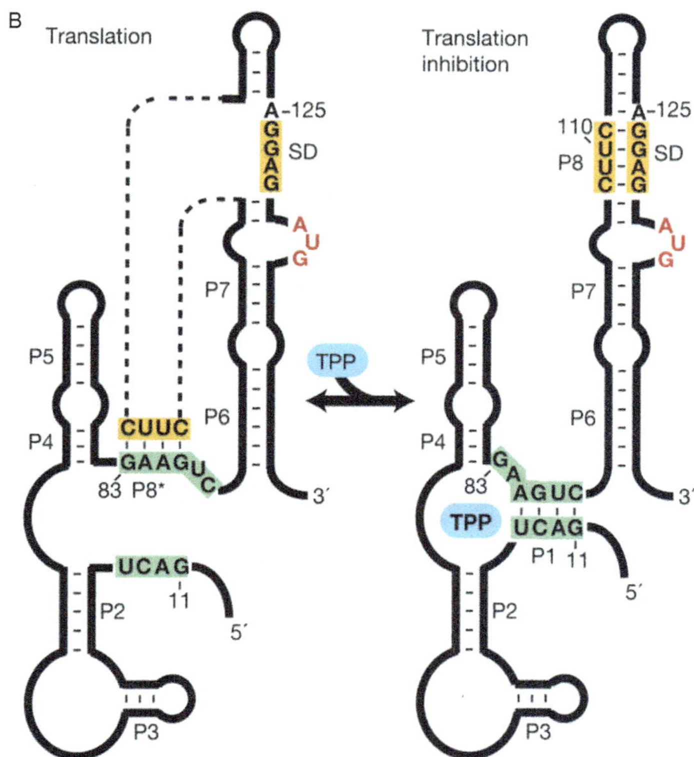

P. Clote, Figure 1 (B) Gene ON and gene OFF structure for the TPP riboswitch from *E. coli*, as determined by Winkler et al. (2002) and Serganov, Polonskaia, Phan, Breaker, and Patel (2006). Note that the Shine–Dalgarno ribosome-binding sequence is sequestered in the gene OFF structure. Panel (B) is a reproduction of Figure 5 of Winkler et al. (2002), reprinted by permission from Macmillan Publishers Ltd.: Nature **419**, 31 October 2002.

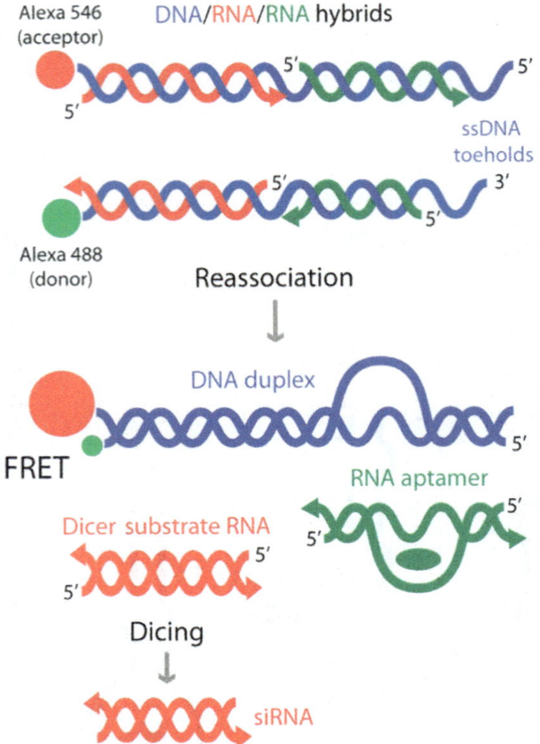

Kirill A. Afonin *et al.*, Figure 1 Schematic representation of reassociation for RNA/DNA hybrids carrying multiple split functionalities (FRET, Dicer Substrate RNA, and RNA aptamer such as malachite green aptamer).

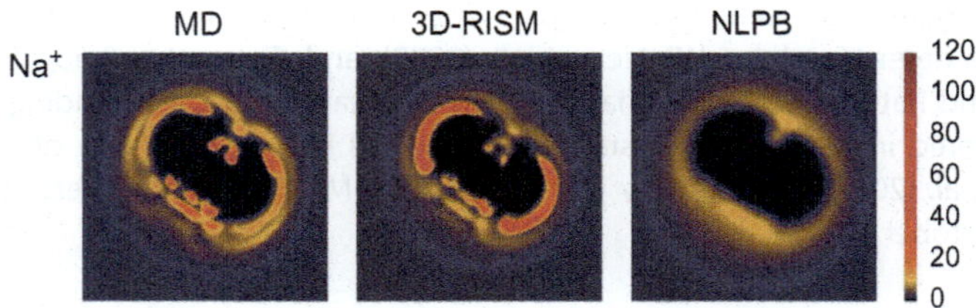

Maria T. Panteva *et al.*, Figure 3 Comparison of ion distributions around a 24-mer of duplex B-DNA from MD simulation, 3D-RISM, and conventional NLPB. Distributions are shown along a rotating "untwisted" coordinate frame along the DNA axis as described in Giambaşu et al. (2014). Shown are the untwisted Na^+ densities from MD, 3D-RISM, and NLPB for 0.17 *M* bulk NaCl concentration. MD and 3D-RISM predict a layered Na^+ density, whereas NLPB is unstructured.

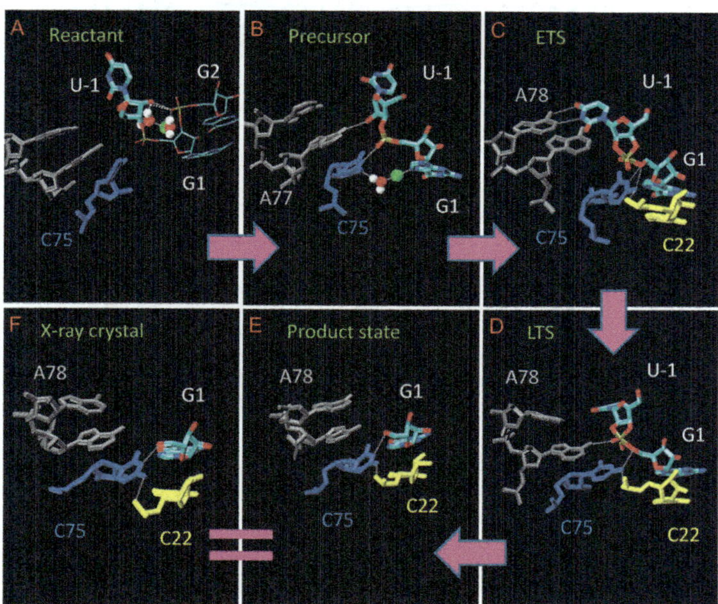

Maria T. Panteva *et al.*, Figure 7 Representative snapshots taken from simulations of HDVr along the reaction path starting from an inactive C75U mutant crystal structure and compared with the crystal structure of the product state (PDB ID: 1CX0).

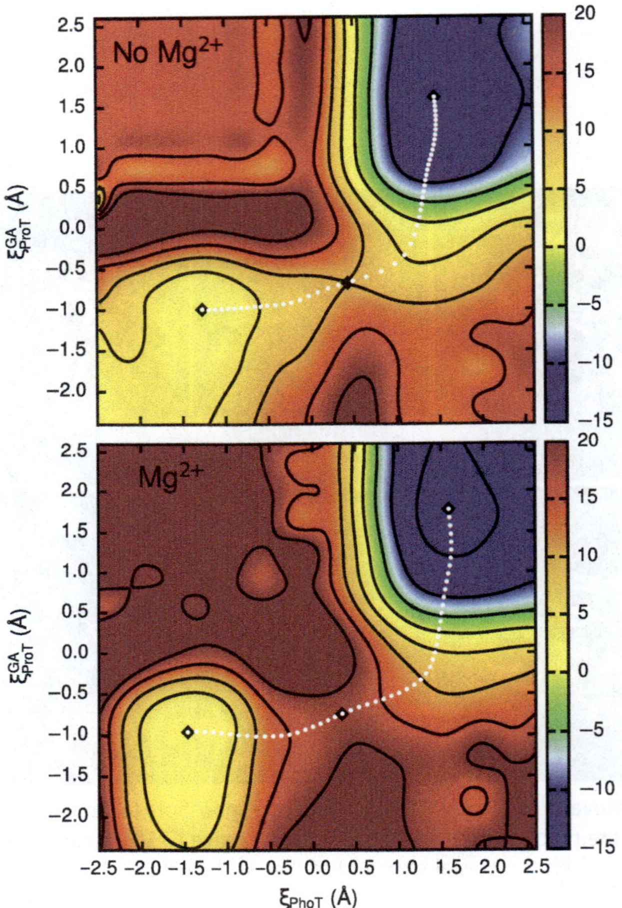

Maria T. Panteva *et al.*, Figure 8 2D free energy surfaces for the general acid step of HDVr catalysis departing from a state where the nucleophile has been activated in a prior step. Shown are simulations in the absence (top) and in the presence (bottom) of a Mg^{2+} bound in the active site. Minima and saddle points (diamonds) and the minimum free energy path (white points) are also indicated.

Printed by Printforce, the Netherlands